T0327423

Computational Intelligence in Bioinformatics

IEEE Press
445 Hoes Lane
Piscataway, NJ 08854

IEEE Computational Intelligence Society, *Sponsor*

IEEE-CIS Liaison to IEEE Press, Gary B. Fogel

Books in the IEEE Press Series on Computational Intelligence

Introduction to Evolvable Hardware: A Practical Guide for Designing Self-Adaptive Systems
Garrison W. Greenwood and Andrew M. Tyrrell
2007 978-0471-71977-9

Evolutionary Computation: Toward a New Philosophy of Machine Intelligence, Third Edition
David B. Fogel
2006 978-0471-66951-7

Emergent Information Technologies and Enabling Policies for Counter-Terrorism
Edited by Robert L. Popp and John Yen
2006 978-0471-77615-4

Computationally Intelligent Hybrid Systems
Edited by Seppo J. Ovaska
2005 0-471-47668-4

Handbook of Learning and Appropriate Dynamic Programming
Edited by Jennie Si, Andrew G. Barto, Warren B. Powell, Donald Wunsch II
2004 0-471-66054-X

Computational Intelligence: The Experts Speak
Edited by David B. Fogel and Charles J. Robinson
2003 0-471-27454-2

Computational Intelligence in Bioinformatics

Edited by

Gary B. Fogel
David W. Corne
Yi Pan

IEEE Computational Intelligence Society, ***Sponsor***

IEEE Press Series on Computational Intelligence
David B. Fogel, *Series Editor*

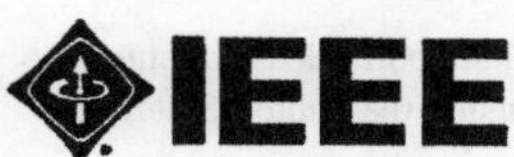

IEEE PRESS

WILEY-INTERSCIENCE
A John Wiley & Sons, Inc., Publication

Wiley Bicentennial Logo: Richard J. Pacifico

Library of Congress Cataloging-in-Publication Data is available.

ISBN 978-0470-10526-9

10 9 8 7 6 5 4 3 2 1

In loving memory of Jacquelyn Shinohara and her extended family, who fought cancer bravely, and of my father Lawrence J. Fogel, whose inspiration continues to guide me.

Gary B. Fogel

Contents

Part Three Molecular Structure and Phylogenetics

Preface

Bioinformatics can be defined as research, development, or application of computational tools and approaches for expanding the use of biological, medical, behavioral, or health data, including those to acquire, store, organize, archive, analyze, or visualize such data. Bioinformatics is a recent scientific discipline that combines biology, computer science, mathematics, and statistics into a broad-based field with profound impacts on all fields of biology and industrial application. Biological data sets have several common problems: They are typically large, have inherent biological complexity, may have significant amounts of noise, and may change with time. These challenges require computer scientists to rethink and adapt modeling approaches. Due to limitations of traditional algorithms, computational intelligence (CI) has been applied to bioinformatics recently and with promising results. This book serves to highlight the importance and recent success of computational intelligence methods over a diverse range of bioinformatics problems. It will encourage others to apply these methods to their research, while simultaneously serving as an introduction to CI methods and their application. The objective of this book is to facilitate collaboration between the CI community, bioinformaticians, and biochemists by presenting cutting edge research topics and methodologies in the area of CI in bioinformatics.

The book is organized as a number of stand-alone chapters developed and edited by leading educators that explain and apply CI methods to problem areas such as gene expression analysis and system biology, sequence analysis and feature detection, RNA and protein structure prediction and phylogenetics, and medicine. This book aims to highlight some of the important applications of CI in bioinformatics and identify how CI can be used to better implement these applications. Therefore, this book represents the unification of two fields (biology and computer science) with CI as a common theme.

Computational intelligence has been defined typically as biologically and linguistically motivated computational paradigms such as artificial neural networks, evolutionary algorithms, fuzzy systems, and hybridizations/extensions thereof. While each of these methods has its own significant history, their hybridization and application to bioinformatics problems remains a recent development. In addition, it is important to note that each problem may require a unique hybridization of CI methods, and it is hoped that part of the education that results from this book is the diversity of ways in which CI methods can be applied for problem solving so that the reader will have a better appreciation of how to solve similar problems in the future.

This book is intended for researchers, professors, students, and engineers in the fields of CI and bioinformatics, who are interested in applying CI to solving problems in bioinformatics applications. The book can also be used as a reference for graduate-level courses, and is divided into four major parts: Gene Expression Analysis and Systems Biology, Sequence Analysis and Feature Detection, Molecular Structure and Phylogenetics, and Medicine. In the following, the key elements of each chapter are briefly summarized.

Among all the challenges encountered in gene expression data analysis, the curse of dimensionality becomes more serious as a result of the availability of overwhelming number of gene expression values relative to the limited number of samples. Commonly used selection methods may result in many highly correlated genes while ignoring truly important ones. Considering this insufficiency of the existing methods in the analysis of more complicated cancer data sets, in Chapter 1, Xu et al. introduce a hybridized computational intelligence technology using neural classifier and swarm intelligence for multiclass cancer discrimination and gene selection.

In Chapter 2, Hong and Cho study two cases of classifying gene expression profiles with evolutionary computation. The first case study presents gene selection using a specialized genetic algorithm. Conventional wrapper methods using the genetic algorithm are applied to feature selection of small or middle-scale feature data sets. It can be difficult to apply such methods directly to large-scale feature data sets such as is typically the case for gene expression profiles due to increased processing time and low efficiency in the number of features used. To deal with these problems, a special genetic algorithm (sGA), which can improve the evolution efficiency, is used to select genes from huge-scale gene expression profiles. The second case study discovers and combines classification rules using genetic programming. Different from conventional approaches, the proposed method directly measures the diversity by comparing the structure of base classifiers, which is free from the side effect of using training samples.

Clustering gene expression data from microarray experiments is one way to identify potentially co-regulated genes. By analyzing the promoter regions of these genes, putative transcription factor binding sites may be identified. In Chapter 3, Ma et al. outline an algorithm, EvoCluster, for the purpose of handling noisy gene expression data. This algorithm is based on the use of an evolutionary computation approach and has several unique characteristics.

Gene networks are not only complex, they tend to be very large. Chapter 4, by Hallinan, considers the role of evolutionary computation in both optimization and the modeling of gene networks. The author examines the application of evolutionary computation to the understanding of network topology and ends with speculation and suggestions as to the future of this exciting field of research. The author concludes that the combination of evolutionary computation and gene networks is a powerful one, and its potential is only beginning to be realized.

In Chapter 5, a new fuzzy-granular gene selection algorithm is proposed by He et al. for reliable gene selection. The performance of this method is evaluated on two microarray expression data sets. A typical gene expression data set is extremely sparse compared to a traditional classification data set. Microarray data typically uses only dozens of tissue samples but contains thousands or even tens of thousands of gene expression values. This extreme sparseness is believed to significantly deteriorate the performance of the resulting classifiers. As a result, the ability to extract a subset of informative genes while removing irrelevant or redundant genes is crucial for accurate classification.

Chapter 6 by Jourdan et al. surveys and investigates the potential of evolutionary algorithms for feature selection in bioinformatics. First, the authors present generalities of evolutionary feature selection and show the common characteristics shared by evolutionary algorithms dedicated to the feature selection problem without considering the application. Then, the feature selection in an unsupervised context is discussed. Feature selection linked with supervised classification is also presented. Later, they present some frameworks and some data sets used in this domain, in order to help the reader to develop and test his own methods. Selecting features from a large set of potential features is clearly a combinatorial problem, and combinatorial optimization methods may be applied to this problem. The aim of this chapter is to evaluate how evolutionary algorithms handle such problems.

In Chapter 7, Sjahputera at al. study two types of microarrays: expression and methylation. A large number of different types of microarrays are available, differing in their basic chip technology, probe type, and experimental methodology. Expression microarrays quantify the level of gene activity, while methylation arrays measure amounts of gene methylation. Using both types of microarrays in the same tumor sample permits investigation of the molecular genetic mechanisms of cancer by detecting genes that are both silenced and inactivated by methylation. The authors introduce two sets of fuzzy algorithms to analyze the relationships between DNA methylation and gene expression. They chose a fuzzy logic framework for their application because it is reportedly better at accounting for noise in the data.

Chapter 8, by Thomsen, contains an overview of protein–ligand docking algorithms currently used by most practitioners as well as a short introduction to the biochemical background and the scoring functions used to evaluate docking solutions. The idea in molecular docking is to design pharmaceuticals computationally by identifying potential drug candidates targeted against proteins. The candidates can be found using a docking algorithm that tries to identify the bound conformation of a small molecule ligand to a macromolecular target at its active site, which is the region of an enzyme where the natural substrate binds. Molecular docking algorithms try to predict the bound conformations of two interacting molecules, such as protein–ligand and protein–protein complexes. The chapter also introduces the author's own research contributions in this area and gives an overview of future work.

The importance of RNA structure to cellular function, such as protein synthesis, or even structure or catalysis is well known. The secondary structure of RNA is described by a list of base pairs formed from the primary sequence. Many algorithms have been written for the prediction of RNA secondary structure. In Chapter 9, Wiese et al. survey various computational approaches for RNA secondary structure prediction, and the authors highlight their own development of evolutionary algorithms for RNA secondary structure prediction. Comparison of these CI approaches is provided relative to existing methods.

In Chapter 10, Huang et al. utilize experimentally determined human mitochondrial proteins to develop a classifier for protein localization using machine learning approaches. Currently, several high-performance computational classification methods are available via the Internet to predict the localization of proteins to the mitochondria for eukaryotes. The authors suggest this is the first development of a mitochondria localization algorithm specific to humans.

Phylogenetic inference is the process of constructing a model of the hypothesized evolutionary relationships among the species or individuals in a data set. The task of inferring a putative evolutionary tree, or phylogeny, is computationally daunting for modest data sets and intractable for realistically large data sets and thus has traditionally proceeded as a heuristic search. As with many domains, well-designed evolutionary computation techniques can assist greatly in the search for phylogenies. In Chapter 11, Congdon describes the general concept of phylogenetic inference, the challenges and opportunities for applying evolutionary computation to the task, some of the contributions made, and some open questions and directions for future work.

Cancer is a complex disease and chemotherapy is its most complex and difficult mode of treatment. In Chapter 12, McCall et al. explore the application of evolutionary algorithms to the design of cancer chemotherapy treatments. They briefly outline the nature of cancer and describe the strategy of chemotherapy in terms of tumor cell reduction. The authors review the major applications of evolutionary algorithms to cancer chemotherapy over the last 10 years, summarizing achievements and identifying challenges that have yet to be overcome.

There is an increasing demand for automatic information extraction (IE) schemes to extract knowledge from scientific documents and store them in a structured form. Ontology-guided IE mechanisms can help in the extraction of information stored within unstructured or semistructured text documents effectively. In Chapter 13, Dey and Abulaish present the design of a fuzzy ontology-based text extraction system that locates relevant ontological concepts within text documents and mines important domain relations from the repository. The system extracts all frequently occurring biological relations among a pair of biological concepts through text mining and natural language processing (NLP). They also propose a mechanism to generalize each relation to be defined at the most appropriate level of specificity, before it can be added to the ontology.

Datasets and other materials associated with these chapters are available through a curated website www.ci-in-bioinformatics.com so that readers can test their own approaches with similar data.

We would like to express our sincere appreciation to all of the authors for their important contributions. We would like to thank the outside referees for their efforts with chapter reviews and providing valuable comments and suggestions to the authors. We would like to extend our deepest gratitude to Catherine Faduska (Acquisitions Editor) from IEEE for her guidance and help in finalizing this book and to David Fogel, the book series editor on behalf of the IEEE Computational Intelligence Society. Yi Pan would like to thank Sherry for her friendship, help, and encouragement during the preparation of the book. David Corne would like to thank his family and Ph.D. students for their patience with him while he has been busy editing this and other books.

GARY B. FOGEL
DAVID W. CORNE
YI PAN

Contributors

Muhammad Abulaish
Department of Mathematics
Jamia Millia Islamia
New Delhi, India

Georgios C. Anagnostopoulos
Department of Electrical & Computer Engineering
Florida Institute of Technology
Melbourne, Florida

Charles W. Caldwell
Ellis Fischel Cancer Center
University of Missouri—Columbia
Columbia, Missouri

Keith C. C. Chan
Department of Computing
Hong Kong Polytechnic University
Hung Hom, Kowloon
Hong Kong

Sung-Bae Cho
Soft Computing Laboratory
Department of Computer Science
Yonsei University
Seoul, Korea

Clare Bates Congdon
Department of Computer Science
University of Southern Maine
Portland, Maine

David W. Corne
School of Mathematical and Computer Sciences
Heriot-Watt University
Edinburgh, United Kingdom

Alain A. Deschênes
Chemical Computing Group, Inc.
Montreal, Quebec

Lipika Dey
Department of Mathematics
Indian Institute of Technology, Delhi
New Delhi, India

Clarisse Dhaenens
INRIA Futurs, LIFL
Villeneuve D'Ascq, France

Gary B. Fogel
Natural Selection, Inc.
San Diego, California

Jennifer Hallinan
School of Computing Science
Newcastle University
Newcastle, United Kingdom

Yuanchen He
Department of Computer Science
Georgia State University
Atlanta, Georgia

Andrew G. Hendriks
School of Computing Science
Simon Fraser University
Surrey, British Columbia
Canada

Jin-Hyuk Hong
Soft Computing Laboratory
Department of Computer Science
Yonsei University
Seodaemoon-Gu, Seoul

Xiaohua Hu
College of Information Science and Technology
Drexel University
Philadelphia, Pennsylvania

Zhong Huang
College of Information Science and Technology
Drexel University
Philadelphia, Pennsylvania

Laetitia Jourdan
INRIA Futurs, LIFL
Villeneuve D'Ascq, France

James M. Keller
Electrical & Computer Engineering
University of Missouri—Columbia
Columbia, Missouri

Patrick C. H. Ma
Department of Computing
Hong Kong Polytechnic University
Hung Hom, Kowloon,
Hong Kong

John McCall
School of Computing
Robert Gordon University
Aberdeen, Scotland, United Kingdom

Yi Pan
Department of Computer Science
Georgia State University
Atlanta, Georgia

Andrei Petrovski
School of Computing
Robert Gordon University
Aberdeen, Scotland, United Kingdom

Mihail Popescu
Department of Health Management and Informatics
University of Missouri—Columbia
Columbia, Missouri

Siddhartha Shakya
School of Computing
Robert Gordon University
Aberdeen, Scotland, United Kingdom

Ozy Sjahputera
Center for Geospatial Intelligence
University of Missouri—Columbia
Columbia, Missouri

Rajshekhar Sunderraman
Department of Computer Science
Georgia State University
Atlanta, Georgia

El-Ghazali Talbi
INRIA Futurs, LIFL
Villeneuve D'Ascq, France

Yuchun Tang
Department of Computer Science
Georgia State University
Atlanta, Georgia

René Thomsen
Molegro ApS
Århus C, Denmark

Kay C. Wiese
School of Computing Science
Simon Fraser University
Surrey, British Columbia, Canada

Donald C. Wunsch II
Department of Electrical and Computer Engineering
University of Missouri—Rolla
Rolla, Missouri

Rui Xu
Department of Electrical and Computer Engineering
University of Missouri—Rolla
Rolla, Missouri

Xuheng Xu
College of Information Science and Technology
Drexel University
Philadelphia, Pennsylvania

Xin Yao
School of Computer Science
University of Birmingham
Edgbaston, Birmingham, United Kingdom

Yan-Qing Zhang
Department of Computer Science
Georgia State University
Atlanta, Georgia

Part One

Gene Expression Analysis and Systems Biology

Chapter 1

Hybrid of Neural Classifier and Swarm Intelligence in Multiclass Cancer Diagnosis with Gene Expression Signatures

Rui Xu, Georgios C. Anagnostopoulos, and Donald C. Wunsch II

1.1 INTRODUCTION

With the rapid advancement of deoxyribonucleic acid (DNA) microarray technologies, cancer classification through gene expression profiles has already become an important means for cancer diagnosis and treatment and attracted numerous efforts from a wide variety of research communities (McLachlan et al., 2004). Compared with the traditional classification methods that are largely dependent on the morphological appearance and clinical parameters, gene expression signature-based methods offer cancer researchers new methods for the investigation of cancer pathologies from a molecular angle, under a systematic framework, and further, to make more accurate prediction in prognosis and treatment.

Although previous research has included binary cancer classification (Alizadeh et al., 2000; Golub et al., 1999; West et al., 2001), it is more common practice to discriminate more than two types of tumors (Khan et al., 2001; Nguyen and Rocke, 2002; Ooi and Tan, 2003; Ramaswamy et al., 2001; Scherf et al., 2000). For example, Khan et al. (2001) used multilayer perceptrons to categorize small round blue-cell tumors (SRBCTs) with 4 subclasses. Scherf et al. (2000) constructed a gene expression database to investigate the relationship between genes and drugs for

Computational Intelligence in Bioinformatics. Edited by G. B. Fogel, D. Corne, and Y. Pan

60 human cancer cell lines originating from 10 different tumors, which provides an important criterion for therapy selection and drug discovery. Although these methods manifest interesting performance for some cancer data sets, their classification accuracy deteriorates dramatically with the increasing number of classes in the data, as shown in a comparative study by Li et al. (2004).

Among all the challenges encountered in gene expression data analysis, the curse of dimensionality becomes more serious as a result of the availability of overwhelming number of gene expression values relative to the limited number of samples. The existence of numerous genes in the data that are completely irrelevant to the discrimination of tumors not only increases the computational complexity but restricts the discovery of truly relevant genes. Therefore, feature selection, also known as *informative gene selection* in this context, becomes critically important (Deng et al., 2004; Golub et al., 1999; Ooi and Tan, 2003). Commonly used selection methods to discover informative genes rank the genes in terms of their expression difference in two different categories (Golub et al., 1999; Jaeger et al., 2003; Nguyen and Rocke, 2002; Ben-Dor et al., 2000). These criteria can provide some meaningful insights for binary classification. However, they may result in many highly correlated genes while ignoring the truly important ones. For multiclass cancer-type discrimination, these same approaches also cause some additional complexity due to the requirement for either one-versus-all or all-pairs comparison.

Considering this insufficiency of the existing methods in the analysis of more complicated cancer data sets, such as the NCI60 data set, in this chapter we propose a hybridized computational intelligence technology of semisupervised ellipsoid ARTMAP (ssEAM) and particle swarm optimization (PSO) (Kennedy et al., 2001) for multiclass cancer discrimination and gene selection. We apply the proposed method on publicly available cancer data sets and investigate the effect of various parameters on system performance. The experimental results demonstrate the effectiveness of ssEAM/PSO in addressing the massive, multidimensional gene expression data and are comparable to, or better than, those obtained by other classifiers. Additional analysis and experimental results on ssEAM/PSO in cancer classification are published elsewhere (Xu et al., 2007).

The rest of the chapter is organized as follows. In Section 1.2, we present the ssEAM/PSO system for multiclass cancer discrimination, together with the experiment design. In Section 1.3, we show the experimental results on two publicly accessible benchmark data sets. We conclude the chapter in Section 1.4.

1.2 METHODS AND SYSTEMS

1.2.1 EAM and Semisupervised EAM

Semisupervised ellipsoid ARTMAP is based on adaptive resonance theory (ART) (Grossberg, 1976), which was inspired by neural modeling research, and developed as a solution to the *plasticity–stability dilemma*. It is designed as an enhancement and generalization of ellipsoid ART (EA) and ellipsoid ARTMAP (EAM)

(Anagnostopoulos, 2001; Anagnostopoulos and Georgiopoulos, 2001), which, in turn, follow the same learning and functional principles of fuzzy ART (Carpenter et al., 1991) and fuzzy ARTMAP (Carpenter et al., 1992).

Ellipsoid ARTMAP accomplishes classification tasks by clustering data that are attributed with the same class label. The geometric representations of these clusters, which are called EA *categories*, are hyperellipsoids embedded in the feature space. A typical example of such a category representation, when the input space is two-dimensional, is provided in Figure 1.1, where each category C_j is described by a center location vector $\mathbf{m}_j$, orientation vector $\mathbf{d}_j$, and Mahalanobis radius M_j, which are collected as the template vector $\mathbf{w}_j = [\mathbf{m}_j, \mathbf{d}_j, M_j]$. If we define the distance between an input pattern $\mathbf{x}$ and a category C_j as

$$D(\mathbf{x}, \mathbf{w}_j) = \max\{\|\mathbf{x} - \mathbf{m}_j\|_{\mathbf{C}_j}, M_j\} - M_j, \tag{1.1}$$

$$\|\mathbf{x} - \mathbf{m}_j\|_{\mathbf{C}_j} = \sqrt{(\mathbf{x} - \mathbf{m}_j)^{\mathrm{T}} \mathbf{S}_j (\mathbf{x} - \mathbf{m}_j)}, \tag{1.2}$$

where $\mathbf{S}_j$ is the category's shape matrix, defined as

$$\mathbf{S}_j = \frac{1}{\mu^2}[\mathbf{I} - (1 - \mu^2)\mathbf{d}_j \mathbf{d}_j^{\mathrm{T}}], \tag{1.3}$$

and μ is the ratio between the length of the hyperellipsoid's minor axes (with equal length) and major axis, the *representation region* of C_j, which is the shaded area in the figure, can be defined as a set of points in the input space, satisfying the condition

$$D(\mathbf{x}, \mathbf{w}_j) = 0 \Rightarrow \|\mathbf{x} - \mathbf{m}_j\|_{\mathbf{S}_j} \leq M_j. \tag{1.4}$$

A category encodes whatever information the EAM classifier has learned about the presence of data and their associated class labels in the locality of its geometric representation. This information is encoded into the location, orientation, and size of the hyperellipsoid. The latter feature is primarily controlled via the baseline vigilance $\bar{\rho} \in [0, 1]$ and indirectly via the choice parameter $a > 0$ and a network parameter $\omega \geq 0.5$ (Anagnostopoulos and Georgiopoulos, 2001). Typically, small values of $\bar{\rho}$ produce categories of larger size, while values close to 1 produce the opposite effect. As a special case, when $\bar{\rho} = 1$, EAM will create solely point categories (one for each

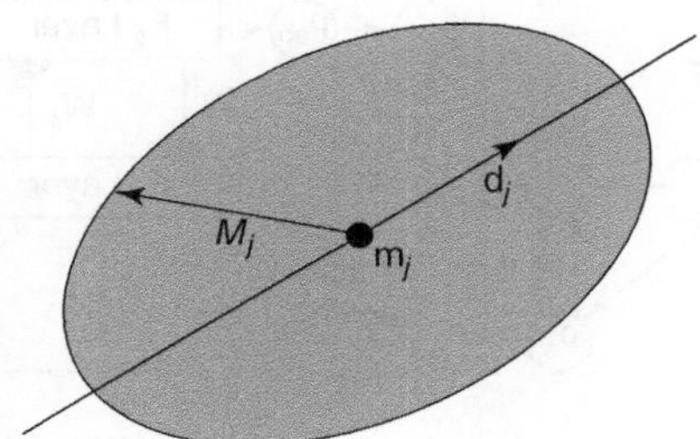

Figure 1.1 Example of geometric representation of EA category C_j in two-dimensional feature space.

training pattern) after completion of training, and it implements the ordinary, Euclidian 1-nearest neighbor classification rule. A category's particular shape (eccentricity of its hyperellipsoid) is controlled via the network parameter $\mu \in (0, 1]$; for $\mu = 1$ the geometric representations become hyperspheres, in which case the network is called *hypersphere ARTMAP* (Anagnostopoulos and Georgiopoulos, 2000).

Figure 1.2 illustrates the block diagram of an EAM network, which consists of two EA modules (ART_a and ART_b) interconnected via an inter-ART module. The ART_a module clusters patterns of the input domain while ART_b clusters patterns of the output domain. The information regarding the input–output associations is stored in the weights $\mathbf{w}_j^{ab}$ of the inter-ART module, while EA category descriptions are contained in the template vectors $\mathbf{w}_j$. These vectors are the top–down weights of F_2-layer nodes in each module.

Learning in EAM occurs by creating new categories or updating already existing ones. If a training pattern $\mathbf{x}$ initiates the creation of a new category C_J, then C_J receives the class label $L(\mathbf{x})$ of $\mathbf{x}$, by setting the class label of C_J to $I(J) = L(\mathbf{x})$. The recently created category C_J is initially a *point category*, meaning that $\mathbf{m}_J = \mathbf{x}$ and $M_J = 0$. While training progresses, point categories are being updated, due to the presentation of other training patterns, and their representation regions may grow. Specifically, when it has been decided that a category C_j must be updated by a training pattern $\mathbf{x}$, its representation region expands, so that it becomes the minimum-volume hyperellipsoid that contains the entire, original representation region and the new pattern. Learning eventually ceases in EAM, when no additional categories are being created and the existing categories have expanded enough to capture all training data. Notice that, if $\mathbf{x}$ falls inside the representation region of C_j, no update occurs since C_j has already taken into account the presence of $\mathbf{x}$.

The procedure of deciding which category C_j is going to be updated, with a training pattern $\mathbf{x}$, involves competition among preexisting categories. Let us define this set of categories as N as well as the set $S \subseteq N$ of all categories that are candidates

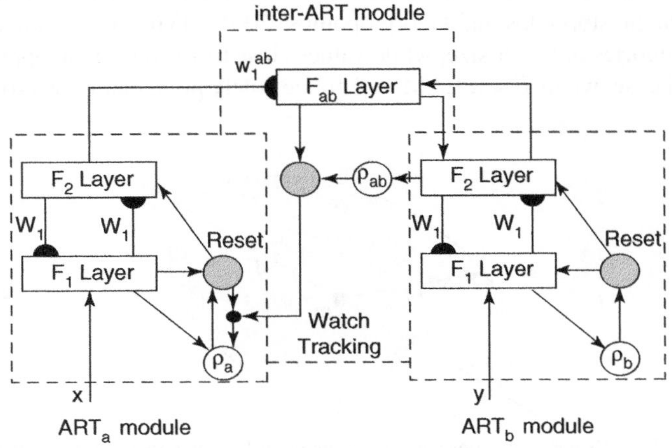

Figure 1.2 Ellipsoid ARTMAP block diagram.

in the competition; initially, $S = N$. Before the competition commences, for each category C_j, two quantities are calculated: the *category match function* (CMF) value

$$\rho(\mathbf{w}_j|\mathbf{x}) = \frac{D_{max} - 2M_j - D(\mathbf{x}, \mathbf{w}_j)}{D_{max}}, \tag{1.5}$$

where D_{max} is a parameter greater than 0, and the *category choice function* (CCF) value

$$T(\mathbf{w}_j|\mathbf{x}) = \frac{D_{max} - 2M_j - D(\mathbf{x}, \mathbf{w}_j)}{D_{max} - 2M_j + a}. \tag{1.6}$$

Next, EAM employs two category-filtering mechanisms: the *vigilance test* (VT) and the *commitment test* (CT). Both tests decide if the match between the pattern and the category's representation region is sufficient to assign that pattern to the cluster represented by the category in question. These tests can function as a novelty detection mechanism as well: If no category in S passes both tests, then $\mathbf{x}$ is not a typical pattern in comparison to the data experienced by the classifier in the past. Categories that do not pass these tests can be subsequently removed from the candidate set S. Next, the competition for $\mathbf{x}$ is won by the category C_J that features the maximum CCF value with respect to the pattern; in case of a tie, the category with minimum index is chosen. The final verdict on whether to allow C_J to be updated with $\mathbf{x}$ or not is delivered by the *prediction test* (PT): C_J is allowed to be updated with $\mathbf{x}$ only if both C_J and $\mathbf{x}$ feature the same class label, that is, if $I(J) = L(\mathbf{x})$. If C_J fails the PT, a *match tracking* process is invoked by utilizing a stricter VT in the hope that another suitable EAM category will be found that passes all three tests. If the search eventually fails, a new point category will be created as described before.

Ellipsoid ARTMAP does not allow categories to learn training patterns of dissimilar class labels. This property is ideal when the individual class distributions of the problem are relatively well separated. However, in the case of high class overlap, or when dealing with increased amount of noise in the feature domain, EAM will be forced to create many small-sized categories, a phenomenon called the *category proliferation problem*. Moreover, when EAM is trained in an offline mode to perfection, its posttraining error will be zero, which can be viewed as a form of data overfitting.

The semisupervised EAM (ssEAM) classifier extends the generalization capabilities of EAM by allowing the clustering into a single category of training patterns not necessarily belonging to the same class (Anagnostopoulos et al., 2002). This is accomplished by augmenting EAM's PT in the following manner: A winning category C_J may be updated by a training pattern $\mathbf{x}$, even if $I(J) \neq L(\mathbf{x})$, as long as the following inequality holds:

$$\frac{w_{J,I(J)}}{1 + \sum_{c=1}^{C} w_{J,c}} \geq 1 - \varepsilon, \tag{1.7}$$

where C denotes the number of distinct classes related to the classification problem at hand and the quantities $w_{J,c}$ contain the count of how many times category C_J was updated by a training pattern belonging to the cth class. In other words, Eq. (1.7) ensures that the percentage of training patterns that are allowed to update category C_J and carry a class label different than the class label $I(J)$ (the label that was initially assigned to C_J, when it was created) cannot exceed 100ε %, where $\varepsilon \in [0,1]$ is a new network parameter, the *category prediction error tolerance*, which is specific only to ssEAM. For $\varepsilon = 1$ the modified PT will allow categories to be formed by clustering training patterns, regardless of their class labels, in an unsupervised manner. In contrast, with $\varepsilon = 0$ the modified PT will allow clustering (into a single category) only of training patterns belonging to the same class, which makes the category formation process fully supervised. Under these circumstances, ssEAM becomes equivalent to EAM. For intermediate values of ε, the category formation process is performed in a semisupervised fashion.

Both EAM and ssEAM feature a common performance phase, which is almost identical to their training phases. However, during the presentation of test patterns, no categories are created or updated. The predicted label for a test pattern $\mathbf{x}$ is determined by the *dominant class label* $O(J)$ of the winning category C_J defined as

$$\hat{L}(\mathbf{x}) = O(J) = \arg\max_{c=1..C} w_{J,c} = 1. \tag{1.8}$$

When $\varepsilon < 0.5$, ssEAM's PT guarantees that throughout the training phase $O(j) = I(j)$ for any category C_j.

For $\varepsilon > 0$ ssEAM will, in general, display a nonzero posttraining error, which implies a departure from EAM's overfitting and category proliferation issues. For classification problems with noticeable class distribution overlap or noisy features, ssEAM with $\varepsilon > 0$ will control the generation of categories representing localized data distribution exceptions, thus, improving the generalization capabilities of the resulting classifier. Most importantly, the latter quality is achieved by ssEAM without sacrificing any of the other valuable properties of EAM, that is, stable and finite learning, model transparency, and detection of atypical patterns.

Semisupervised EAM has many attractive properties for classification or clustering. First, ssEAM is capable of both online (incremental) and offline (batch) learning. Using *fast learning* in offline mode, the network's training phase completes in a small number of epochs. The computational cost is relatively low, and it can cope with large amounts of multidimensional data, maintaining efficiency. Moreover, ssEAM is an *exemplar-based model*, that is, during its training the architecture summarizes data via the use of exemplars in order to accomplish its learning objective. Due to its exemplar-based nature, responses of an ssEAM architecture to specific test data are easily explainable, which makes ssEAM a *transparent learning model*. This fact contrasts other, *opaque* neural network architectures, for which it is difficult, in general, to explain why an input $\mathbf{x}$ produced a particular output $\mathbf{y}$. Another important feature of ssEAM is the capability of detecting atypical patters during either its training or performance phase. The detection of such patterns is accomplished via the employment of a match-based criterion that decides to which degree a particular pattern matches the characteristics of an already formed category in ssEAM. Additionally, via the utilization of hyperellipsoidal categories, ssEAM can learn complex decision boundaries, which arise frequently in gene expression classification problems. Finally, ssEAM is far simpler to implement, for example, than backpropagation for feedforward neural networks and the training algorithm of support vector machines. Many of these advantages are inherited, general properties of the ART family of neural networks including fast, exemplar-based, match-based, learning (Grossberg, 1976), transparent learning (Healy and Caudell, 1997), capability to handle massive data sets (Healy et al., 1993), and implementability in software and hardware (Serrano-Gotarredona et al., 1998; Wunsch et al., 1993; Wunsch, 1991). Also, ART neural networks dynamically generate clusters without specifying the number of clusters in advance as the classical *k*-means algorithm requires (Xu and Wunsch, 2005).

1.2.2 Particle Swarm Optimization

Particle swarm optimization (PSO) is a heuristic, global, stochastic maximization meta-algorithm, motivated by the complex social behavior and originally intended to explore optimal or near-optimal solutions in sophisticated continuous spaces (Kennedy et al., 2001). Like most evolutionary computation meta-algorithms, PSO is also population based, consisting of a swarm of particles, each of which represents a candidate solution $\mathbf{x}_i$ and is associated with a random velocity $\mathbf{v}_i$. The basic idea of PSO is that each particle randomly searches through the problem space by updating itself with its own memory and the social information gathered from other particles. Specifically, at each time step, each particle is accelerated toward two best locations, based on the fitness value: p_{best} for the previous best solution for this particle and g_{best} for the best overall value in the entire swarm. Accordingly, the canonical PSO velocity and position update equations are written as

$$v_{ij}(t+1) = W_I \times v_{ij}(t) + c_1 \operatorname{rand}_1[p_{\text{best}_{ij}} - x_{ij}(t)] + c_2 \operatorname{rand}_2[g_{\text{best}_{ij}} - x_{ij}(t)], \quad (1.9)$$

$$x_{ij}(t+1) = x_{ij}(t) + v_{ij}(t+1), \quad (1.10)$$

where W_I is an inertia weight, c_1 and c_2 are acceleration constants, and rand_1 and rand_2 are samples of random variables uniformly distributed in the range of [0, 1]. PSO has many desirable characteristics, such as a memory mechanism of keeping track of previous best solutions, the easiness to implement, fast convergence to high-quality solutions, and the flexibility in balancing global and local exploration.

Since our goal is to select important genes from a large gene pool with *M* genes in total, we employ a discrete binary version of PSO (Kennedy and Eberhart, 1997). The major change of the binary PSO lies in the interpretation of the meaning of the particle velocity. Given a set of *N* particles $\mathbf{X} = \{\mathbf{x}_1, \mathbf{x}_2, \ldots, \mathbf{x}_N\}$, each of which corresponds to a subset of genes, the velocity for the *i*th particle $\mathbf{x}_i = (x_{i1}, x_{i2}, \ldots, x_{iM})$ is represented as $\mathbf{v}_i^* = (v_{i1}^*, v_{i2}^*, \ldots, v_{iM}^*)$. The possible values for each bit x_{ij} ($1 \leq i \leq N$, $1 \leq j \leq M$) are either one or zero, indicating the corresponding genes are selected or not. The velocity v_{ij}^* associated with it is defined as the probability that x_{ij} takes the value of 1 and is calculated by the logistic probability law

$$v_{ij}^*(t+1) = \frac{1}{1+\exp[-v_{ij}(t+1)]}, \tag{1.11}$$

where v_{ij} is calculated using Eq. (1.9). Accordingly, the position update equation is given as

$$x_{ij}(t+1) = \begin{cases} 1 & \text{if } \text{rand}_3 + \delta < v_{ij}^*(t+1), \\ 0 & \text{otherwise}, \end{cases} \tag{1.12}$$

where rand_3 is a sample of random variable uniformly distributed in the range of [0, 1], and δ is a parameter that limits the total number of genes selected to a certain range. Compared to the original binary PSO by Kennedy and Eberhart (1997), we add the parameter δ to obtain more flexibility in controlling the number of informative genes. If the value of δ is large, the number of genes selected becomes small and vice versa.

Now, we summarize the basic procedure of binary PSO for informative gene selection as follows:

1. Initialize a population of N particles with random positions and velocities. The dimensionality M of the problem space equals the number of genes in the data.
2. Evaluate the classification performance of ssEAM and calculate the optimization fitness function for each particle. The definition of PSO's fitness function aims to minimize the classification error while favoring the subset with fewer genes; it is defined as

$$f(\mathbf{x}_i) = \text{Acc}_{\text{LOOCV}} + \frac{1}{M_i}, \tag{1.13}$$

   where $\text{Acc}_{\text{LOOCV}}$ is the *leave-one-out cross-validation* (LOOCV) (Kohavi, 1995) classification accuracy defined in Eq. (1.16) and M_i is the number of informative genes selected.
3. Compare the fitness value of each particle with its associated p_{best} value. If the current value is better than p_{best}, set both p_{best} and the particle's attained location to the current value and location.
4. Compare p_{best} of the particles with each other and update g_{best} with the greatest fitness.
5. Update the velocity and position of the particles using Eqs. (1.9), (1.11), and (1.12).
6. Return to step 2 until the stopping condition is met, which, usually, is reaching the maximum number of iterations or the discovery of high-quality solutions.

The inertial weight is similar to the momentum term in the backpropagation algorithm for multilayer perceptrons (Bergh and Engelbrecht, 2004). It specifies the trade-off between the global and local search (Shi and Eberhart, 1998). Larger values of W_I facilitate the global exploration while lower values encourage local search. W_I can take on a fixed value, or more commonly, decreases linearly during a PSO run (Shi and Eberhart, 1999),

$$W_I = (W_{I1} - W_{I2})\frac{T-t}{T} + W_{I2}, \tag{1.14}$$

where W_{I1} and W_{I2} are the initial and final values, respectively, T is the maximum number of epochs allowed, and t is the current epoch number. Alternately, the inertial weight can be set to change randomly within a range (Eberhart and Shi, 2001), for instance,

$$W_I = W_{I0} + \frac{\text{rand}}{2}, \tag{1.15}$$

where rand is a uniformly distributed random function in the range of [0, 1]. As an example, if W_{I0} is set as 0.5, Eq. (1.15) makes W_I vary between 0.5 and 1, with a mean of 0.75. In this chapter, these three methods are referred to as PSO-FIXEW, PSO-LTVW, and PSO-RADW, respectively. c_1 and c_2 are known in the PSO literature as cognition and social components, respectively, and are used to adjust the velocity of a particle toward p_{best} and g_{best}. Typically, they are both set to a value of 2, although different values may achieve better performance (Eberhart and Shi, 2001).

1.2.3 Experiment Design

Since the data sets consist of only a small number of samples for each cancer type, it is important to use an appropriate method to estimate the classification error of the classifier. In the experiment below, we perform a double cross validation [*10-fold cross validation* (CV10) with LOOCV], instead of just the commonly used LOOCV, to examine the performance of ssEAM/PSO. This is because, although the LOOCV is an unbiased point estimate of the classification error, it has high variance, which is not preferred in cancer classification. On the contrary, resampling strategies such as bootstrap (Efron, 1982), may become largely biased in some cases (Ambroise and McLachlan, 2002; Kohavi, 1995) despite having lower variance. During the double cross-validation procedure, the data set with Q samples is divided into 10 mutually exclusive sets of approximately equal size, with each subset consisting of approximately the same proportions of labels as the original data set, known as stratified cross validation (Kohavi, 1995). The classifier is trained 10 times, with a different subset left out as the test set and the other samples are used to train the classifier each time. During the training phase, gene selection is performed based on the 9 out of the 10 data subsets (without considering the test data), for which LOOCV classification accuracy is used as the fitness function in Eq. (1.13). The prediction performance of the classifier is estimated by considering the average classification accuracy of the 10 cross-validation experiments, described as

$$\mathrm{Acc}_{\mathrm{CV10}} = \left(\frac{1}{Q} \sum_{i=1}^{10} A_i \right) \times 100\%, \tag{1.16}$$

where A_i is the number of correctly classified samples. Previous studies have shown that CV10 is more appropriate when a compromise between bias and variance is preferred (Ambroise and McLachlan, 2002; Kohavi, 1995).

We also compare our approach with four other classifiers, that is, multilayer perceptrons (MLPs) (Haykin, 1999), probabilistic neural networks (PNNs) (Specht, 1990), learning vector quantization (LVQ) (Kohonen, 2001), and *k*-nearest-neighbor (kNN) (Duda et al., 2001), together with Fisher's discriminant criterion (Hastie et al., 2003), which is used for informative genes selection and defined as

$$F(i) = \frac{|\mu_+(i) - \mu_-(i)|^2}{\sigma_+^2(i) + \sigma_-^2(i)}, \tag{1.17}$$

where $\mu_+(i)$ and $\mu_-(i)$ are the mean values of gene i for the samples in class +1 and class −1, and $\sigma_+^2(i)$ and $\sigma_-^2(i)$ are the variances of gene i for the samples in class +1 and −1. The score aims to maximize the between-class difference and minimize the within-class spread. Currently, other proposed rank-based criteria with these considerations show similar performance (Jaeger et al., 2003). Since our ultimate goal is to classify multiple types of cancer, we utilize a one-versus-all strategy to seek gene predictors. In order to overcome *selection bias*, which is caused by including the test samples in the process of feature selection and which leads to an overoptimistic estimation of the performance for the classifier (Ambroise and McLachlan, 2002; Nguyen and Rocke, 2002; West et al., 2001), we utilize the strategy that separates gene selection from cross-validation assessment. Note in this case that the subsets of genes selected at each stage tend to be different.

1.3 EXPERIMENTAL RESULTS

1.3.1 NCI60 Data

The NCI60 data set includes 1416 gene expression profiles for 60 cell lines in a drug discovery screen by the National Cancer Institute (Scherf et al., 2000). These cell lines belong to 9 different classes: 8 breast (BR), 6 central nervous system (CNS), 7 colorectal (CO), 6 leukemia (LE), 9 lung (LC), 8 melanoma (ME), 6 ovarian (OV), 2 prostate (PR), and 8 renal (RE). Since the PR class only had two samples, these were excluded from further analysis. There were 2033 missing gene expression values in the data set, which were imputed by the method described by Berrar et al. (2003). This process left the final matrix in the form of $E = \{e_{ij}\}_{58\times1409}$, where $e_{i,j}$ represents the expression level of gene j in tissue sample i.

We first investigated the effect of the three different strategies for the inertia weight (i.e., PSO-FIXEW, PSO-LTVW, and PSO-RADW) on the performance of PSO. The values of c_1 and c_2 were both set to 2 in this study. We fixed the W_I at 0.8 for PSO-FIXEW and linearly decreased W_I from 0.9 to 0.4 for PSO-LTVW. For the random method PSO-RADW, we chose W_{I0} equal to 0.5, so that W_I varied in the range of [0.5, 1]. The performance of the three methods was compared in terms of the number of iterations required to reach a prespecified classification accuracy, say, 81% (47 out of 58 samples) in this case. We further set the maximum number of iterations to 100. If PSO reached 81% classification accuracy within 100 iterations, we considered PSO to have converged. The results over 20 runs are summarized in Table 1.1, which consists of the number of times that the iteration exceeded the allowed maximum and the average number of epochs if PSO converged. As indicated in Table 1.1, the most effective performance was achieved when PSO-FIXEW was used, where PSO achieves the expected accuracy within 39 iterations on average, except for 2 runs that did not converge. The result for PSO-LTVW was slightly inferior to PSO-FIXEW, as the average number of iterations was 42.9 for 16 out of 20 converging cases. PSO-RADW did not perform well in this problem and was more dependent on its initialization. In the following discussion, we used PSO-FIXEW with W_I at 0.8, and set both c_1 and c_2 to 2.

We set the parameters for ssEAM as follows: $\mu = 0.3$, $\rho = 0.4$, and $\alpha = 2.5$, learning rate = 0.8, and adjusted the value of ε, which controlled the amount of misclassification allowed in the training phase. The parameters of ssEAM were determined based on a simple selection procedure in which the data set was randomly divided into training and validation sets. We compared the different parameter combination and chose the ones that lead to relatively high performance. The parameter δ controlled the total number of genes selected in the subsets, and we evaluated the program with δ at 0.5, 0.45, 0.4, 0.3, 0.2, 0.1, and 0.0. Each time, evolution was processed for 300 generations with a swarm population of 50 particles. The algorithm was iterated 20 times using different partitions of the data set and performance was reviewed relative to mean performance. The mean and standard deviation of the classification accuracies from the 20 runs are summarized in Table 1.2, and the best results are depicted in Figure 1.3(*a*). For the purpose of comparison, we also show the results of PNN, MLP, kNN, and LVQ1, in which the Fisher criterion was

Table 1.1 Comparison of PSO-FIXEW, PSO-LTVW, and PSO-RADW on PSO Performance

W_I	Performance	
	>100 Iterations	Average Number of Iterations
PSO-FIXEW (at 0.8)	2	38.3
PSO-LTVW (0.9–0.4)	3	42.9
PSO-RADW (0.5–1)	5	45.5

Table 1.2 Classification Accuracy for NCI60 Data Set[a]

		Number of Features (Genes)						
NCI60		10	79	135	252	385	555	695
EAM ($\varepsilon = 0$)	PSO	65.52 (1.86)	83.02 (1.51)	79.40 (1.42)	76.64 (2.47)	71.21 (1.59)	68.10 (1.90)	67.76 (1.49)
ssEAM ($\varepsilon = 0.1$)	PSO	65.78 (2.19)	84.66 (1.36)	81.12 (1.98)	78.62 (1.89)	75.26 (1.79)	73.10 (2.27)	72.50 (2.34)
PNN	Fisher criterion	24.05 (2.27)	71.12 (2.23)	72.24 (2.09)	74.65 (2.02)	76.81 (2.27)	76.03 (1.24)	76.12 (1.40)
MLP	Fisher criterion	16.81 (3.21)	39.14 (5.82)	39.40 (4.38)	44.91 (5.08)	45.43 (5.46)	45.17 (4.25)	47.59 (7.51)
kNN	Fisher criterion	41.90 (2.86)	69.22 (2.64)	69.74 (1.63)	72.59 (1.57)	73.71 (1.28)	72.59 (1.10)	71.81 (1.28)
LVQ1	Fisher criterion	42.67 (2.73)	71.81 (2.87)	72.76 (2.48)	73.10 (1.80)	73.88 (1.28)	72.33 (1.98)	72.24 (1.47)

[a]Given are the mean and standard deviation (in parentheses) of percent of correct classification of 58 tumor samples with CV10 ($\rho = 0.4$, $\mu = 0.3$, learning rate = 0.8, $\alpha = 2.5$).

used for gene selection. For PNN, the smoothing parameter of the Gaussian kernel was set to 1. The MLP consisted of a single hidden layer featuring 20 nodes and was trained with the one-step secant algorithm for fast learning. We varied the number of prototypes in LVQ1 and the value of k in kNN from 8 to 17 and 1 to 10, respectively. Both methods were evaluated based on average classification accuracy. From Table 1.2 it should be noted that ssEAM/PSO is superior to the other methods used in these experiments or additional ones found in the literature. Specifically, the best result attained with ssEAM/PSO is 87.9% (79 genes are selected by PSO), which is better than other results reported in the literature. We performed a formal t test to compare the difference between the best overall results of ssEAM and other methods. All p values are less than 10^{-15}, which indicates the classification accuracy for ssEAM is statistically better than those of other methods at a 5% significance level. The same conclusion can also be reached via a nonparametric Wilcoxon rank test and a Kruskal–Wallis test. Another interesting observation from the experimental results is that the introduction of the error tolerance parameter ε could provide an effective way to increase generalization capability and decrease overfitting, which is encountered frequently in cancer classification. The performance of ssEAM can usually be improved with the appropriate selection of ε (0.1 for this data set). However, the overrelaxing of the misclassification tolerance criterion during the category formation process in ssEAM training could cause the degradation of the classifier performance.

We further compared the top 100 genes selected by the Fisher criterion with those selected by PSO. This comparison shows that there is only a small fraction of overlaps between the genes chosen by these two methods. For example, for the 79

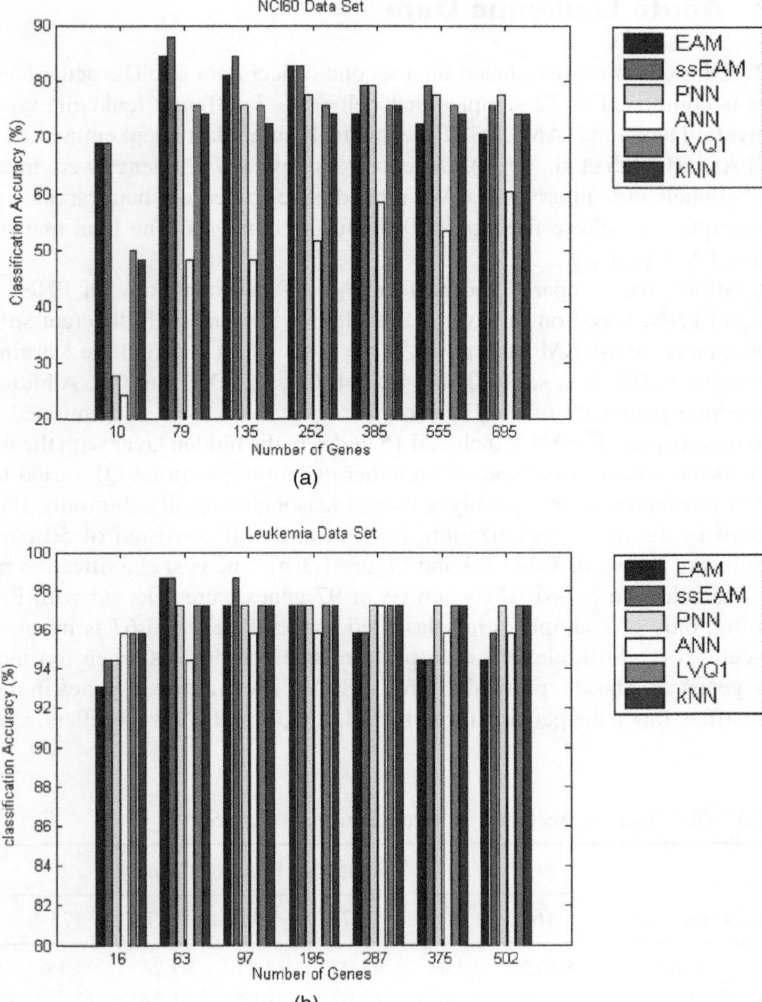

Figure 1.3 Best classification accuracy of 20 runs for the (*a*) NCI60 and (*b*) leukemia data set. Order for bars is EAM, ssEAM, PNN, ANN, LVQ1, and kNN, from left to right.

genes that lead to the best classification result, only 7 were also selected by the Fisher criterion. Although the Fisher criterion can be effective in some binary classification problems, as shown in Xu and Wunsch (2003), it does not achieve effective performance in cases of multiclass discrimination. The reason for this may lie in the fact that the Fisher criterion tends to choose many highly correlated genes, ignoring those genes that are really important for classification. In addition, the use of the Fisher discriminant criterion is justified when the data follow an approximately Gaussian distribution. This may not be true for this data set.

1.3.2 Acute Leukemia Data

The ssEAM method was evaluated on a second cancer data set. The acute leukemia data set is comprised of 72 samples that belong to 3 different leukemia types: 25 acute myeloid leukemia (AML), 38 B-cell acute lymphoblastic leukemia (ALL), and 9 T-cell ALL (Golub et al., 1999). Gene expressions for 7129 genes were measured using oligonucleotide microarrays. We ranked genes based on their variance across all the samples and chose the top 1000 for further analysis. The final matrix is in the form of $E = \{e_{ij}\}_{72\times1000}$.

As before, we compared the performance of our method with PNN, MLP, LVQ1, and kNN, based on the average results for 20 runs with different splitting. The parameters for ssEAM were $\mu = 0.9$, $\rho = 0.45$, and $a = 4$, and the learning rate was set equal to 0.8; δ is set to 0.5, 0.45, 0.4, 0.3, 0.2, 0.1, and 0.0. Additionally, the smoothing parameter of the Gaussian kernel was set to 1, as mentioned previously in this chapter. The MLP included 15 nodes in the hidden layer with the logistic function as the transfer function. The number of prototypes in LVQ1 varied from 3 to 12. For this data set, we typically achieved reasonable results after only 100 generations of evolutionary optimization. Each swarm still consisted of 50 particles. The results are shown in Table 1.3 and Figure 1.3(*b*). The best classification performance was achieved by ssEAM when 63 or 97 genes were selected with PSO. In this setting, only one sample is misclassified (i.e., a T-cell ALL67 is misclassified as a B-cell ALL). Still, classification performance deteriorates when too many or too few genes are chosen, particularly for ssEAM. The number of genes in the data does not affect much the performance of PNN, LVQ1, and kNN classifiers, although

Table 1.3 Classification Accuracy for Acute Leukemia Data Set[a]

Acute Leukemia Data		Number of Features (Genes)						
		16	63	97	195	287	375	502
EAM ($\varepsilon = 0$)	PSO	89.72 (2.08)	94.44 (2.67)	95.07 (1.65)	94.31 (1.68)	93.26 (1.58)	93.13 (1.23)	92.36 (1.39)
ssEAM ($\varepsilon = 0.1$)	PSO	91.60 (1.23)	97.15 (0.95)	97.50 (0.73)	95.83 (0.90)	94.65 (0.82)	93.68 (0.71)	92.64 (1.43)
PNN	Fisher criterion	90.00 (2.61)	96.32 (0.68)	96.74 (0.68)	96.46 (0.71)	96.39 (0.70)	96.25 (0.91)	96.18 (0.99)
MLP	Fisher criterion	93.61 (2.27)	92.50 (1.23)	93.47 (2.07)	91.60 (4.19)	91.74 (3.94)	91.60 (2.90)	91.67 (3.60)
LVQ1	Fisher criterion	94.86 (0.65)	95.83 (0.45)	96.45 (0.84)	95.97 (0.77)	96.04 (0.93)	96.10 (0.86)	95.90 (0.71)
kNN	Fisher criterion	95.07 (0.71)	95.83 (0.45)	96.18 (0.62)	96.11 (0.73)	95.90 (0.84)	95.69 (1.00)	95.69 (0.77)

[a]Given are the mean and standard deviation (in parentheses) of percent of correct classification for 72 tumor samples with CV10 ($\rho = 0.45$, $\mu = 0.9$, learning rate = 0.8, $\alpha = 4$).

there is some slight decrease when more genes are included. In contrast to the performance results obtained with the NCI60 data set, kNN and LVQ1 work well for this data set. The p values of the associated t tests comparing performances of ssEAM, PNN, ANN, LVQ1, and kNN classifiers are 0.0015, 9.9×10^{-9}, 1.6×10^{-4}, and 3.1×10^{-7}, respectively, which, again, shows the significantly better performance of ssEAM models (at a 5% significance level).

Among all examples, sample AML66 and T-cell ALL67 cause the most misclassification (e.g., when $\delta = 0.45$, 23 out of 50 particles misclassified AML66 as ALL, and 32 out of 50 particles misclassified T-cell ALL67 either as B-cell ALL or AML). This is similar to the results from other analyses (Golub et al., 1999; Nguyen and Rocke, 2002). For the acute leukemia data set, the effect of introducing the error tolerance parameter ε ($\varepsilon > 0$) is not as pronounced as in the NCI60 data set. This may be the result of reduced overlap among the Golub data in comparison to the NCI60 data set. The improvement of performance is more effective for semisupervised training when applied to higher overlapped data sets.

It is interesting to examine whether the genes selected by PSO are really meaningful in a biological sense. Among the top 50 genes selected, many of them have already been identified as the important markers for the differentiation of the AML and ALL classes. Specifically, genes such as *NME4, MPO, CD19, CTSD, LTC4S, Zyxin*, and *PRG1*, are known to be useful in AML/ALL diagnosis (Golub et al., 1999). Also, some new genes are selected that previously were not reported as being relevant to the classification problem. Additional investigation is required for these genes. Moreover, we found that the Fisher discriminant criterion can also identify genes that contribute to the diagnosis of these three leukemia types, such as with genes *Zyxin, HoxA9*, and *MB-1*. The reason for this may be due to the underlying biology represented in the Golub data: Genes express themselves quite differently under different tumor types in the AML/ALL data relative to the NCI60 data. Furthermore, we observe that different feature selection methods usually lead to different subsets of selected, informative genes with only very small overlap, although the classification accuracy does not change greatly. Genes that have no biological relevance still can be selected as an artifact of the feature selection algorithms themselves. This suggests that feature selection may provide effective insight in cancer identification; however, careful evaluation is critical due to the problems caused by insufficient data.

1.4 CONCLUSIONS

Classification is critically important for cancer diagnosis and treatment. Microarray technologies provide an effective way to identifying different kinds of cancer types, while simultaneously spawning many new challenges. Here, we utilized semisupervised ellipsoid ARTMAP and particle swarm optimization to address the multitype cancer identification problem, based on gene expression profiling. The proposed method achieves qualitatively good results on two publicly accessible benchmark data sets, particularly, with the NCI60 data set, which is not effectively dealt with by previous methods. The comparison with four other important machine learning techniques shows that the combined ssEAM/PSO scheme can outperform them on both data sets and the difference in classification accuracy is statistically significant.

With all the improvement we obtain, we also note that there are still many problems that remain to be solved in gene expression profiles-based cancer classification, particularly, the curse of dimensionality, which becomes more serious due to the rapidly and persistently increasing capability of gene chip technologies, in contrast to the limitations in sample collections. Thus, questions, such as how many genes are really needed for disease diagnosis, and whether these gene subsets selected are really meaningful in a biological sense, still remain open. Without any doubt, larger data sets would be immensely useful in effectively evaluating different kinds of classifiers and constructing cancer discrimination systems. In the meantime, more advanced feature selection approaches are required in order to find informative genes that are more efficient in prediction and prognosis.

Acknowledgments

Partial support for this research from the National Science Foundation and from the M.K. Finley Missouri endowment is gratefully acknowledged.

REFERENCES

ALIZADEH, A., M. EISEN, R. DAVIS, C. MA, I. LOSSOS, A. ROSENWALD, J. BOLDRICK, H. SABET, T. TRAN, X. YU, J. POWELL, L. YANG, G. MARTI, T. MOORE, J. HUDSON, Jr., L. LU, D. LEWIS, R. TIBSHIRANI, G. SHERLOCK, W. CHAN, T. GREINER, D. WEISENBURGER, J. ARMITAGE, R. WARNKE, R. LEVY, W. WILSON, M. GREVER, J. BYRD, D. BOSTEIN, P. BROWN, and L. STAUDT (2000). "Distinct types of diffuse large B-cell lymphoma identified by gene expression profiling," *Nature*, Vol. 403, pp. 503–511.

AMBROISE, C. and G. MCLACHLAN (2002). "Selection bias in gene extraction on the basis of microarray gene-expression data," *Proc. Natl. Acad. Sci., USA*, Vol. 99, pp. 6562–6566.

ANAGNOSTOPOULOS, G. C. (2001). Novel Approaches in Adaptive Resonance Theory for Machine Learning, Doctoral Thesis, University of Central Florida, Orlando, Florida.

ANAGNOSTOPOULOS, G. C. and M. GEORGIOPOULOS (2000). "Hypersphere ART and ARTMAP for unsupervised and supervised incremental learning," Proceedings of the IEEE-INNS-ENNS International Joint Conference on Neural Networks (IJCNN'00), pp. 59–64.

ANAGNOSTOPOULOS, G. C. and M. GEORGIOPOULOS (2001). "Ellipsoid ART and ARTMAP for incremental unsupervised and supervised learning," Proceedings of the IEEE-INNS-ENNS Intl. Joint Conf. on Neural Networks (IJCNN'01), pp. 1221–1226.

ANAGNOSTOPOULOS, G. C., M. GEORGIOPOULOS, S. VERZI, and G. HEILEMAN (2002). "Reducing generalization error and category proliferation in ellipsoid ARTMAP via tunable misclassification error tolerance: Boosted ellipsoid ARTMAP," Proceedings of the IEEE-INNS-ENNS International Joint Conference on Neural Networks (IJCNN'02), pp. 2650–2655.

BEN-DOR, A., L. BRUHN, N. FRIEDMAN, I. NACHMAN, M. SCHUMMER, and Z. YAKHINI (2000). "Tissue classification with gene expression profiles," Proceedings of the Fourth Annual International Conference on Computational Molecular Biology, pp. 583–598.

BERGH, F. and A. ENGELBRECHT (2004). "A cooperative approach to particle swarm optimization," *IEEE Trans. Evol. Computat.*, Vol. 8, pp. 225–239.

BERRAR, D., C. DOWNES, and W. DUBITZKY (2003). "Multiclass cancer classification using gene expression profiling and probabilistic neural networks," *Pacific Symp. Biocomput.*, Vol. 8, pp. 5–16.

CARPENTER, G., S. GROSSBERG, N. MARKUZON, J. REYNOLDS, and D. ROSEN (1992). "Fuzzy ARTMAP: A neural network architecture for incremental supervised learning of analog multidimensional maps," *IEEE Trans. Neural Net.*, Vol. 3, pp. 698–713.

CARPENTER, G, S. GROSSBERG, and D. ROSEN (1991). "Fuzzy ART: Fast stable learning and categorization of analog patterns by an adaptive resonance system," *Neural Net.*, Vol. 4, pp. 759–771.

DENG, L., J. PEI, J. MA, and D. LEE (2004). "A Rank Sum Test Method for Informative Gene Discovery," Proceedings of the 2004 ACM SIGKDD International Conference on Knowledge Discovery and Data Mining, pp. 410–419.

DUDA, R., P. HART, and D. STORK (2001). *Pattern Classification* 2nd ed., Wiley, New York.

EBERHART, R. and Y. SHI (2001). "Particle swarm optimization: Developments, applications and resources," Proceedings of the 2001 Congress on Evolutionary Computation, pp. 81–86.

EFRON, B. (1982). *The Jackknife, the Bootstrap and Other Resampling Plans*. Society for Industrial and Applied Mathematics (SIAM), Philadelphia, PA.

GOLUB, T., D. SLONIM, P. TAMAYO, C. HUARD, M. GAASENBEEK, J. MESIROV, H. COLLER, M. LOH, J. DOWNING, M. CALIGIURI, C. BLOOMFIELD, and E. LANDER (1999). "Molecular classification of cancer: class discovery and class prediction by gene expression monitoring," *Science*, Vol. 286, pp. 531–537.

GROSSBERG, S. (1976). "Adaptive pattern recognition and universal encoding II: Feedback, expectation, olfaction, and illusions," *Biol. Cybern.*, Vol. 23, pp. 187–202.

HASTIE, T., R. TIBSHIRANI, and J. FRIEDMAN (2003). *The Elements of Statistical Learning: Data Mining, Inference, and Prediction*. Springer, New York.

HAYKIN, S. (1999). *Neural Networks: A Comprehensive Foundation*, 2nd ed. Prentice Hall, Upper Saddle River, NJ.

HEALY, M. and T. CAUDELL (1997). "Acquiring rule sets as a product of learning in the logical neural architecture LAPART," *IEEE Trans. Neural Net.*, Vol. 8, pp. 461–474.

HEALY, M., T. CAUDELL, and S. SMITH (1993). "A neural architecture for pattern sequence verification through inferencing," *IEEE Trans. Neural Net.*, Vol. 4, pp. 9–20.

JAEGER, J., R. SENGUPTA, and W. RUZZO (2003). "Improved gene selection for classification of microarrays," *Pacific Symp. Biocomput.*, Vol. 8, pp. 53–64.

KENNEDY, J. and R. EBERHART (1997). "A discrete binary version of the particle swarm optimization," *Proc. IEEE Intl. Conf. System, Man, Cybern.*, Vol. 5, pp. 4104–4108.

KENNEDY, J., R. EBERHART, and Y. SHI (2001). *Swarm Intelligence*. Morgan Kaufmann, San Francisco.

KHAN, J., J. WEI, M. RINGNÉR, L. SAAL, M. LADANYI, F. WESTERMANN, F. BERTHOLD, M. SCHWAB, C. ANTONESCU, C. PETERSON, and P. MELTZER (2001). "Classification and diagnostic prediction of cancers using gene expression profiling and artificial neural networks," *Nature Med.*, Vol. 7, pp. 673–679.

KOHAVI, R. (1995). "A study of cross-validation and bootstrap for accuracy estimation and model selection," Proceedings of the 14th International Joint Conference Artificial Intelligence, Morgan Kaufman, San Francisco, pp. 1137–1145.

KOHONEN, T. (2001). *Self-Organizing Maps*, 3rd ed., Springer, Berlin, Heidelberg.

LI, T., C. ZHANG, and M. OGIHARA (2004). "A comparative study of feature selection and multiclass classification methods for tissue classification based on gene expression," *Bioinformatics*, Vol. 20, pp. 2429–2437.

MCLACHLAN, G., K. DO, and C. AMBROISE (2004). *Analyzing Microarray Gene Expression Data*. Wiley, Hoboken, NJ.

NGUYEN, D. and D. ROCKE (2002). "Multi-class cancer classification via partial least squares with gene expression profiles," *Bioinformatics*, Vol. 18, pp. 1216–1226.

OOI, C. and P. TAN (2003). "Genetic algorithms applied to multi-class prediction for the analysis of gene expression data," *Bioinformatics*, Vol. 19, pp. 37–44.

RAMASWAMY, S., P. TAMAYO, R. RIFKIN, S. MUKHERJEE, C. YEANG, M. ANGELO, C. LADD, M. REICH, E. LATULIPPE, J. MESIROV, T. POGGIO, W. GERALD, M. LODA, E. LANDER, and T. GOLUB (2001).

"Multiclass cancer diagnosis using tumor gene expression signatures," *Proc. Natl. Acad. Sci., USA*, Vol. 98, pp. 15149–15154.

SCHERF, U., D. ROSS, M. WALTHAM, L. SMITH, J. LEE, L. TANABE, K. KOHN, W. REINHOLD, T. MYERS, D. ANDREWS, D. SCUDIERO, M. EISEN, E. SAUSVILLE, Y. POMMIER, D. BOTSTEIN, P. BROWN, and J. WEINSTEIN (2000). "A gene expression database for the molecular pharmacology of cancer," *Nat. Genet.*, Vol. 24, pp. 236–244.

SERRANO-GOTARREDONA, T., B. LINARES-BARRANCO, and A. ANDREOU (1998). *Adaptive Resonance Theory Microchip*. Kluwer Academic, Norwell, MA.

SHI, Y. and R. EBERHART (1998). "Parameter selection in particle swarm optimization," Proceedings of the 7th Annual Conference on Evolutionary Programming, pp. 591–601.

SHI, Y. and R. EBERHART (1999). "Empirical study of particle swarm optimization," Proceedings of the 1999 Congress on Evolutionary Computation, pp. 1945–1950.

SPECHT, D. (1990) "Probabilistic neural networks," *Neural Net.*, Vol. 3, pp. 109–118.

WEST, M., C. BLANCHETTE, H. DRESSMAN, E. HUANG, S. ISHIDA, R. SPANG, H. ZUZAN, J. OLSON, J. MARKS, and J. NEVINS (2001). "Prediction the clinical status of human breast cancer by using gene expression profile," *Proc. Natl. Acad. Sci., USA*, Vol. 98, pp. 11462–11467.

WUNSCH II, D. (1991) An Optoelectronic Learning Machine: Invention, Experimentation, Analysis of First Hardware Implementation of the ART1 Neural Network. Ph.D. Dissertation, University of Washington.

WUNSCH II, D., T. CAUDELL, D. CAPPS, R. MARKS, and A. FALK (1993). "An optoelectronic implementation of the adaptive resonance neural network," *IEEE Trans. Neural Net.*, Vol. 4, pp. 673–684.

XU, R., G. ANAGNOSTOPOULOS, and D. WUNSCH II (2006). "Multi-class cancer classification using semi-supervised ellipsoid artmap and particle swarm optimization with gene expression data," *IEEE/ACM Trans. Computat. Biol. Bioinform.*, Vol. 4, pp. 65–77.

XU, R. and D. WUNSCH II (2003). "Probabilistic neural networks for multi-class tissue discrimination with gene expression data," Proceedings of the International Joint Conference on Neural Networks (IJCNN'03), pp. 1696–1701.

XU, R. and D. WUNSCH II (2005). "Survey of clustering algorithms," *IEEE Trans. Neural Net.*, Vol. 16, pp. 645–678.

Chapter 2

Classifying Gene Expression Profiles with Evolutionary Computation

Jin-Hyuk Hong and Sung-Bae Cho

2.1 DNA MICROARRAY DATA CLASSIFICATION

Tumor classification is a major research area in medicine and is important for proper prognosis and treatment (Deutsch, 2003). Conventional cancer diagnosis relies on morphological and clinical analysis. Unfortunately, different types of cancers often show morphologically similar symptoms making diagnosis difficult. The recent advent of the DNA (deoxyribonucleic acid) microarray technology (often called gene chips) provides a means for cancer diagnosis that utilizes gene expression data (Ben-Dor et al., 2000). DNA microarrays measure the transcription activities of thousands of genes simultaneously and produce large-scale gene expression profiles that include valuable information on cellular organization as well as cancer. Although there still remain unsettled issues, the potential applications of the technique are numerous, and many systematic approaches have been actively investigated to classify tumor tissues and to discover useful information on tumors.

The DNA microarrays consist of spatially ordered cDNA (complementary DNA) probes or oligonucleotides (Futschik et al., 2003). Ribonucleic acid (RNA) is extracted from a tissue sample, amplified, and reverse transcribed to dyed cDNA, then hybridized to the probes where the cDNA binds only to specific, complementary, oligonucleotides. A laser is used to excit the dye so that the amount of hybridized cDNA can be measured in terms of fluorescence intensity. The expression value is calculated after subtracting the fluorescence background for each gene monitored (Schmidt and Begley, 2003).

Various DNA microarray data sets can be downloaded from the Internet. Several representative two-class data sets are explored as follows:

Computational Intelligence in Bioinformatics. Edited by G. B. Fogel, D. Corne, and Y. Pan

- A data set consisting of 62 samples of colon epithelial cells taken from colon cancer patients. Each sample was represented as 2000 gene expression levels, which were reduced from the original data of 6000 gene expression levels according to the confidence in the measured expression levels. Forty of 62 samples were labeled as colon cancer and the remaining samples were labeled as normal. Each sample was taken from tumor and normal tissues from the same colon patient and measured using high-density oligonucleotide arrays. For this data set, 31 out of the 62 samples are used as training data and the remaining are used as test data. The data set can be downloaded at http://www.sph.uth.tmc.edu:8052/hgc/default.asp (Alon et al., 1999).
- A data set consisting of 72 samples obtained from 47 acute lymphoblastic leukemia (ALL) patients and 25 acute myeloid leukemia (AML) patients. Each sample had 7129 gene expression levels. Researchers usually use 38 samples for training and the remaining 34 samples for testing. The data are available for download at http://www-genome.wi.mit.edu/cgi-bin/cancer/publications/pub_paper.cgi?mode=view&paper_id=43 (Golub et al., 1999).
- A data set composed of 47 samples from patients with diffuse large B-cell lymphoma (DLBCL), the common subtype of non-Hodgkin's lymphoma. Twenty-four samples were grouped into the germinal center B-like, and the remaining samples were grouped as the activated B-like. Each sample was represented by 4026 gene expression levels. Twenty-two samples are often used as training samples, and 25 samples are used as test samples. The data set is accessed at http://llmpp.nih.gov/lymphoma (Alizadeh et al., 2000).
- A data set consisting of 31 malignant pleural mesothelioma (MPM) and 150 adenocarcinoma (ADCA) samples obtained from lung tissue. Each tissue has 12,533 gene expressions. Thirty-two samples are typically used for training and the remaining for testing. The data set can be downloaded at http://www.chestsurg.org (Gordon et al., 2002).
- A data set consisting of 97 samples, 46 of which are from patients who developed metastases within 5 years of first indication of cancer while the remaining samples are collected from patients who remained healthy after 5 years. Each sample was composed of 24,481 gene expression levels. Usually, the data set is separated into 78 training samples and 19 testing samples. It can be downloaded from http://www.rii.com/publications/2002/vantveer.htm (van't Veer et al., 2002).
- A data set consisting of 102 samples collected from 52 prostate tumor patients and 50 nonprostate people by using oligonucleotide microarrays containing probes for 12,600 genes. The raw data is available at http://www-genome.wi.mit.edu/MPR/prostate (Singh et al., 2002).
- A data set including 60 tissues to analyze the outcome of cancer treatment, containing 21 survivors and 39 patients who succumbed to the disease. Each

tissue was represented as 7129 gene expression levels. It can be accessed through the website http://www-genome.wi.mit.edu/mpr/CNS (Pomeroy et al., 2002).

Recently, many authors have investigated multiclass cancer classification using gene expression profiles. Several benchmark data sets include:

- NCI60 data set consisting of 61 tissues of 9 classes according to the site of origin such as breast (7), central nervous system (6), colon (7), leukemia (6), melanoma (8), non-small-cell lung carcinoma (9), ovarian (6), renal (8) and reproductive (4). Each sample was represented by using 9703 spotted cDNA sequences. It is available at http://genome-www.stanford.edu/sutech/download/nci60 (Ross et al., 2000).
- A leukemia data set can be used as the multiclass cancer data set by dividing acute lymphoblastic leukemia samples into two subtypes of B cell and T cell. There are 38 ALL-B-cell tissues, 9 ALL-T-cell tissues, and 25 AML tissues (Golub et al., 1999).
- A small, round blue-cell tumor (SRBCT) data set consists of 83 samples of four childhood malignancies including neuroblastoma (NB), rhabdomyosarcoma (RMS), non-Hodgkin lymphoma (NHL), and the Ewing family of tumors (EWS). Each sample has 2308 gene expression levels from cDNA experiments. Researchers often compose the data set into 63 training samples and 20 testing samples. The data set can be downloaded at http://research.nhgri.nih.gov/microarray/Supplement (Khan et al., 2001).
- A lung carcinoma data set consisting of 203 samples including 139 lung adenocarcinomas (AD), 21 squamous (SQ) cell carcinomas, 20 pulmonary carcinoid (COID) tumors, 6 small-cell lung cancers (SCLC) and 17 normal lung tissues. Each sample contains 12,600 mRNA (messenger RNA) expression levels. It can be downloaded at http://research.dfci.harvard.edu/meyersonlab/lungca.html (Bhattacherjee et al., 2001).
- A global cancer map (GCM) data set composed of 198 samples of 14 different tumor types including breast adenocarcinoma, prostate, lung adenocarcinoma, colorectal adenocarcinoma, lymphoma, bladder, melanoma, uterine adenocarcinoma, leukemia, renal cell carcinoma, pancreatic adenocarcinoma, ovarian adenocarcinoma, pleural mesothelioma, and central nervous system. Each sample has 16,063 gene expression levels. Researchers often use 144 samples for the training and 54 for the test. It is available at http://www-genome.wi.mit.edu/mpr/publications/projects/Global_Cancer_Map (Ramaswamy et al., 2001).

The DNA microarray data often consists of a large number of genes but few samples. Since not all the genes in the data are related with cancer classification, many data sets include significant noise as well as useful information on cancer classification. Moreover, the low availability of samples makes the classification task more difficult (Brazma and Vilo, 2000).

Research for cancer classification based on gene expression profiles aims to find the functionality of genes related to cancers. The usual focus of previous studies in this area can be summarized as follows:

1. *Selecting features*: Determine informative genes for the target cancer classification to reduce the dimensionality of the data.
2. *Learning classifiers*: Construct a classifier with the selected genes by using various techniques including neural networks, decision trees, support vector machines, and the like.
3. *Discovering knowledge*: Understand relationships among genes and cancer from the result and classifiers learned.

In selecting features, it is important to identify a small subset of genes sufficiently informative to distinguish cancers for diagnostic purposes (Inza et al., 2001; Li and Yang, 2002). Two representative approaches such as filter and wrapper approaches are used widely for selecting genes. The filter approach selects informative features (genes) regardless of classifiers by independently measuring the importance of features (Cho and Ryu, 2002), while the wrapper approach selects features with classifiers by simultaneously using some features with training classifiers to produce the optimal combination of features and classifiers (Deutsch, 2003; Inza et al., 2001).

Most filter methods consider the features as being independent to avoid additional computational overhead. In doing so, filter methods are fraught with information loss. On the other hand, wrapper methods consider mutual dependency among features during model development. Conventional wrapper methods using evolutionary algorithms are applied to feature selection of small- or middle-scale feature data sets (Deb and Goldberg, 1989; Kohavi and John, 1997). It is hard to apply these approaches directly to huge-scale feature data sets due to significant processing time and low efficiency to optimize the number of features used (Bins and Draper, 2001).

Various machine learning techniques such as neural networks, Bayesian approaches, support vector machines, and decision trees have been actively investigated as cancer classifiers. Neural networks and support vector machines show a good performance in cancer classification (Ramaswamy et al., 2001), while Bayesian approaches and decision trees provide interpretability for the obtained classifier. Evolutionary techniques have also been used to analyze gene expression data by discovering classification rules or constructing an optimal ensemble classifier (Ando and Iba, 2004; Langdon and Buxton, 2004; Hong and Cho, 2006b). Usually, learning classifiers often involves the feature selection as a preceding step (Cho and Ryu, 2002).

Clinicians sometimes demand that classifiers be not only highly accurate but also interpretable. Therefore, it is important to discover useful knowledge from data sets such as a relationship between genes and diseases. Bayesian networks, decision trees, and evolutionary methods have been applied for that purpose.

2.2 EVOLUTIONARY APPROACH TO THE PROBLEM

Evolution is a remarkable problem-solving method proposed by several research groups such as Rechenberg and Schewefel (Schwefel, 1981), Fogel (Fogel et al., 1966), and Holland (Holland, 1975), which mimics natural evolution to solve problems in a wide variety of domains (Goldberg, 1998). Evolutionary algorithms (EAs) utilize solutions in a multidimensional space containing a hypersurface known as the fitness surface. In the case of genetic algorithms (GAs), a population of solutions is maintained in the form of chromosomes encoded as strings. The strings are converted into problem solutions to be evaluated according to an objective scoring function. Following fitness evaluation, a new population of chromosomes is produced by applying a set of genetic operators such as selection, crossover, and mutation to the original population. A new population is composed of parent chromosomes directly proportional to their fitness. Crossover is used to recombine genetic material of two parent chromosomes to produce offspring. Provided two chromosomes $a =$ (1 1 1 1) and $b =$ (0 0 0 0), one-point crossover at the third point produces two new chromosomes, $a' =$ (1 1 0 0) and $b' =$ (0 0 1 1). Finally, mutation can be used at a random position of a random string and negates the bit value. For instance, if mutation is applied to the fourth bit of string a', the transformed string becomes (1 1 0 1). Typical evolutionary algorithms repeat evolution until satisfying a terminal condition as the following.

```
t = 0
Initialization Population P(t);
Evaluate P(t);
while not done do
    t = t + 1;
    P' = SelectParents P(t);
    Recombine P'(t);
    Mutate P'(t);
    Evaluate P'(t);
    P = Survive P, P'(t);
end_while
```

In summary, evolutionary approaches comprise a set of individual elements (the population) and a set of biologically inspired operators defined over the population itself. According to evolutionary theories, only the most suited elements in a population are likely to survive and generate offspring, thus offering the information they represent to new generations. Table 2.1 summarizes the state-of-the-art of evolutionary computation in cancer classification using DNA microarray data.

Evolutionary approaches were often used to select informative features working together with a classifier like support vector machines. Evolutionary neural networks have been applied to classifying colon cancers, while recently Langdon and Buxton (2004) and Hong and Cho (2006b) proposed a novel method for learning classification rules using genetic programming (GP).

Table 2.1 State-of-the-Art Evolutionary Computation in Analyzing Gene Expression Data

Author	Method	Object	Data Set
Deb and Reddy (2003)	EA	Feature selection	Leukemia, lymphoma, colon
Deutsch (2003)	EA	Feature selection	SRBCT, leukemia, lymphoma, GCM
Ando and Iba (2004)	GA	Feature selection	Leukemia, colon
Liu et al. (2005)	GA	Feature selection	NCI60, Brown
Ooi and Tan (2003)	GA	Feature selection	NCI60, GCM
Paul and Iba (2005)	GA	Feature selection	Lung, brain, prostate
Peng et al. (2003)	GA	Feature selection	NCI60, GCM, leukemia, colon
Hong and Cho (2006a)	GA	Feature selection	Leukemia, lymphoma, GCM
Futschik et al. (2003)	EFuNN	Classifier learning	Colon
Kim and Cho (2004)	ENN	Classifier learning	Colon
Hong and Cho (2006b)	GP	Rule learning	Lymphoma, lung, ovarian
Langdon and Buxton (2004)	GP	Rule learning	Central nervous system (CNS)

Deb and Reddy (2003) applied a multiobjective evolutionary algorithm to minimizing gene subset size. In order to identify a proper gene subset, three objectives were defined such as minimizing the size of gene subset in the classifier, minimizing the number of mismatches in the training samples, and minimizing the number of mismatches in the test samples. A simple binary string whose length is the number of genes was employed. Elitist selection and an explicit diversity preserving mechanism were used to enhance the optimization process. Experiments were conducted for three cancer data sets such as leukemia, lymphoma, and colon cancer, where some informative genes are selected.

Deutsch (2003) proposed a replication algorithm that evolved an ensemble of predictors using different subsets of genes. Initializing an ensemble of different genes, an evolutionary algorithm was used to produce a better set of predictors by means of the scoring function. A *k*-nearest-neighbor approach was used as a baseline measure of fitness. Four representative cancer data sets including multiclass data were used to verify the approach, and the proposed method worked well as compared with conventional approaches.

Wrapper approaches that use GAs have been often used to select useful genes. Ando and Iba (2004) compared the simple GA, artificial immune system, correlation metric, and random selection for feature selection, and used *k*-nearest-neighbor, support vector machines, and naïve Bayes classifier as base classifiers. On a set of benchmark data, the combination of naïve Bayes classifiers and artificial immune system yielded the best results.

Liu et al. (2005) applied the GAs to multiclass cancer classification. They selected informative genes using a GA and constructed a classifier using all paired support vector machines. The method selected only 40 genes from the NCI60 data set and produced a leave-one-out cross-validation (LOOCV) accuracy of 88.5%. On the other hand, Ooi and Tan (2003) used a maximum-likelihood classification method instead of support vector machines. They defined the fitness function with a cross-validation error rate and independent test error rate, which might produce a biased result. The approach was validated on NCI60 and GCM data sets and showed a comparative result with previous studies. In addition, Peng et al. (2003) generated similar experiments to Ooi and Tan (2003), but recursive feature elimination was employed to yield a compact gene subset.

Paul and Iba (2005) recently proposed a gene selection algorithm using the GA. Instead of applying crossover and mutation operators, they generated new possible individuals by sampling a probability distribution. Three representative microarray data sets were used to verify the method, and it showed improved performance compared to a filter approach with signal-to-noise ratio for ranking genes.

Hong and Cho (2006a) proposed a wrapper approach that uses a speciated GA. To improve the performance of the GA, the explicit fitness sharing was used to maintain the diversity of the population. They demonstrated that the method was superior to feature selection using the simple GA.

In contrast to the above examples, some researchers have attempted to apply EAs to optimize classifiers or to discover classification rules. For example, Futschik et al. (2003) proposed evolving fuzzy neural networks to classify cancer tissues. The evolving connectionist system paradigm that was recently introduced by Kasabov was imposed into the method (Kasabov, 2001). Moreover, to provide knowledge about the classification process, fuzzy logic rules that present genes strongly associated with specific types of cancer were extracted easily from the trained networks. The approach was validated on leukemia and colon cancer data sets and not only produced a high accuracy but also provided useful information on those cancers.

Kim and Cho (2004) applied an evolutionary neural network to predict colon cancer. After selecting genes by utilizing information gain, they evolved neural networks with the genes selected. A matrix and the corresponding genetic operators were defined to represent a neural network. Through the experiment, it produced a neural network with an accuracy of 94% on classifying the colon cancer data set.

Hong and Cho (2006b) used GP to produce classification rules and proposed the structural diversity among them for ensembling. A subset of diverse classification rules obtained by GP was combined using a fusion scheme. The method was verified on three representative data sets and produced better performance than conventional approaches.

Langdon and Buxton (2004) also proposed a rule discovering method by using GP. Various arithmetic functions were employed to construct a rule and the size of rule was limited at 5 or 9. The experiment on central nervous systems showed the usefulness of mining genes from genetic programming.

2.3 GENE SELECTION WITH SPECIATED GENETIC ALGORITHM

The first case study in this chapter focuses on gene selection using the speciated GA (sGA). As mentioned before, the GA is often used to select informative genes working together with classifiers to consider the mutual dependency among them. However, conventional wrapper methods using the GA are applied to feature selection of small- or middle-scale feature data set, so it is hard to apply them directly to huge-scale feature data sets like gene expression profiles due to much processing time and low efficiency in reducing the number of features used (Hong and Cho, 2006a).

To deal with these problems, an sGA, which can improve the evolution efficiency, was used to select genes from huge-scale gene expression profiles. Speciation might help to obtain diverse gene subsets for classifying cancers rather than conventional evolution that does not consider the diversity among populations (Deb and Goldberg, 1989). Moreover, the representation of chromosomes was modified to be suitable for huge-scale feature selection.

Since many applications have various suboptimal solutions in search space as shown in Figure 2.1, speciation is useful to obtain multiple species in evolutionary methods by restricting an individual to mate only with similar ones or manipulating its fitness with niching pressure at selection. Crowding, local mating, fitness sharing, and island GAs are representative speciation methods in an EA. In particular, explicit fitness sharing limits the number of individuals residing in one region of the fitness

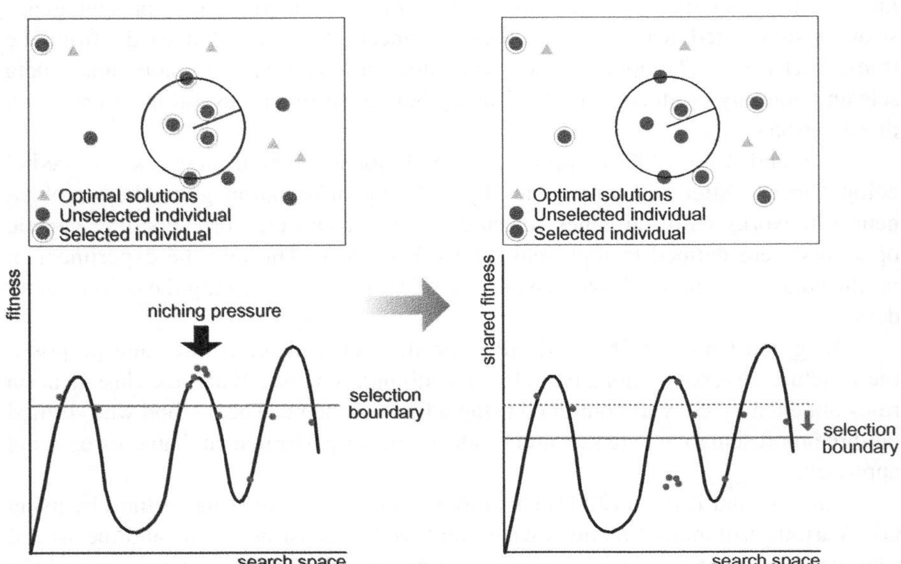

Figure 2.1 Example of explicit fitness sharing in fitness landscape.

landscape by sharing fitness (resource) with similar individuals (species). This leads to avoid converging on a point, which is often appeared when using the simple GA.

As shown in Figure 2.1, the selection boundary is moved by sharing the fitness so as to give more chances to various individuals to be selected. It reduces the fitness of individuals in densely populated regions to the amount divided into the number of similar individuals as follows:

$$\mathrm{sf}_i = \frac{f_i}{m_i}, \tag{2.1}$$

where sf_i is the shared fitness of an individual i and m_i is the number of individuals in the population within a fixed distance. The niche count m_i is calculated as follows:

$$m_i = \sum_{j=1}^{N} \mathrm{sh}(d_{ij}), \quad \mathrm{sh}(d_{ij}) = \begin{cases} 1-\left(\dfrac{d_{ij}}{\sigma_s}\right)^{\alpha} & \text{for } 0 \le d_{ij} < \sigma_s, \\ 0 & \text{for } d_{ij} \ge \sigma_s, \end{cases} \tag{2.2}$$

where N denotes the population size and d_{ij} indicates the distance between the individuals i and j. The sharing function (sh) returns "1" when the members are regarded as identical or "0" when the distance is greater than a threshold σ_s (sharing radius); α is a constant to regulate the shape of the sharing function, where the performance of speciation is not sensitive to the value of σ_s (set as 5.0 for the work presented in this chapter).

The distance d_{ij} can be determined by a similarity metric based on either genotypic or phenotypic similarity. Genotypic similarity is obtained with Hamming distance based on bit-string representation, while phenotypic similarity is determined from real parameters of the search space such as Euclidian distance.

Conventional encoding schemes that encode all possible features are not suitable to large-scale feature selection. Therefore, a modified representation as shown in Figure 2.2 is used, which fixes the number of features involved. The number of selected features n_s can vary according to the number of all features n_f, where n_s is set as 25 for the work presented in this chapter. The fitness of an individual is

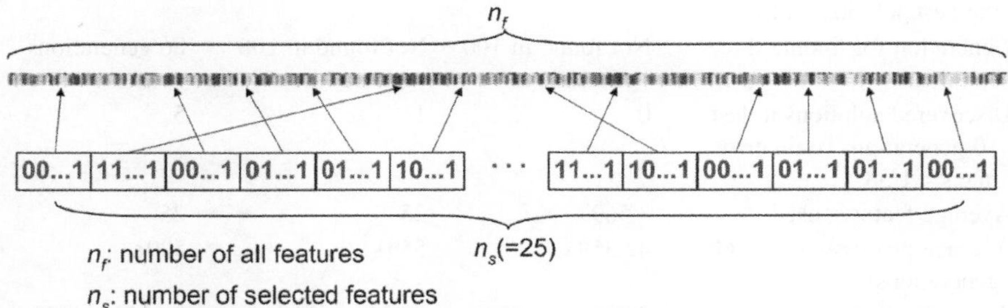

Figure 2.2 Chromosome representation used in work presented in this chapter.

measured by classifying training samples with neural networks that use features selected according to the chromosome representation.

To verify this proposed method, we utilized a leukemia cancer data set. As described in Section 2.1, this data set consisted of 72 samples including 7129 gene expression levels, among which 38 samples are used for training and the remaining 34 samples for testing. The parameters of the experiment were as shown in Table 2.2, and the experiment was repeated 10 times.

Table 2.3 shows the performance of the proposed method in comparison with two conventional GA-based feature selection methods, one of which uses a traditional encoding scheme (method 1) and the other uses the proposed encoding scheme but not speciation (method 2). The proposed method shows improved performance relative to the other methods in discovering optimal solutions, while method 2 yields a larger deviation than the proposed method due to the characteristics of speciation. Speciation might be useful to get diverse solutions and it helps to improve the performance of classification. Despite much processing time, the simple GA with the traditional encoding scheme does not show any effect on reducing the dimensionality of the data set as shown in Figure 2.3.

Table 2.2 Parameters of Experimental Environment

Genetic Operator	Value	Neural Network	Value
Population size	50	Learning rate	[0.1, 03]
Selection rate	[0.6, 0.8]	Momentum	[0.5, 0.8]
Crossover rate	[0.6, 0.8]	Maximum iteration	[500, 50,000]
Mutation rate	[0.01, 0.2]	Minimum error	0.02
		Hidden node number	[2, 5]

Table 2.3 Comparison of Performance for Leukemia (10 runs on average)

Measure	Method 1	Method 2	Proposed Method
Average training error rate (std.)	0.342 (±0.02)	0.210 (±0.03)	0.236 (±0.03)
Average training error rate of the best solution (std.)	0.211 (±0.06)	0.008 (±0.07)	0.026 (±0.08)
Generation for finding three solutions	Not found in 100 generations	Not found in 100 generations	86 generations
Discovered solutions in last 10 generations (train error rate <= 1/38)	0	1	5
Average features used	3,569	25	25
Average processing time (10 generations)	42,359 s	559 s	590 s
Average test error rate (std.)	0.3006 (±0.03)	0.3401 (±0.03)	0.2252 (±0.09)

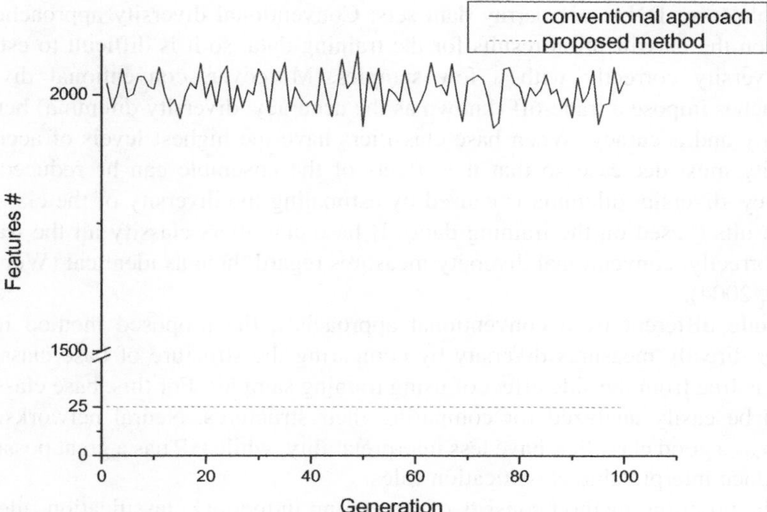

Figure 2.3 Number of features used for leukemia cancer classification.

2.4 CANCER CLASSIFICATION BASED ON ENSEMBLE GENETIC PROGRAMMING

A second case study was used to discover and combine classification rules using GP. Constructing a classifier system and discovering useful knowledge from data sets are very important issues in bioinformatics. Machine learning techniques are apt to be overfitted for gene expression profiles that consist of only a few samples with a large number of genes. Therefore, ensemble approaches offer increased accuracy and reliability when dealing with such problems, and various methods for combining multiple classifiers have been investigated in the past decade (Kuncheva, 2002). One of main issues in ensembling is how best to generate diverse base classifiers since performance improvements can be obtained only when the base classifiers are complementary.

Estimating the diversity among base classifiers has been actively investigated in the field of ensemble techniques to select most diverse ones for constructing the ensemble classifier. Various diversity measures from pairwise to nonpairwise such as the *Q* statistic, the correlation coefficient, the Kohavi–Wolpert variance, and the entropy measure have been proposed. Zenobi and Cunningham (2001) used diversity based on different feature subsets. Shipp and Kuncheva (2002) analyzed relationships between fusion methods and diversity measures, and Kuncheva and Whitaker (2003) compared various diversity measures in classifier ensembles. Windeatt (2005) conducted an empirical analysis on diversity measures for the multiple classifier system.

Even though there are many diversity measures, it is very difficult to apply them to cancer classification using gene expression profiles since only a few samples are available from DNA microarray data sets. Conventional diversity approaches are based on the classification results for the training data, so it is difficult to estimate the diversity correctly with a few samples. Moreover, conventional diversity approaches impose a trade-off (known as the accuracy–diversity dilemma) between diversity and accuracy. When base classifiers have the highest levels of accuracy, diversity must decrease so that the effects of the ensemble can be reduced. The accuracy–diversity dilemma is caused by estimating the diversity of the classification results (based on the training data). If base classifiers classify all the training data correctly, conventional diversity measures regard them as identical (Webb and Zheng, 2004).

Quite different from conventional approaches, the proposed method in this chapter directly measures diversity by comparing the structure of base classifiers, which is free from the side effect of using training samples. For this, base classifiers should be easily analyzed for comparing their structures. Neural networks, well known as a good classifier, have less interpretability, while GP has a great possibility to produce interpretable classification rules.

The proposed method consists of generating individual classification rules and combining them to construct an ensemble classifier. Individual classification rules are generated by using GP after reducing the dimensionality of data sets with Pearson correlation (Radonić et al., 2005). According to their diversity, a set of diverse classification rules is selected to consist of an ensemble classifier. The overall procedure of the proposed method is as follows:

```
Original data set
```

$$\mathrm{OD} = \{\{d_1^1, d_2^1, d_3^1, \ldots, d_f^1\}, \{d_1^2, d_2^2, d_3^2, \ldots, d_f^2\}, \ldots, \{d_1^n, d_2^n, d_3^n, \ldots, d_f^n\}\}$$

```
(f: # of genes, n: # of samples)
A set of extracted rules R = {r1, r2, r3, ... , rm}
(m: # of rules in the pool)
Diversity for DIV = {div1,div2,div3, ... ,divm}
A set of selected rules S = {s1, s2, s3, ... , sk} (k: #
of rules used in ensembling)

D = feature_selection(OD);
for(i=0; i<m; i++) ri = rule_generation(D);
for(i=0; i<m; i++)
for(j=0; j<m; j++)
{
divi += different_feature_number(ri, rj) - common_
feature_number(ri, rj) - α × edit_distance(ri, rj);
divj += different_feature_number(ri, rj) - common_
feature_number(ri, rj) - α × edit_distance(ri, rj);
}
sort(R, DIV);
for(i=0; i<k; i++) si = ri;
```

Two ideal markers are defined to select informative genes by scoring the respective similarity with each ideal marker. The first marker consists of 1 for all the samples in class A and 0 for all the samples in class B, while the second marker is composed of 0 for all the samples in class A and 1 for all the samples in class B.

Pearson correlation is employed to estimate the similarity between genes and ideal markers as follows:

$$\mathrm{PC} = \frac{\sum_{i=1}^{n}(\mathrm{ideal}_i g_i) - \frac{\left(\sum_{i=1}^{n}\mathrm{ideal}_i \sum_{i=1}^{n} g_i\right)}{n}}{\sqrt{\left[\sum_{i=1}^{n}\mathrm{ideal}_i^2 - \frac{\left(\sum_{i=1}^{n}\mathrm{ideal}_i\right)^2}{n}\right]\left[\sum_{i=1}^{n} g_i^2 - \frac{\left(\sum_{i=1}^{n} g_i\right)^2}{n}\right]}}, \tag{2.3}$$

After selecting informative genes, GP generates cancer classification rules using arithmetic operations employed to figure out the regulation of genes. For an individual that means a classification rule is represented as a tree including the function set {+, −, ×, ÷} and the terminal set $\{f_1, f_2, \ldots, f_n, \text{constant}\}$, where n is the number of features. Finally, the rule set P is defined as follows:

```
G={V={EXP, OP, VAR}, T={+, −, ×, ÷, f1, f2, ..., fn,
constant}, P, {EXP}}
EXP→EXP OP EXP | VAR
OP→+|−|×|÷
VAR→ f1|f2| ... |fn|constant
```

Since GP uses trees as the basic representation of individuals, operators such as crossover, mutation, and permutation are conducted (Fig. 2.4). Crossover randomly selects and changes subtrees from two individuals, mutation changes a subtree into new one, and permutation exchanges two subtrees of an individual. All genetic operations are conducted according to predefined probabilities.

Design of the fitness function is based on the classification accuracy and simplicity of a classification rule as the following formula, where an instance is

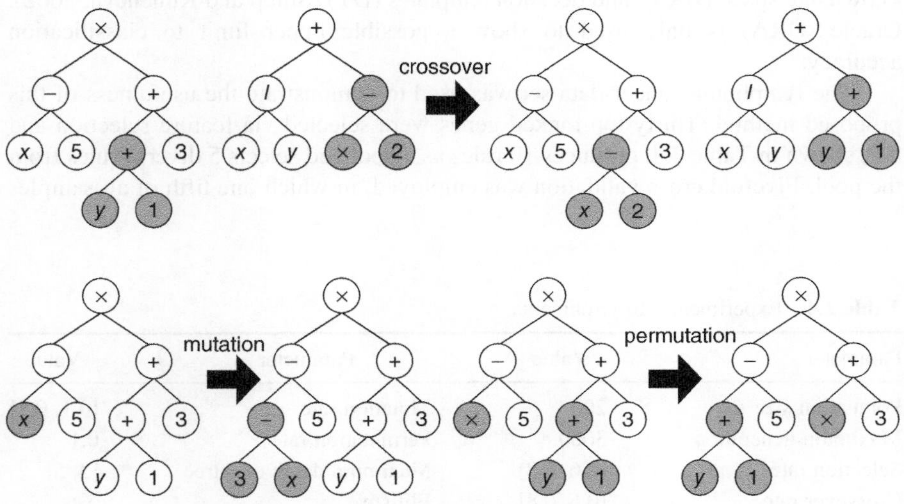

Figure 2.4 Variation operations used in GP.

classified into class 1 if the evaluated value for the rule is larger than 0, otherwise it is classified into class 2. The notion of Occam's razor also supports the introduction of simplicity (Brameier and Banzhaf, 2001).

$$\text{Fitness of individual}_i = \frac{\text{number of correct samples}}{\text{number of total train data}} \times w_1 + \text{simplicity} \times w_2, \qquad (2.4)$$

where simplicity = number of nodes/number of maximum nodes

w_1 = weight for training rate

w_2 = weight for simplicity

As mentioned before, the diversity among base classification rules is measured by comparing their structures, so the proposed method does not require any training data in estimating the diversity. Among several diversity measures used in genetic programming such as pseudo-isomorphs and edit distance, a simplified edit distance is used to calculate the diversity between two classification rules r_i and r_j as follows:

$$\text{distance}\,(r_i, r_j) = \begin{cases} d(p, q) & \text{if neither } r_i \text{ nor } r_j \text{ have any children,} \\ d(p, q) + \text{distance(RS of } r_i\text{, RS of } r_j) + \text{distance(LS of } r_i\text{,} \\ \quad \text{LS of } r_j) \quad \text{otherwise,} \end{cases}$$

where RS means right subtree and LS means left subtree and

$$d(p, q) = \begin{cases} 1 & \text{if } p \text{ and } q \text{ overlap,} \\ 0 & \text{if } p \text{ and } q \text{ do not overlap.} \end{cases}$$

Once it selects the most diverse subset of classification rules, a fusion method such as majority vote (MAJ), maximum (MAX), average (AVG), behavior-knowledge space (BKS), and decision templates (DT) (Shipp and Kuncheva, 2002). Oracle (ORA) is only used to show a possible upper limit to classification accuracy.

The lymphoma cancer data set was used to demonstrate the usefulness of this proposed method. Thirty top-ranked genes were selected via feature selection and GP, shown in Table 2.4, produces 10 rules as a pool and selects 5 diverse rules from the pool. Fivefold cross validation was employed, in which one fifth of all samples

Table 2.4 Experimental Environments

Parameter	Value	Parameter	Value
Population size	200	Mutation rate	[0.1, 0.3]
Maximum generation	3000	Permutation rate	0.1
Selection rate	[0.6, 0.8]	Maximum depth of a tree	[3, 5]
Crossover rate	[0.6, 0.8]	Elitism	Yes

was used as test data and the others as training data. This process was repeated 10 times to obtain the average results with 50 (5 × 10) experiments in total.

Table 2.5 summarizes the predictive accuracy of the proposed method; the highlighted values represent high accuracy. "10 classifiers" and "5 classifiers" are based on ensembling from random forest, while "5 diverse classifiers" is the result of the proposed method. The results show that the ensemble classifier performs better than the individual classifier. In most cases, about 5% increments are observed when using ensemble techniques. In addition, the proposed method shows superior classification performance when combining the 10 rules and the 5 rules. This signifies that considering diversity improves the performance of the ensemble.

The 10-rule ensemble is better than the 5-rule ensemble, and the proposed approach not only supports the same degree of useful information with the ensemble that uses 10 classification rules but also minimizes the increment of the error.

The relationship between the number of base classification rules and the performance of the ensemble is examined and shown in Figure 2.5. Performance

Table 2.5 Test Accuracy on Lymphoma Cancer Data Set (%)

1. Fusion Method	2. 10 Rules	3. 5 Rules	4. 5 Diverse Rules	5. Individual Rule
6. MAJ	7. 95.0	8. 93.2	9. 97.1	
11. MAX	12. 96.7	13. 95.2	14. 97.6	
15. AVG	16. 96.7	17. 94.4	18. 96.9	10. 91.3
19. BKS	20. 95.0	21. 93.2	22. 97.1	
23. DT	24. 95.8	25. 95.1	26. 97.0	
27. ORA	28. 100	29. 100	30. 100	

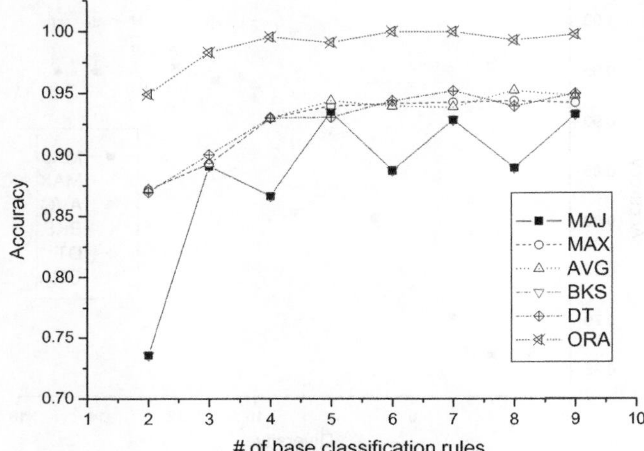

Figure 2.5 Test accuracy according to number of base classification rules.

increases with the number of base classification rules, but it is almost converged when using 4–6 rules. BKS shows an oscillation between even and odd numbers. When the ensemble uses the odd number of rules, it shows better performance than when using even numbers.

The relationship between diversity and performance is also analyzed and shown in Figure 2.6. The results indicate that classification accuracy increases according to the increment of diversity in most cases. A decline in accuracy occasionally appears because diversity is apt to increase when there is a peculiar rule. This can be solved by a nonpairwise approach for estimating diversity in ensemble GP.

Experiments on conventional diversity measurements such as the Q statistic, correlation coefficient, disagreement measure, double-fault measure, Kohavi–Wolpert variance, measurement of interrater agreement, entropy measure, measure of difficulty, generalized diversity, and coincident failure diversity were conducted to compare with the proposed method. These diversity measurements are explained in Shipp and Kuncheva, (2002).

As mentioned before, conventional diversity measurements have limitations in applying cancer classification when using gene expression data. They require a number of samples to measure diversity correctly, and there is less enhancement when base classifiers are highly accurate. Most gene expression data sets provide only a few samples, and the classification rules obtained by genetic programming produce high accuracy as shown in Table 2.5. In most cases, the individual classification rules obtain 100% accuracy for training data, so conventional diversity measurements output only 0 for all possible combinations of classification rules. Instead, since the proposed method does not consider the training data, it produces improved performance in the ensemble. Figure 2.7 shows the proposed method compared to conventional diversity measurements. In most cases, classification rules classify all training data correctly. The proposed method obtained higher accuracy than the

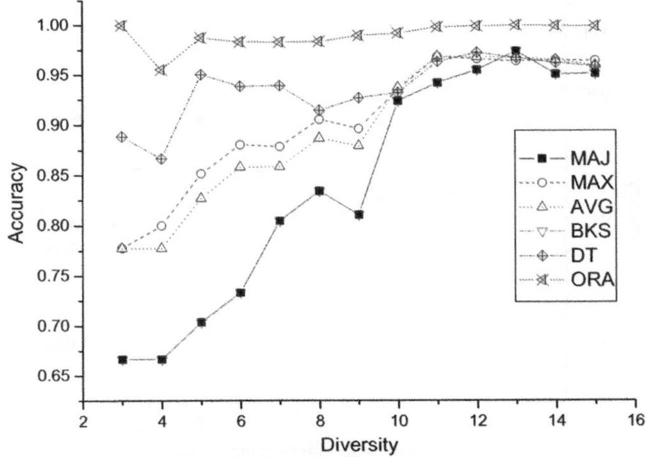

Figure 2.6 Predictive accuracy according to diversity.

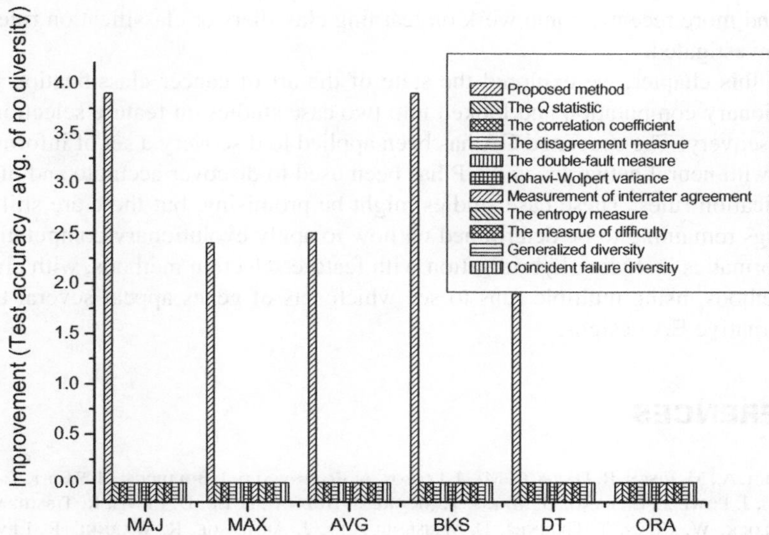

Figure 2.7 Improvement of accuracy according to diversity measurement.

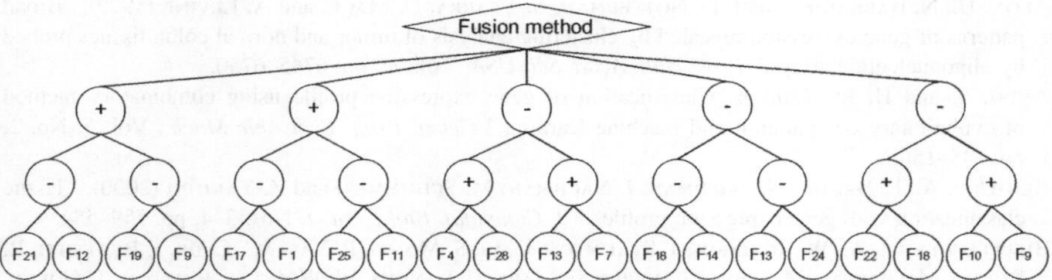

Figure 2.8 Ensemble classifier obtained by proposed method.

ensemble without diversity while the other diversity measurements did not provide any improved accuracy. Figure 2.8 shows an example of a set of rules developed by the proposed method.

2.5 CONCLUSION

Cancer classification using DNA microarray data has been regarded as an important issue in bioinformatics, and many researchers have developed a method for analyzing the data. Reducing the dimensionality of data sets by selecting informative genes, optimizing a classifier of a high accuracy, and discovering useful knowledge on the cancer and genes are important issues. Computational intelligence approaches have a great ability of not only producing a high accuracy but also providing useful knowledge in cancer classification. There are many studies on gene selection using

GAs and more recently some work on learning classifiers or classification rules has been investigated.

In this chapter, we explored the state of the art of cancer classification using evolutionary computation and looked into two case studies on feature selection and rule discovery. The speciated GA has been applied to discovery a set of informative genes with neural networks, and GP has been used to discover accurate and diverse classification rules. These case studies might be promising, but there are still a lot of things remaining to be determined on how to apply evolutionary computation in bioinformatics including hybridization with feature selection methods, with clustering methods, using multiple runs to see which sets of genes appear several times, or alternative EA designs.

REFERENCES

Alizadeh, A., M. Eisen, R. Davis, C. Ma, I. Lossos, A. Rosenwald, J. Boldrick, H. Sabet, T. Tran, X. Yu, J. Powell, L. Yang, G. Marti, T. Moore, J. Hudson, L. Lu, D. Lewis, R. Tibshirani, G. Sherlock, W. Chan, T. Greiner, D. Weisenburger, J. Armitage, R. Warnke, R. Levy, W. Wilson, M. Grever, J. Byrd, D. Botstein, P. Brown, L. Staudt (2000). "Distinct types of diffuse large B-cell lymphoma identified by gene expression profiling," *Nature*, Vol. 403, No. 6769, pp. 503–511.

Alon, U., N. Barkai, K. Gish, D. Notterman, S. Ybarra, D. Mack, and A. Levine (1999). "Broad patterns of gene expression revealed by clustering analysis of tumor and normal colon tissues probed by oligonucleotide arrays," *Proc. Natl. Acad. Sci. USA*, Vol. 96, pp. 6745–6750.

Ando, S. and H. Iba (2004). "Classification of gene expression profile using combinatory method of evolutionary computation and machine learning," *Genet. Prog. Evolvable Mach.*, Vol. 5, No. 2, pp. 145–156.

Ben-Dor, A., L. Bruhn, N. Friedman, I. Nachman, M. Schummer, and Z. Yakhini (2000). "Tissue classification with gene expression profiles," *J. Computat. Biol.*, Vol. 7, Nos. 3–4, pp. 559–584.

Bhattacherjee, A., W. Richards, J. Staunton, C. Li, S. Monti, P. Vasa, C. Ladd, J. Beheshti, R. Bueno, M. Gillete, M. Loda, G. Weber, E. Mark, E. Lander, W. Wong, B. Johnson, T. Golub, D. Sugarbaker, and M. Meyerson (2001). "Classification of human lung carcinomas by mRNA expression profiling reveals distinct adenocarcinoma subclasses," *Proc. Natl. Acad. Sci. USA*, Vol. 98, pp. 13790–13795.

Bins J. and B. Draper (2001). "Feature selection from huge feature sets," *Proc. Int. Conf. Computer Vision*, Vol. 2, pp. 159–165.

Brameier, M. and W. Banzhaf (2001). "A comparison of linear genetic programming and neural networks in medical data mining," *IEEE Trans. Evolut. Comput.*, Vol. 5, No. 1, pp. 17–26.

Brazma, A. and J. Vilo (2000). "Gene expression data analysis," *Fed. Europ. Biochemi. Soc. Lett.*, Vol. 480, No. 1, pp. 17–24.

Cho, S.-B. and J.-W. Ryu (2002). "Classifying gene expression data of cancer using classifier ensemble with mutually exclusive features," *Proc. the IEEE*, Vol. 90, No. 11, pp. 1744–1753.

Deb, K. and D. Goldberg (1989). "An investigation of niche and species formation in genetic function optimization," in *Proc. 3rd Int. Conf. Genetic Algorithms*, J. D. Schaffer, Ed. Morgan Kaufmann, San Mateo, CA, pp. 42–50.

Deb, K. and A. Reddy (2003). "Reliable classification of two-class cancer data using evolutionary algorithms," *BioSystems*, Vol. 72, No. 1, pp. 111–129.

Deutsch, J. (2003). "Evolutionary algorithms for finding optimal gene sets in microarray prediction," *Bioinformatics*, Vol. 19, No. 1, pp. 45–52.

Fogel, L., A. Owens, and M. Walsh (1966). *Artificial Intelligence through Simulated Evolution*. Wiley, New York.

FUTSCHIK, M., A. REEVE, and N. KASABOV (2003). "Evolving connectionist systems for knowledge discovery from gene expression data of cancer tissue," *Artif. Intell. Med.*, Vol. 28, No. 2, pp. 165–189.

GOLDBERG, D. (1989). *Genetic Algorithms in Search, Optimization and Machine Learning*, Addison Wesley, Reading, MA, 14.

GOLUB, T., D. SLONIM, P. JAMAYO, C. HUARD, M. GASSENBEEK, J. MESIROV, H. COLLER, M. LOH, J. DOWNING, and M. LANDER, (1999). "Molecular classification of cancer: Class discovery and class prediction by gene-expression monitoring," *Science*, Vol. 286, pp. 531–537.

GORDON, G., R. JENSEN, L. HSIAO, S. GULLANS, J. BLUMENSTOCK, S. RAMASWAMY, W. RICHARDS, D. SUGARBAKER, and R. BUENO (2002). "Translation of microarray data into clinically relevant cancer diagnostic tests using gene expression ratios in lung cancer and mesothelioma," *Can. Res.*, Vol. 62, No. 17, pp. 4963–4967.

HAYKIN, S. (1999). *Neural Networks*. Prentice Hall, Upper Saddle River, NJ.

HOLLAND, J. (1975). *Adaptation in Natural and Artificial Systems*. University of Michigan Press, Ann Arbor, MI.

HONG, J.-H. and S.-B. CHO (2006a). "Efficient huge-scale feature selection with speciated genetic algorithm," *Pattern Recog. Lett.*, Vol. 27, No. 2, pp. 143–150.

HONG, J.-H. and S.-B. CHO (2006b). "The classification of cancer based on DNA microarray data that uses diverse ensemble genetic programming," *AI Med.*, Vol. 36, No. 1, pp. 43–58.

INZA, I., M. MERINO, P. LARRAÑAGA, J. QUIROGA, B. SIERRA, and M. GIRALA (2001). "Feature subset selection by genetic algorithms and estimation of distribution algorithms. A case study in the survival of cirrhotic patients treated with TIPS," *Artif. Intell. Med.*, Vol. 23, No. 2, pp. 187–205.

KASABOV, N. (2001). "Evolving fuzzy neural networks for on-line learning, reasoning and rule extraction," *IEEE Trans. Syst., Man, Cyberns.—Part B: Cybernetics*, Vol. 31, No. 6, pp. 902–918.

KHAN, J., J. WEI, M. RINGNER, L. SAAL, M. LADANYI, F. WESTERMANN, F. BERTHOLD, M. SCHWAB, C. ANTONESCU, C. PETERSON, and P. MELTZER (2001). "Classification and diagnostic prediction of cancers using gene expression profiling and artificial neural networks," *Nature Med.*, Vol. 7, No. 6, pp. 673–679.

KIM, K.-J. and S.-B. CHO (2004). "Prediction of colon cancer using an evolutionary neural network," *Neurocomputing*, Vol. 61, pp. 361–379.

KOHAVI, R. and G. JOHN (1997). "Wrappers for feature subset selection," *Artif. Intell.*, Vol. 97, Nos. 1–2, pp. 273–324.

KUNCHEVA, L. (2002). "A theoretical study on six classifier fusion strategies," *IEEE Trans. Pattern Anal. Mach. Intell.*, Vol. 24, No. 2, pp. 281–286.

KUNCHEVA, L. and C. WHITAKER (2003). "Measures of diversity in classifier ensembles and their relationship with the ensemble accuracy," *Mach. Learn.*, Vol. 51, No. 2, pp. 181–207.

LANGDON, W. and B. BUXTON (2004). "Genetic programming for mining DNA chip data from cancer patients," *Genet. Program. Evolvable Mach.*, Vol. 5, No. 3, pp. 251–257.

LI, W. and Y. YANG (2002). "How many genes are needed for a discriminant microarray data analysis?" in *Methods of Microarray Data Analysis*, Kluwer Academic, Boston, pp. 137–150.

LIU, J., G. CUTLER, W. LI, Z. PAN, S. PENG, T. HOEY, L. CHEN, and X. LING (2005). "Multiclass cancer classification and biomarker discovery using GA-based algorithms," *Bioinformatics*, Vol. 21, No. 11, pp. 2691–2697.

OOI, C. and P. TAN (2003). "Genetic algorithms applied to multi-class prediction for the analysis of gene expression data," *Bioinformatics*, Vol. 19, No. 1, pp. 37–44.

PAUL, T. and H. IBA (2005). "Gene selection for classification of cancers using probabilistic model building genetic algorithm," *BioSystems*, Vol. 82, No. 3, pp. 208–225.

PENG, S., Q. XU, X. LING, X. PENG, W. DU, and L. CHEN (2003). "Molecular classification of cancer types from microarray data using the combination of genetic algorithms and support vector machines," *FEBS Lett.*, Vol. 555, No. 2, pp. 358–362.

POMEROY, S., P. TAMAYO, M. GAASENBEEK, L. STURLA, M. ANGELO, M. MCLAUGHLIN, J. KIM, L. GOUMNEROVA, P. BLACK, C. LAU, J. ALLEN, D. ZAGZAG, J. OLSON, T. CURRAN, C. WETMORE, J. BIEGEL, T. POGGIO, S. MUKHERJEE, R. RIFKIN, A. CALIFANO, G. STOLOVITZKY, D. LOUIS, J. MESIROV, E. LANDER, and T. GOLUB (2002). "Prediction of central nervous system embryonal tumour outcome based on gene expression," *Nature*, Vol. 415, No. 6870, pp. 436–442.

RADONIĆ, A., S. THULKE, H. BAE, M. MÜLLER, W. SIEGERT, and A. NITSCHE (2005). "Reference gene selection for quantitative real-time PCR analysis in virus infected cells: SARS corona virus, Yellow fever virus, Human Herpesvirus-6, Camelpox virus and Cytomegalovirus infections," *Virology J.*, Vol. 2, No. 1, pp. 7, 2005.

RAMASWAMY, S., P. TAMAYO, R. RIFKIN, S. MUKHERJEE, C. YEANG, M. ANGELO, C. LADD, M. REICH, E. LATULIPPE, J. MESIROV, T. POGGIO, W. GERALD, M. LODA, E. LANDER, and R. GOLUB (2001). "Multiclass cancer diagnosis using tumor gene expression signatures," *Proc. Natl. Acad. Sci., USA*, Vol. 98, pp. 15149–15154.

ROSS, D., U. SCHERF, M. EISEN, C. PEROU, P. SPELLMAN, V. IYER, S. JEFFREY, M. VAN DE RIJN, M. WALTHAM, A. PERGAMENSCHIKOV, J. LEE, D. LASHKARI, D. SHALON, T. MYERS, J. WEINSTEIN, D. BOTSTEIN, and P. BROWN (2000). "Systematic variation in gene expression patterns in human cancer cell lines," *Nature Genet.*, Vol. 24, No. 3, pp. 227–234.

SCHMIDT, U. and C. BEGLEY (2003). "Cancer diagnosis and microarrays," *Int. J. Biochem. Cell Biol.*, Vol. 35, No. 2, pp. 119–124.

SCHWEFEL, H. (1981). *Numerical Optimization of Computer Models*. Wiley, Chichester, UK.

SHIPP, C. and L. KUNCHEVA (2002). "Relationships between combination methods and measures of diversity in combining classifiers," *Info. Fusion*, Vol. 3, No. 2, pp. 135–148.

SINGH, D., P. FEBBO, K. ROSS, D. JACKSON, J. MANOLA, C. LADD, P. TAMAYO, A. RENSHAW, A. D'AMICO, J. RICHIE, E. LANDER, M. LODA, P. KANTOFF, T. GOLUB, and W. SELLERS (2002). "Gene expression correlates of clinical prostate cancer behavior," *Can. Cell*, Vol. 1, No. 2, pp. 203–209.

VAN'T VEER, L., H. DAI, M. VAN DE VIJVER, Y. HE, A. HART, M. MAO, H. PETERSE, K. VAN DE KOOY, M. MARTON, A. WITTEVEEN, G. SCHREIBER, R. KERKHOVEN, C. ROBERTS, P. LINSLEY, R. BEMARDS, and S. FRIEND (2002). "Gene expression profiling predicts clinical outcome of breast cancer," *Nature*, Vol. 415, No. 6871, pp. 530–536.

WEBB, G. and Z. ZHENG (2004). "Multistrategy ensemble learning: Reducing error by combining ensemble learning techniques," *IEEE Trans. Knowl. Data Eng.*, Vol. 16, No. 8, pp. 980–991.

WINDEATT, T. (2005). "Diversity measures for multiple classifier system analysis and design," *Info. Fusion*, Vol. 6, No. 1, pp. 21–36.

ZENOBI, G. and P. CUNNINGHAM (2001). "Using diversity in preparing ensembles of classifiers based on different feature subsets to minimize generalization error," *Lect. Notes Comput. Sci.* (ECML), Vol. 2167, pp. 576–587.

Chapter 3

Finding Clusters in Gene Expression Data Using EvoCluster

Patrick C. H. Ma, Keith C. C. Chan, and Xin Yao

3.1 INTRODUCTION

Given a database of records each characterized by a set of attributes, the clustering problem is concerned with the discovery of interesting record groupings based on the attribute values. Many algorithms have been developed to tackle clustering problems in a variety of application domains (Ward, 1963; MacQueen, 1967; Kohonen, 1989). In particular, some of them have been used to uncover hidden groupings in gene expression data (Eisen et al., 1998; Tavazoie et al., 1999; Tamayo et al., 1999).

Gene expression is the process by which a gene's coded information is converted into the structures present and operating in a cell. This information flow from gene to protein occurs via a ribonucleic acid (RNA) intermediate in two stages: transcription and translation. If one would like to prevent the expression of undesirable genes such as those related to cancer, prevention of transcription initiation is important (Hartl and Jones, 2001; Cox and Sinclair, 1997). To prevent transcription initiation, a set of transcription factor binding sites must be identified. These sites consist of untranscribed nucleotide sequences located typically in the promoter regions of genes. Once identified, protein repressors can be developed to bind at these sites and block transcription (Cox and Sinclair, 1997). To locate transcription factor binding sites, coexpressed genes (i.e., genes that have similar transcriptional responses to the external environment such as temperature, pH value, pressure, etc.) need to be identified. Coexpression can be an indication of coregulation by the same transcription factors to common binding sites. Clustering gene expression data from microarray experiments (Fernandes et al., 2004; Lapointe et al., 2004; Spellman

Computational Intelligence in Bioinformatics. Edited by G. B. Fogel, D. Corne, and Y. Pan

et al., 1998; Cho et al., 1998, 2001; DeRisi et al., 1997; Lashkari et al., 1997) is one way to identify potentially coregulated genes. By analyzing the promoter regions of these genes, we may be able to discover patterns, which have relatively high occurring frequencies compared to other sequences, and may be possible binding sites of these genes (Zheng et al., 2003).

Various types of noise can be introduced during the collection of microarray data. They can arise when deoxyribonucleic acid (DNA) arrays are being produced or from the preparation of biological samples, extraction of results, and the like (Berrar et al., 2003; Sherlock, 2000). For the purpose of handling very noisy gene expression data, we have developed an algorithm called EvoCluster (Ma et al., 2006). This algorithm is based on the use of an evolutionary approach and has several characteristics: (1) It encodes the entire cluster grouping in a *chromosome*[1] so that each *gene* encodes one cluster and has a set of crossover and mutation operators that facilitates the exchange of grouping information between two *chromosomes* on one hand and allows variation to be introduced to avoid entrapment at local optima on the other; (2) it makes use of a fitness function that measures the quality of a particular grouping of data records encoded in a *chromosome*; (3) unlike similarity measures that are based on local pairwise distances that may not give very accurate measurements in the presence of very noisy data (Jain and Dubes, 1998), the EvoCluster fitness measure is probabilistic and it takes into consideration global information contained in a particular grouping of data; (4) it is able to distinguish between relevant and irrelevant feature values in the data during the clustering process; (5) it is able to explain clustering results discovered by explicitly revealing hidden patterns in each cluster in an easily understandable if–then rule representation, and (6) there is no requirement for the number of clusters to be decided in advance. For performance evaluation, we have tested EvoCluster using both simulated and real data. Experimental results show that it can be very effective and robust even in the presence of noisy and missing values. In Ma et al. (2006), some of the details are not presented. Here, we would like to put more discussions on the implementation and present results of further experiments done to evaluate the performance of EvoCluster.

The rest of this chapter is organized as follows. In Section 3.2, we provide an overview of existing algorithms used for clustering gene expression data. In Section 3.3, EvoCluster is described in detail. In Section 3.4, we discuss how EvoCluster can be evaluated and compared with some existing clustering algorithms using both simulated and real data. The evaluation results and the biological interpretation of the clusters discovered by EvoCluster are then presented and discussed. Finally, in Section 3.5, we provide a summary of the chapter.

[1] The development of evolutionary algorithms has, in some instances, taken great liberty with biological terms and theory. The terms such as *chromosomes* and *genes*, when used in a computational context, may not have the same meanings as their biological counterparts. In order to avoid possible confusion, when referring to these terms in the contexts of evolutionary computation, they are made italic.

3.2 RELATED WORK

Let us assume that we are given a set of gene expression data, **G**, consisting of the data collected from N genes in M experiments each carried out under different sets of conditions. Let us represent the data set as a set of N records/genes, $\mathbf{G} = \{\mathbf{g}_1, \ldots, g_i, \ldots, g_N\}$, with each record, $\mathbf{g}_i$ where $i = 1, \ldots, N$, characterized by M attributes, $E_1, \ldots, E_j, \ldots, E_M$, whose values, $e_{i1}, \ldots, e_{ij}, \ldots, e_{iM}$ where $e_{ij} \in$ domain(E_j), represents the value of the ith record under the jth attribute.

To discover clusters in the data, the hierarchical agglomerative clustering algorithm (Ward, 1963) can be used to perform a series of successive fusions of records into clusters. The fusion process is guided by a measure of similarity between clusters so that clusters that are similar to each other are merged. This fusion process is repeated until all clusters are merged into a single cluster. The results of the fusion process are normally presented in the form of a two-dimensional hierarchical structure, called a dendrogram. The records falling along each branch in a dendrogram form a cluster. Depending on user preferences, a specific number of clusters can be obtained from the dendrogram by cutting across the branches at a specific level. Compared to the hierarchical agglomerative clustering algorithm that does not require users to specify the number of clusters ahead of time, users of the k-means algorithm (MacQueen, 1967) are required to do so. Given a data set **G**, the k-means algorithm can group the records, $\mathbf{g}_1, \ldots, g_i, \ldots, \mathbf{g}_N$ into k clusters by initially selecting k records as centroids. Each record is then assigned to the cluster associated with its closest centroid. The centroid for each cluster is then recalculated as the mean of all records belonging to the cluster. This process of assigning records to the nearest clusters and recalculating the position of the centroids is then performed iteratively until the positions of the centroids remain unchanged.

The self-organizing map (SOM) algorithm (Kohonen, 1989) is one of the best-known artificial neural network algorithms. It can be considered as a mapping from M-dimensional input data space onto a map so that every neuron of the map is associated with an M-dimensional reference vector. The reference vectors together form a codebook. The neurons of the map are connected to adjacent neurons by a neighborhood relation, which dictates the topology of the map. The most common topologies used are rectangular and hexagonal. In the basic SOM algorithm, the topology and the number of neurons remain fixed from the beginning. The number of neurons determines the granularity of the mapping, which has an effect on the accuracy and generalization of the SOM. During the training phase, the reference vectors in the codebook drift to the areas where the density of the input data is high. Eventually, only few codebook vectors lie in areas where the input data is sparse. After the training is over, the map should be topologically ordered. This means that n topologically close input data vectors map to n adjacent map neurons or even to the same single neuron.

In should be noted that existing clustering algorithms in gene expression data analysis (Eisen, 1998; Tavazoie, 1999; Tamayo et al., 1999; Quackenbush, 2001; Ben-Dor et al., 1999; Herrero et al., 2001) do not provide very accurate measurements when the data concerned are noisy and contain missing values. This may be a result of their use of such metrics and functions (e.g., Euclidian distance measure or the Pearson correlation coefficient, etc.) that do not differentiate between the importance of different attributes when measuring similarities. Since these metrics and functions measure only pairwise distances, the measurements obtained could be too local (Jain and Dubes, 1998). Clustering algorithms based only on the local pairwise information may, therefore, miss important global information. In addition to these deficiencies, clustering results obtained with the use of many clustering algorithms can be difficult to interpret. For example, although one can visualize the result of hierarchical clustering as a treelike dendrogram, and note correlations between genes, it is the users' responsibilities to discover the similarities and differences between various clusters and to decide on the number of clusters and the cluster boundaries to form. To do so, users need to have prior knowledge about the data. Similarly, for the *k*-means algorithm and SOM, users have to decide on the number of clusters into which to partition a data set. They also have to use a separate technique to uncover underlying patterns in the clusters. Given the increased popularity of DNA microarray technologies in the study of gene variations and given that more and more microarray data are generated as a result of advances in these technologies, there is a great need to develop clustering algorithms that can overcome these limitations. Toward this goal, we have developed a novel clustering algorithm that is based on evolutionary computation.

3.3 EVOLUTIONARY CLUSTERING ALGORITHM

Evolutionary algorithms (EAs) have been successfully used to solve different data mining problems (Muni et al., 2004; Chi et al., 2003; Cano et al., 2003; Parpinelli et al., 2002). They have, in particular, been used for clustering (Murthy and Chowdhury, 1996; Chan and Chung, 1999; Park and Song, 1998; Hall et al., 1999; Maulik et al., 2000; Babu and Murty, 1994; Franti et al., 1997; Krishna and Murty, 1999). For example, in Murthy and Chowdhury (1996) and Chan and Chung (1999), the data records were encoded as *genes* in a *chromosome* and are given a label from 1 to *k*, where *k* was the maximum number of clusters to be discovered. Such an approach was relatively easy to implement given it does not require special evolutionary operators. Unfortunately, they are not very scalable. As the length of each *chromosome* is exactly the size of the training set, these algorithms are not very practical when handling large data sets. An alternative data and cluster representation was proposed in Park and Song (1998) where the clustering problem is formulated as a graph-partitioning problem. Each data record was represented as a node in a graph and each node was mapped to a certain position in a *chromosome* and was encoded as a *gene*. The indices of other records were encoded as *alleles* so that if a *gene* *i* contains value *j*, an edge is created in the graph to link the nodes *i* and *j*. The *alleles* in each *gene* *i* were therefore the nearest neighbors of *i*, and the users were required to specify the number of nearest neighbors as input parameters ahead of time. With this representation, an evolutionary algorithm was used to find clusters that were represented as connected subgraphs. This approach was again not very scalable. Other than the length of the *chromosomes* being again the same as the size of a data set, there was an additional need for the nearest neighbors of data records to be computed. This approach also suffers from the same problems as other clustering algorithms that are based on the need to compute pairwise distance measures.

One other popular use of evolutionary algorithms in clustering is to use them to identify the best cluster centers. In Hall et al. (1999) and Maulik et al. (2000), each *chromosome* encoded the coordinates of k centers and a standard genetic algorithm was used to find the best center coordinates. A similar approach to identifying the best centers is to use an EA to search for optimal initial seed values for cluster centroids (Babu and Murty, 1994). As with other problems, for clustering, domain knowledge can be used in several ways to try to improve algorithm performance. For example, specialized evolutionary operators can be developed when hybridizing an evolutionary algorithm with a conventional clustering algorithm such as k means. In Babu and Murty (1994) and Krishna and Murty (1999), each *chromosome* represents the coordinates of the cluster centroids and different crossover methods are used to generate the offspring. After crossover, each *chromosome* underwent several iterations of the k-means clustering algorithm. The authors observed that adding the k-means iterations was crucial for obtaining good results, although there could be a considerable increase in computation time if many generations of evolution were used. This kind of hybridization raises the question of how to allocate the computing time (e.g., using many generations of the EAs and a few iterations of the local methods or running the EAs for a few generations and using the local methods to improve the solutions). In principle, the centroid-based representation has the advantage of short *chromosomes* because the representation is only required to encode the coordinates of k centroids. This means that the length of the *chromosome* is proportional to the dimensionality of the problem and not the size of the training set. However, just like many EA-based clustering methods, the drawback of the centroid-based representation is that the number of clusters needs to be specified in advance. Moreover, the similarity functions used such as Euclidean distance or correlation coefficient for measuring the similarity of the records do not differentiate between the relative importances of different attributes. Therefore, they do not give accurate measurements when the data concerned are noisy and contain missing values. In addition, these similarity functions measure only pairwise distances, and the measurements obtained could be too local.

Clustering gene expression data as a new area of research poses new challenges due to its unique problem nature with which the previous EA-based clustering algorithms were not originally designed to deal. As discussed in Lu and Han (2003), there are some new challenges in dealing with gene expression data. For example, the presence of both biological and technical noise inherent in the data set, the presence of large number of irrelevant attributes, and the explanatory capability of an algorithm to help biologists in gaining more understanding of the underlying biological process. And also, the clustering structure of gene expression data is usually unknown. Like other evolutionary algorithms (Goldberg, 1989; Ashlock, 2006; Falkenauer, 1998; De Jong, 2006; Baeck et al., 2006b), the EvoCluster algorithm consists of the following steps:

1. Initialization of a population of *chromosomes* with each representing a unique cluster grouping.
2. Evaluate the fitness of each *chromosome*.
3. Select *chromosomes* for reproduction using the roulette wheel selection scheme.
4. Apply crossover and mutation operators.
5. Replace the least-fit *chromosomes* in the existing population by the newly generated offspring.
6. Repeat steps 2 to 5 until the stopping criteria are met.

3.3.1 Cluster Encoding in Chromosomes and Population Initialization

To evolve the best cluster grouping, EvoCluster encodes different grouping arrangements in different *chromosomes* so that one *chromosome* encodes one particular cluster grouping. In each such *chromosome*, each *gene* encodes one cluster (Falkenauer, 1998). Hence, if a particular *chromosome* encodes k clusters, $C_1, \ldots, C_i, \ldots, C_k$, it has k *genes*. Since each cluster contains a number of data records, a *gene* encoding a cluster can be considered as being made up of the labels of a number of data records in gene expression data. For example, assume that C_i contains n_i records, $g_{(i1)}, \ldots, g_{(ij)}, \ldots, g_{(in_i)}$, where $g_{(ij)} \in \mathbf{G} = \{g_1, \ldots, g_i, \ldots, g_N\}$, the unique labels of these records, $g_{(i1)}, \ldots, g_{(ij)}, \ldots, g_{(in_i)}$, can be encoded in each *gene*, C_i, so that a *chromosome* that encodes a particular cluster grouping can then be represented diagrammatically as shown in Figure 3.1.

For the initial population, each *chromosome* is generated randomly by EvoCluster in such a way that the number of clusters, k, to be encoded in a *chromosome* is generated within an acceptable range. Each of the records in $\mathbf{G} = \{g_1, \ldots, g_i, \ldots, g_N\}$ is also randomly assigned to one of the k clusters.

3.3.2 Selection and Reproduction

Reproduction in EvoCluster consists of the application of both the crossover and mutation operations (Fig. 3.2). As the evolutionary process enters into reproduction, two *chromosomes* are selected as parents for crossover using roulette wheel selection (Ashlock, 2006) so that each parent's chance of being selected is directly proportional to its fitness. Since it is the cluster grouping encoded in each *chromosome* that

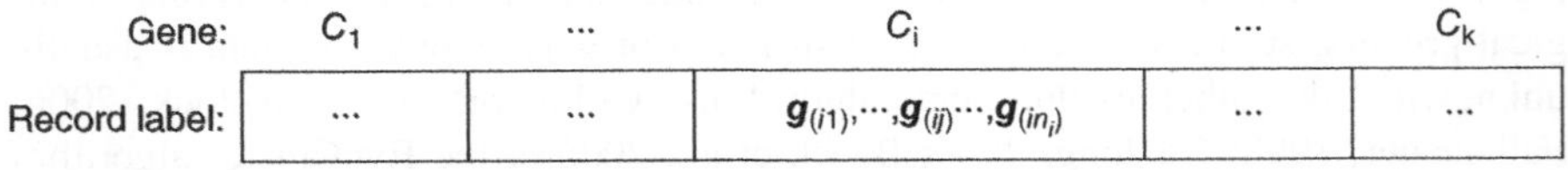

Gene:	C_1	...	C_i	...	C_k
Record label:	...	...	$g_{(i1)}, \cdots, g_{(ij)} \cdots, g_{(in_i)}$	...	...

Figure 3.1 *Chromosome* encoding scheme.

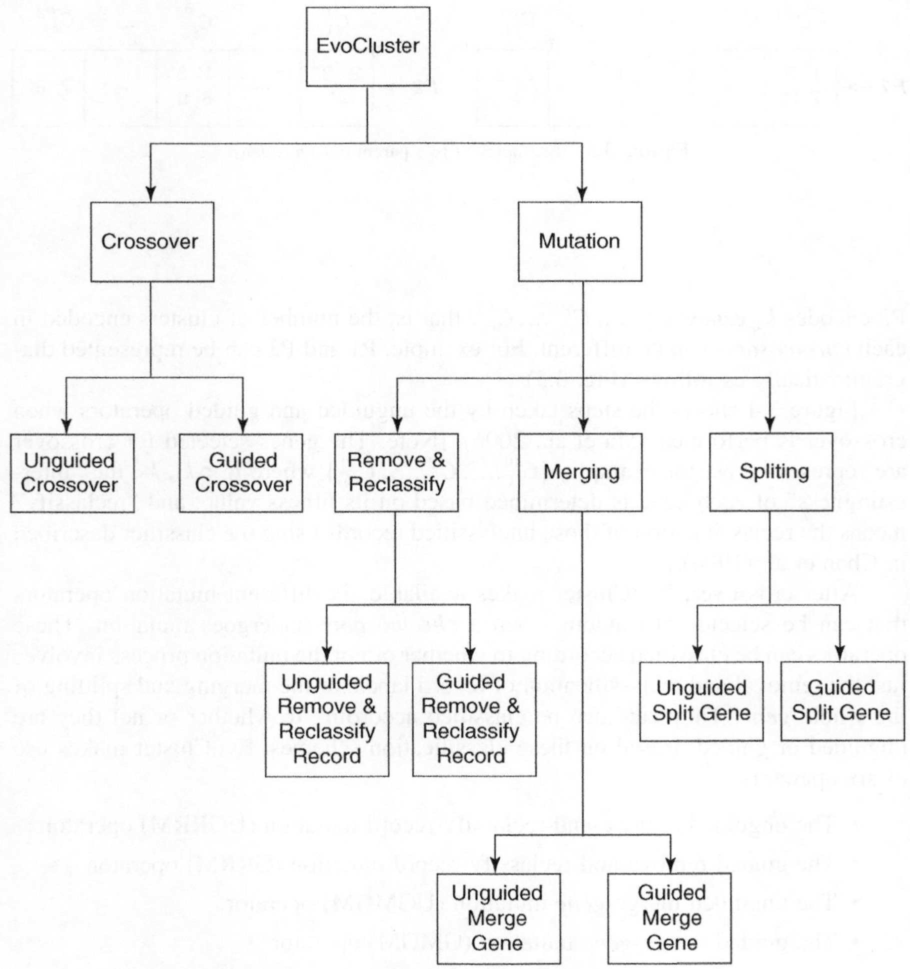

Figure 3.2 Reproduction operators of EvoCluster.

conveys the most important information, our crossover operators are designed to facilitate the exchange of grouping information. And since this process can be "unguided" or "guided," our crossover operators are also classified in the same way. We have both an unguided operator and a guided operator. For the unguided crossover (UGC) operator, the exchange of the grouping information between clusters takes place randomly.

For the guided crossover (GC) operator, the exchange of grouping information is not totally random in the sense that the grouping information of the "best formed" clusters is preserved during the crossover process. Assume that two parent *chromosomes*, P1 and P2 are chosen so that P1 encodes k_1 *genes*, $C_1^{P1}, \ldots, C_i^{P1}, \ldots, C_{k_1}^{P1}$, and

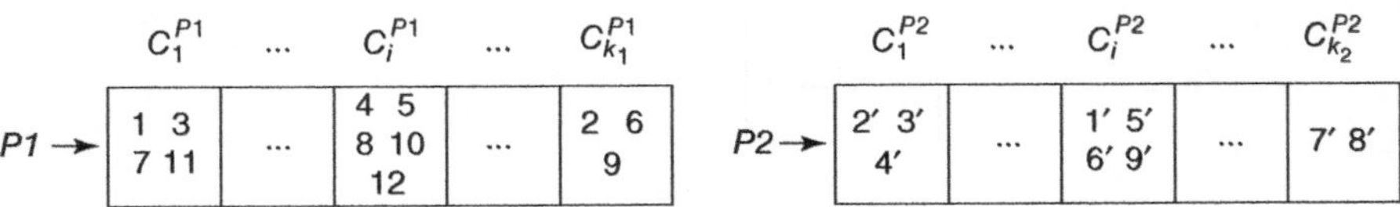

Figure 3.3 Examples of two parent *chromosomes*.

P2 encodes k_2 *genes*, $C_1^{P2},\ldots, C_i^{P2} \ldots, C_{k_2}^{P2}$, that is, the number of clusters encoded in each *chromosome* can be different. For example, P1 and P2 can be represented diagrammatically as follows (Fig. 3.3):

Figure 3.4 shows the steps taken by the unguided and guided operators when crossover is performed (Ma et al., 2006). [Note: The genes selected for crossover are represented as, for example, $\{C_{(1)}^{P1},\ldots, C_{(i)}^{P1},\ldots, C_{(l_1)}^{P1}\}$ where $l_1 < k_1, k_2$, the "interestingness" of each gene is determined based on its fitness value, and "reclassify" means the reclassification of those unclassified records using the classifier described in Chan et al. (1994).]

After crossover, EvoCluster makes available six different mutation operators that can be selected at random when a *chromosome* undergoes mutation. These operators can be classified according to whether or not the mutation process involves just the removal and reclassification of record labels or the merging and splitting of the whole *gene*. They can also be classified according to whether or not they are unguided or guided. Based on these classification schemes, EvoCluster makes use of six operators:

- The unguided remove-and-reclassify-record mutation (UGRRM) operator
- The guided remove-and-reclassify-record mutation (GRRM) operator
- The unguided merge-*gene* mutation (UGMGM) operator
- The guided merge-*gene* mutation (GMGM) operator
- The unguided split-*gene* mutation (UGSGM) operator
- The guided split-*gene* mutation (GSGM) operator

The merge and split mutation operators were specifically designed to allow the length of *chromosomes* to be changed dynamically as the evolutionary process progresses. The advantage with this feature is that the number of clusters that need to be formed does not need to be specified by the users ahead of time. The steps taken by each of these operators are described in Figures 3.5 to 3.7 (Ma et al., 2006). (Note: Assume that MIN is a user-defined minimum number of clusters encoded in a *chromosome* and MAX is a user-defined maximum number of clusters encoded in a *chromosome*.)

For generational replacement, EvoCluster adopts a steady-state reproduction approach (Baeck et al., 2000a) so that only two least-fit *chromosomes* are replaced whenever two new children are generated after each reproduction.

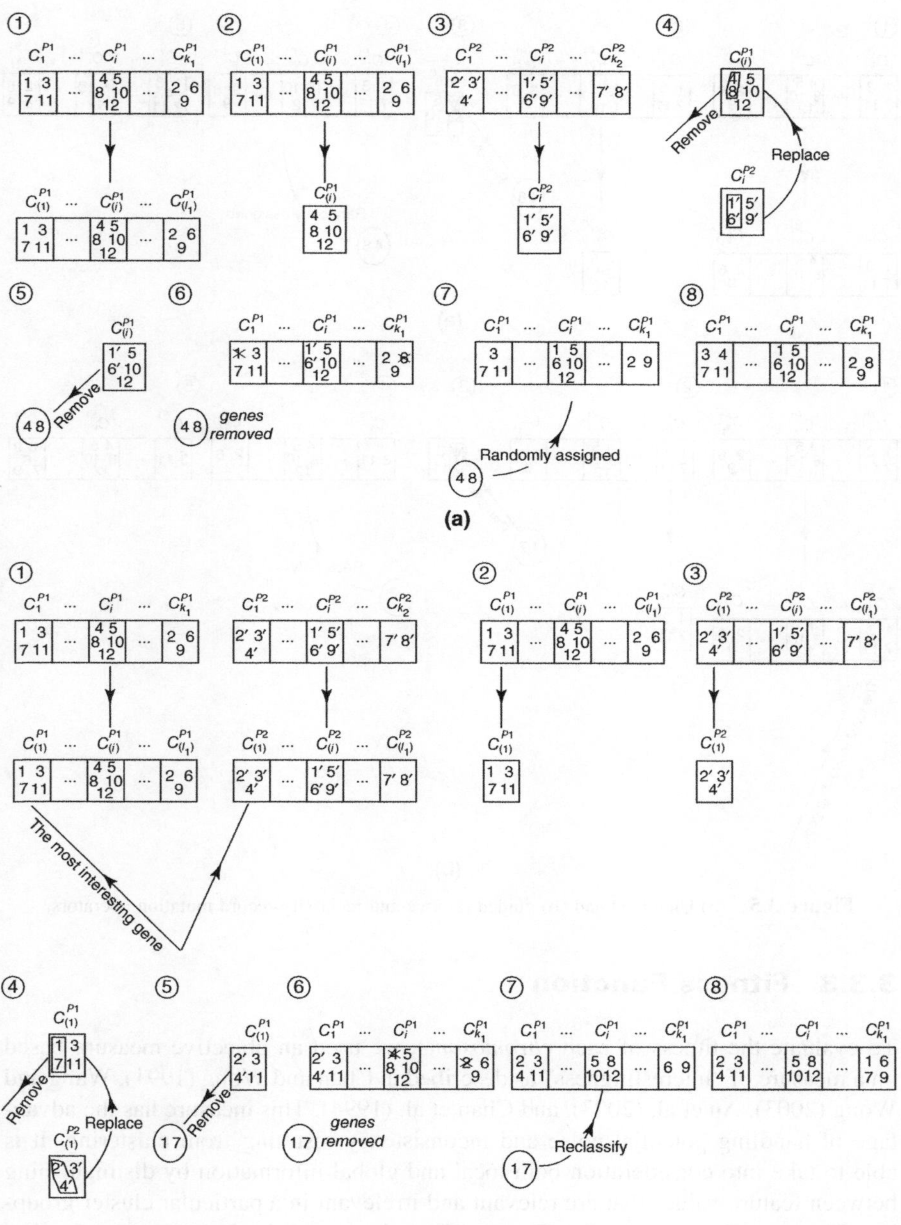

Figure 3.4 (a) Unguided crossover and (b) guided crossover operators.

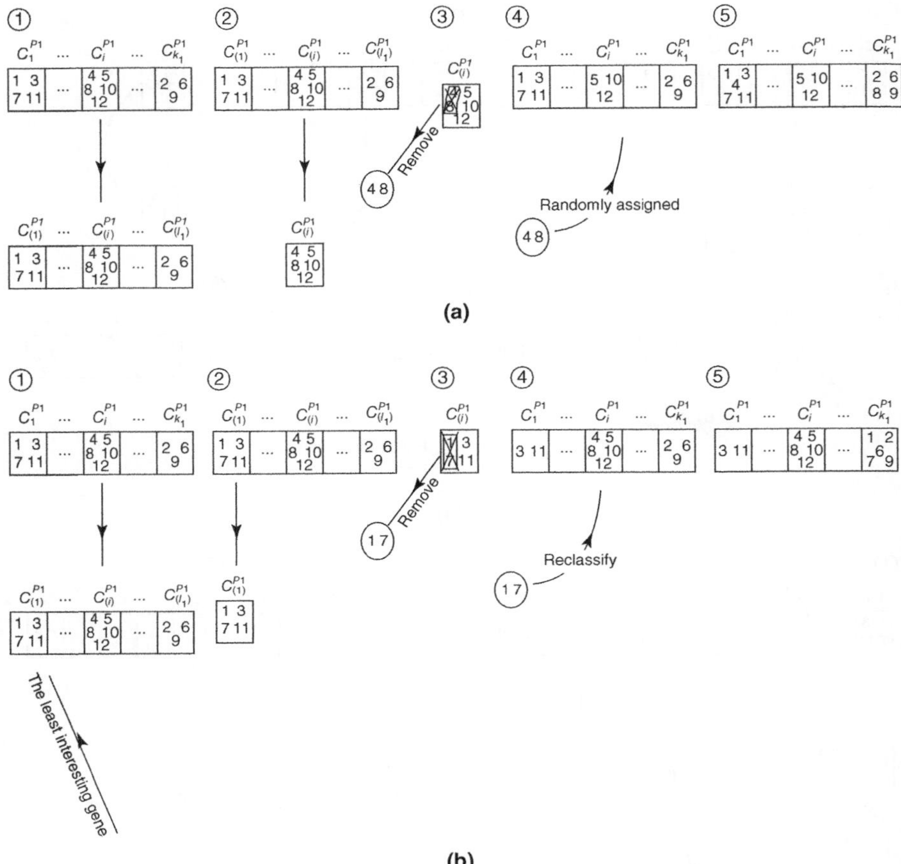

Figure 3.5 (**a**) Unguided and (**b**) guided remove-and-reclassify-record mutation operators.

3.3.3 Fitness Function

To evaluate the fitness of each *chromosome*, we used an objective measure based on a measure of "interestingness" as described in Chan and Wong (1991), Wang and Wong (2003), Au et al. (2003), and Chan et al. (1994). This measure has the advantage of handling potential noise and inconsistency resulting from clustering. It is able to take into consideration both local and global information by distinguishing between feature values that are relevant and irrelevant in a particular cluster grouping. These attributes make EvoCluster very robust even in the presence of noise. The fitness evaluation procedure is invoked after new *chromosomes* are formed. The fitness function accepts a *chromosome* as a parameter and its fitness is evaluated in three steps. In step 1, interesting association patterns in the cluster grouping encoded in the *chromosome* are discovered by using the adjusted residual. To do so, a subset of records from different clusters encoded in a *chromosome* is selected randomly to

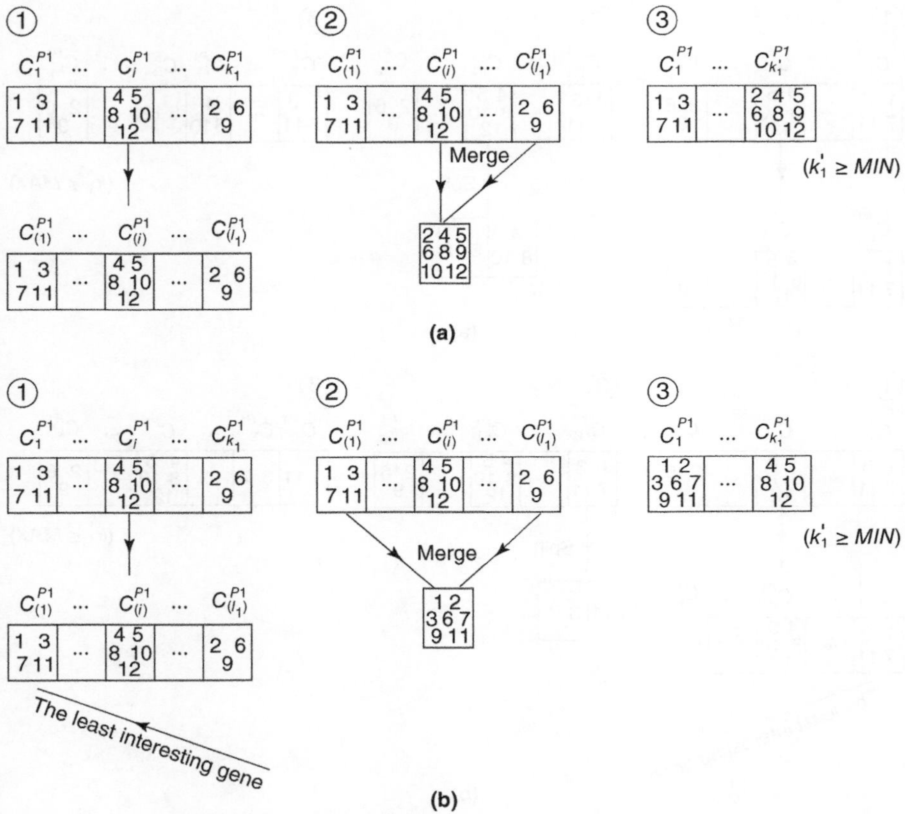

Figure 3.6 (a) Unguided and (b) guided merge-*gene* mutation operators.

form a training set for pattern discovery. In step 2, the weight is assigned to each discovered pattern using the weight of evidence measure. In step 3, those records not selected in step 1 are then reclassified into one of the clusters based on the discovered patterns in order to determine the reclassification accuracy of the *chromosome*. If a clustering algorithm is effective, the clusters that are discovered should contain hidden patterns that can be used to accurately reclassify the records in the testing data. And if this is the case, the reclassification accuracy measure is an indication of how good the quality of the cluster grouping is. For this reason, the reclassification accuracy is then taken to be the fitness of each *chromosome*.

3.4 EXPERIMENTAL RESULTS

To evaluate the effectiveness of EvoCluster, we have tested it on both simulated and real data. In this section, we describe the experiments we carried out and present the results of these experiments.

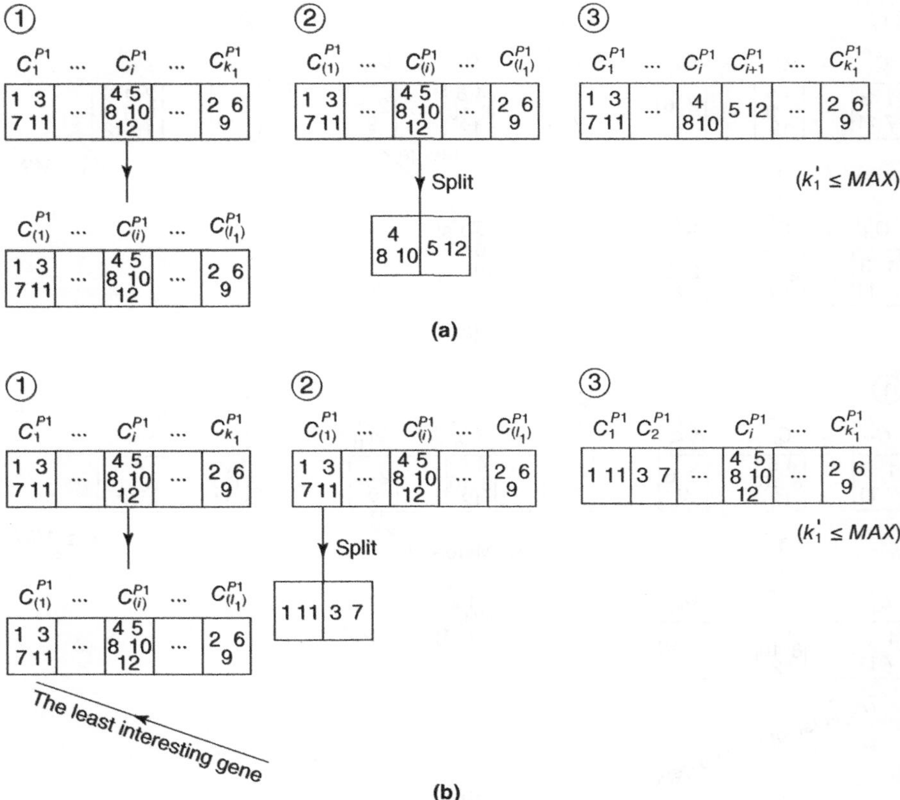

Figure 3.7 (a) Unguided and (b) guided split-*gene* mutation operators.

3.4.1 Experimental Data

For performance evaluation, we used a set of simulated data consisting of 300 records each characterized by 50 different attributes that takes on values from [0.0, 1.0]. Initially, all these records were sampled from a uniform distribution and they were preclassified into one of three clusters so that each cluster contains 100 records. To embed hidden patterns in the data, 10% of the attributes in each cluster were selected randomly. For each selected attribute, 40% of the values in that cluster were generated randomly from the range $[L, U]$, where $0.0 \le L \le U \le 1.0$ so that L was selected uniformly from [0.0, 1.0] first, and U was then also selected uniformly from $[L, 1.0]$. In addition to the simulated data, to test the effectiveness of EvoCluster, we also used two different sets of real gene expression microarray data given in Spellman et al. (1998) and Cho et al. (1998), respectively. Data set 1 contains about 800 cell cycle regulated genes measured under 77 different experimental conditions. According to Spellman et al. (1998), we tried in our experiments to partition the data set also into different clusters from 6 to 10. Data set 2 contains a subset of 384 genes measured under 17 different experimental conditions. For this data set, we also tried to partition it into different clusters from 4 to 8 (Cho et al., 1998).

With the above data sets, the effectiveness of EvoCluster was evaluated objectively based on three objective measures: (1) the Davies–Bouldin validity index (DBI) measure (Davies and Bouldin, 1979), (2) the *F* measure (Larsen and Aone, 1999), and (3) a predictive power measure. The DBI measure is a function of the inter- and intracluster distances. These distances are considered good indicators of the quality of a cluster grouping as a good grouping should be reflected by a relatively large intercluster distance and a relatively small intracluster distance. In fact, many optimization clustering algorithms are developed mainly to maximize intercluster and minimize intracluster distances. The DBI measure combines these two distances into a function to measure the average similarity between a cluster and its most similar one. For the experiments described below, the Euclidean distance was chosen as the distance metric when computing the DBI measure. A low value of DBI therefore indicates good cluster grouping. The *F* measure that is typically used for cluster evaluation combines the "precision" and "recall" ideas from information retrieval (Kowalski, 1997). When the correct classification of the data is known, the *F* measure is useful in the sense that it would provide objective information on the degree to which a clustering algorithm is able to recover the original clusters. The *F*-measure values are in the interval [0,1] and the larger its values, the better the clustering quality is. The predictive power measure is actually a measure of classification accuracy. If the clusters discovered are valid, we should expect patterns to be discovered in them. If these patterns are used to classify some testing data, the classification accuracy can reflect how valid and how good the qualities of the discovered clusters are. In order to determine the classification accuracy, a set of training samples can be randomly selected from each cluster to construct a decision tree classifier using the algorithm known as C4.5 (Quinlan, 1993). C4.5 is a greedy algorithm that recursively partitions a set of training samples by selecting attributes that yield a maximum information gain measure at each step in the tree construction process. After a decision tree is built, a postpruning procedure is used to compensate for any overfitting of training samples. Based on the pruned tree, the cluster memberships of those records that were not selected for training are then predicted. The percentage of accurate predictions can then be determined as classification accuracy. This accuracy measure is also referred to as the predictive power measure. If a clustering algorithm is effective, the clusters that are discovered should contain hidden patterns that can be used to accurately predict the cluster membership of the testing data. And if this is the case, the predictive power of a cluster grouping should be high. Otherwise, if a clustering algorithm is ineffective, the clusters it discovers are not expected to contain too many hidden patterns and the grouping is more or less random. And if this is the case, the predictive power is expected to be low. Hence, the greater the predictive power, the more interesting a cluster grouping is and vice versa. In our experiments, the predictive power measure was computed based on a 10-fold cross-validation approach. For each fold, 90% of the data records in each cluster were selected randomly for training and the remaining 10% used for testing. After 10 experiments corresponding to the 10-folds of data were performed, the average predictive power of the discovered clusters was computed as the average classification accuracy over the 10 experiments.

3.4.2 Results

The effectiveness of EvoCluster was compared with a number of different clustering algorithms using both simulated and real data. For EvoCluster, we performed 10 trials in our experiments. For each trial, we randomly generated different initial populations of size fixed at 50. Using a steady-state reproduction scheme, the evolutionary process was terminated when the maximum number of reproductions reached 5000 or the population converged to a stable fitness value. Moreover, to ensure that the best results for SOM and the k-means algorithm were obtained, 100 runs were performed for each of them with each run using different randomly generated initial cluster centroids. The best result from among these 100 runs was then recorded. Similar to EvoCluster, such 100-run tests were repeated 10 times. The 10 best results obtained from each 100-run test were then selected for comparisons. In addition to the traditional clustering algorithms, we also compared its performance with a clustering algorithm that represents one of the most successful attempts to use EA in clustering (Maulik and Bandyopadhyay, 2000). For this EA-based clustering algorithm, each *gene* in a *chromosome* encodes one dimension of a cluster center and each *chromosome* encodes a fixed number of clusters. For our experiment with it, the population size and the number of reproductions were set exactly the same as that with EvoCluster. And also, the crossover and mutation rates used by this EA-based clustering algorithm were set to different values and the rates that gave us the best clustering result in terms of the objective evaluation measures were selected. Since one of the desirable features of EvoCluster is its ability to distinguish relevant from irrelevant feature values during the evolutionary process, we also compared its performance against various hybrid clustering algorithms that use a feature selection technique in combination with a clustering algorithm. Among different feature selection techniques that can be used for this purpose (Li and Yang, 2001; Xiong et al., 2001; Xing and Karp, 2001; Jaeger et al., 2003; Golub, 1999; Alon et al., 1999; Ding, 2002; Su et al., 2003), we chose to consider the one described in Su et al. (2003), which makes use of the t-statistic measure (DeGroot and Schervish, 2002). Given an initial cluster grouping, the feature selection was performed in several steps as follows: (1) A cluster grouping is first determined using the algorithm, say the k-means algorithm, with which the feature filtering method is hybridizing. (2) Given the initial cluster grouping, a t-statistic measure is then computed for each attribute to determine how well it is able to distinguish one cluster from the others. (3) Based on the t statistic, a new subset of attributes with the largest t-statistic values is obtained by first selecting 10% of the attributes that has the largest t-statistic values. With this new attribute subset, a classifier is then generated using C4.5 and its classification accuracy is measured using 10-fold cross validation. Afterward, the process of adding another 5% of the attributes with the largest t-statistic values to this new attribute subset and measuring the accuracy of the resulting new classifier was repeated. The final attribute subset is determined when the performance of the classifier converges. (4) With this final attribute subset, a new and improved cluster grouping is then determined.

3.4.2.1 Simulated Data

Since the number of clusters ($k = 3$) to discover was known in advance for the simulated data, the length of the *chromosome* was fixed in our experiment to be 3. Other parameter settings of EvoCluster used in the simulated data are as follows: MIN = 3, MAX = 3, $P_c = 0.5$, $P_m = 0.5$, $L_g = 0.2$, $U_g = 0.8$, $L_r = 0.2$, and $U_r = 0.8$ [Note: $P_c(P_m)$ is the probability of a crossover(mutation) operator to be selected, $L_g(L_r)$ is the lower bound on the probability of a *gene*(record label) to be selected, and $U_g(U_r)$ is the upper bound on the probability of a *gene*(record label) to be selected.] As discussed in the previous section, EvoCluster has a set of unguided and guided operators. For the unguided operators, the exchange of the grouping information between clusters takes place randomly. For the guided operators, the exchange of grouping information is not totally random in the sense that the grouping information of the best formed clusters is preserved during the evolutionary process. To determine if there is a real need for both types of operators, three separate experiments were carried out. In the first experiments, a 50:50 mixture of unguided or guided operators were used, whereas in the second and third, only unguided operators and only guided operators were used, respectively. The average number of reproductions performed by each algorithm until the population converges to a stable fitness value, the DBI measure, the F measure, and the average predictive power measure are given in Table 3.1. As expected, when only guided operators were used alone, it appeared that the results converged only to some local optima, and when only unguided operators were used, not only a longer evolutionary process was required, but the results obtained were unsatisfactory. The performance of EvoCluster was best when both unguided and guided operators were used together even though it required more reproductions to converge. Based on these results, we concluded that both the unguided and guided operators have a role to play in the evolutionary process. When they are used together, they can facilitate the exchange of grouping

Table 3.1 Comparison of Clustering Performance Using "Unguided + Guided" Operators, "Unguided" Operators, or "Guided" Operators[a]

	No. of Reproduction	DBI	F-measure	Predictive Power (%)
Unguided + Guided operators	3830	1.84	0.91	87.60
Unguided operators	4561	1.92	0.63	61.18
Guided operators	3050	1.86	0.8	76.93

[a]Simulated data, $k = 3$.

Table 3.2 Comparison of Clustering Performance[a]

	DBI	F-measure	Predictive Power (%)
Evo Cluster	1.84	0.91	87.60
EA	1.9	0.69	67.21
EA + FS	1.87	0.79	76.96
k means	1.94	0.61	58.73
k means + FS	1.89	0.75	72.85
SOM	1.95	0.57	55.61
SOM + FS	1.92	0.72	69.70
Hierarchical	1.96	0.52	44.37
Hierarchical + FS	1.93	0.65	57.64

[a]Simulated data, $k = 3$, FS represents feature selection.

Table 3.3 Association Patterns Discovered in Each Cluster[a]

C1	C2	C3
A3 = [0.011, 0.257] (0.98)	A47 = [0.003, 0.252] (1.0)	A5 = [0.004, 0.252] (1.0)
A15 = [0.251, 0.5] (1.0)	A12 = [0.253, 0.501] (1.0)	A18 = [0.253, 0.501] (1.0)
A31 = [0.498, 0.747] (1.0)	A26 = [0.499, 0.749] (1.0)	A22 = [0.501, 0.75] (1.0)
A40 = [0.75, 0.999] (1.0)	A34 = [0.748, 0.996] (1.0)	A28 = [0.742, 0.989] (0.98)
A46 = [0.001, 0.247] (1.0)	A39 = [0.006, 0.253] (1.0)	A36 = [0.015, 0.261] (0.96)

[a]Simulated data, A represents an attribute, and C represents a cluster.

information in a way that such information in the best formed clusters is preserved as much as possible during the evolutionary process on one hand, but variations can be introduced at the same time on the other so as to avoid trapping at local optima too early. The performance of EvoCluster was compared with other clustering algorithms and the results are given in Table 3.2. In the simulated data, the best result obtained by the EA-based clustering algorithm were when the crossover rate was set to 0.7 and the mutation rate was set to 0.003.

As shown in Tables 3.1 and 3.2, compared with other clustering algorithms, EvoCluster performs better in terms of all measures. Moreover, it seems that none of the existing algorithms is particularly effective when handling very noisy data such as the simulated data. This shows that EvoCluster is very robust in the presence of a very noisy environment. Being able to effectively discover hidden patterns, it should be noted that EvoCluster has the additional advantage of being able to provide a "justification" of the cluster grouping it discovers. Examples of such discovered patterns are given in Table 3.3. For example, the pattern "A3 = [0.011, 0.257] (0.98)" was discovered in C1. This pattern should be understood as: "If the gene expression value of a gene under the experimental condition, A3, is within the interval from 0.011 to 0.257, then there is a probability of 0.98 that this gene belongs to the cluster, C1."

3.4.2.2 Gene Expression Data

The performance of EvoCluster was also evaluated using real expression data. The minimum and maximum number of clusters considered for both data sets 1 and 2 were set at (MIN = 6, MAX = 10) and (MIN = 4, MAX = 8), respectively. Other parameter settings of EvoCluster used in both data sets 1 and 2 are as follows: $P_c = 0.5$, $P_m = 0.17$, $L_g = 0.2$, $U_g = 0.8$, $L_r = 0.2$, and $U_r = 0.8$. As with the simulated data, the experiments with data sets 1 and 2 were repeated three times with a mixture of unguided and guided operators, unguided operators alone, and guided operators alone, respectively. Based on the results showed in Tables 3.4 and 3.5, we found that using both unguided and guided operators together, once again, gave us the best clustering results. The performances of EvoCluster in comparison with other algorithms are given in Tables 3.6 and 3.7. For the EA-based clustering algorithm, the best results were obtained by using the following parameter settings: (1) data set 1, the crossover rate = 0.8 and the mutation rate = 0.001, and (2) data set 2, the crossover rate = 0.8 and the mutation rate = 0.004. The F measure was not computed for the experiments with real data sets as the original, correct clustering results remained unknown.

According to Tables 3.4 to 3.7, EvoCluster again performs better than others even with the combination of the feature selection method. Based on the clustering results obtained by EvoCluster, we are able to discover some interesting patterns that may have great biological significance. For example, for data set 1, when k = 6 (which gives the best results), and for data set 2, when k = 5 (which gives the best results), some discovered patterns are shown in Tables 3.8 and 3.9, respectively.

Table 3.4 Comparison of Clustering Performance Using "Unguided + Guided" Operators, "Unguided" Operators, or "Guided" Operators (data set 1)

	k	No. of Reproduction	DBI	Predictive Power (%)
Unguided + guided operators	6	3325	1.59	84.88
	7	3701	1.64	81.25
	8	4088	1.67	78.19
	9	4717	1.67	74.26
	10	4779	1.68	73.26
Unguided operators	6	4493	1.69	67.28
	7	4989	1.69	65.36
	8	4777	1.79	57.19
	9	4815	1.79	58.34
	10	4622	1.78	59.71
Guided operators	6	2089	1.64	73.23
	7	3385	1.65	71.02
	8	2876	1.68	67.49
	9	4002	1.68	66.33
	10	4747	1.7	68.51

Table 3.5 Comparison of Clustering Performance Using "Unguided + Guided" Operators, "Unguided" Operators, or "Guided" Operators (data set 2)

	k	No. of Reproduction	DBI	Predictive Power (%)
Unguided + guided operators	4	3104	1.52	87.10
	5	3631	1.48	89.86
	6	4703	1.57	80.90
	7	4464	1.58	83.42
	8	4405	1.59	77.33
Unguided operators	4	4852	1.62	67.68
	5	4197	1.61	70.41
	6	4763	1.65	62.33
	7	4945	1.64	64.75
	8	4572	1.65	62.59
Guided operators	4	2670	1.57	76.13
	5	1978	1.57	78.47
	6	2404	1.59	75.94
	7	4204	1.6	72.16
	8	3978	1.61	70.08

Table 3.6 Comparison of Average DBI and Predictive Power (data set 1)

	DBI					Predictive Power (%)				
	$k = 6$	$k = 7$	$k = 8$	$k = 9$	$k = 10$	$k = 6$	$k = 7$	$k = 8$	$k = 9$	$k = 10$
Evo Cluster	1.59	1.64	1.67	1.67	1.68	84.88	81.25	78.19	74.26	73.26
EA	1.65	1.7	1.69	1.72	1.73	72.26	61.37	67.12	60.51	64.86
EA + FS	1.62	1.69	1.68	1.69	1.7	78.43	69.55	72.38	65.43	70.24
k means	1.67	1.7	1.71	1.7	1.72	67.23	63.19	61.40	64.77	62.59
k means + FS	1.63	1.68	1.7	1.68	1.7	75.86	72.37	68.88	70.12	71.11
SOM	1.74	1.77	1.78	1.79	1.85	62.88	63.75	56.21	56.87	48.38
SOM + FS	1.69	1.72	1.75	1.75	1.79	69.14	70.04	66.56	68.38	59.48
Hierarchical	1.71	1.76	1.73	1.76	1.79	65.30	59.86	63.11	62.80	58.21
Hierarchical + FS	1.7	1.74	1.72	1.75	1.76	71.75	66.32	68.29	67.74	68.86

These patterns can be interpreted as follows. In Table 3.8, the pattern discovered in C1 "alpha21 = [0.45, 2.12] (0.86)" states that if the gene expression value of a gene, under the experimental condition, alpha21, was within the interval from 0.45 to 2.12, then there is a probability of 0.86 that it belongs to cluster C1. In Table 3.9, the pattern discovered in C3 "Cond3 = [−2.874, −1.413] (0.94)" means that if the gene expression value of a gene, under the experimental condition, Cond3, is within

Table 3.7 Comparison of Average DBI and Predictive Power (data set 2)

	DBI					Predictive Power (%)				
	$k = 4$	$k = 5$	$k = 6$	$k = 7$	$k = 8$	$k = 4$	$k = 5$	$k = 6$	$k = 7$	$k = 8$
Evo Cluster	1.52	1.48	1.57	1.58	1.59	87.10	89.86	80.90	83.42	77.33
EA	1.61	1.54	1.61	1.61	1.68	71.47	75.82	68.96	71.08	63.43
EA + FS	1.56	1.51	1.59	1.6	1.63	79.06	81.33	74.42	76.66	68.79
k means	1.59	1.56	1.62	1.66	1.68	76.43	70.68	66.98	64.77	63.12
k means + FS	1.56	1.52	1.6	1.61	1.64	80.02	77.48	75.92	72.37	69.55
SOM	1.62	1.64	1.62	1.7	1.79	72.38	68.53	64.26	63.84	61.39
SOM + FS	1.58	1.55	1.59	1.67	1.72	78.21	75.51	69.23	70.06	70.84
Hierarchical	1.61	1.59	1.58	1.68	1.69	74.39	75.22	73.23	64.66	62.30
Hierarchical + FS	1.6	1.56	1.57	1.61	1.67	75.37	76.82	75.16	66.79	64.23

Table 3.8 Association Patterns Discovered in Each Cluster (data set 1)

C1	alpha21 = [0.45, 2.12] (0.86)
	cdc28_50 = [0.56, 3.35] (0.88)
C2	cdc15_270 = [0.43, 1.96] (0.83)
	alpha35 = [0.04, 0.24] (0.86)
C3	elu360 = [−1.27, −0.29] (0.92)
	cdc15_30 = [0.53, 2.54] (0.84)
C4	cdc15_120 = [−2.5, −0.42] (0.92)
	alpha42 = [−2.24, −0.21] (0.88)
C5	cdc28_150 = [−1.83, −0.32] (0.9)
	cdc28_20 = [0.56, 3.35] (0.85)
C6	elu270 = [0.32, 1.32] (0.9)
	alpha49 = [−2.17, −0.33] (0.84)

Table 3.9 Association Patterns Discovered in Each Cluster (data set 2)

C1	Cond4 = [−2.564, −1.275] (0.96)
	Cond10 = [1.481, 2.974] (0.92)
C2	Cond12 = [0.872, 1.953] (0.9)
	Cond11 = [0.995, 2.258] (0.86)
C3	Cond3 = [−2.847, −1.413] (0.94)
	Cond17 = [−0.23, 0.861] (0.86)
C4	Cond11 = [−1.53, −0.267] (0.9)
	Cond3 = [0.021, 1.454] (0.92)
C5	Cond8 = [−0.284, 1.01] (0.9)
	Cond1 = [−0.139, 1.247] (0.88)

the interval from −2.874 to −1.413, then there is a probability of 0.94 that it belongs to cluster C2.

With the fact that the gene expression patterns discovered in each cluster are different, we attempted to see if there are any well-known binding sites in each discovered cluster. To do so, we looked at the corresponding promoter regions for the genes in each cluster by downloading the sequences from the Saccharomyces Genome Database (SGD) (Ball et al., 2001). We used a popular motif discovery algorithm described in Helden et al. (1998) to try to search for transcription factor binding sites in the DNA sequences. Since the essential element of most binding sites found is in the form of a six-character motif such as ATTCGT (Helden et al., 1998), we also tried to detect six-character patterns in the DNA sequences of each cluster. All discovered sites in each cluster were then checked against the well-known binding sites reported in the published literature (Spellman et al., 1998; Helden et al., 2000). As shown in Table 3.10, we did discover the patterns that are well-known transcription factor binding sites for being involved in the cell cycle. Moreover, in addition to identifying known binding sites, we were able to discover some other potentially important sites (Helden et al., 1998) as shown in Table 3.11. The validity of these sites awaits future confirmation by biologists using different biochemical methods (Suzuki et al., 1995).

Comparing against EvoCluster, other clustering algorithms are only able to discover some of the known binding sites in some of the clusters they discovered (Table 3.12). This is an indication that the cluster groupings discovered by EvoCluster are more biologically meaningful and significant than the groupings discovered by others. The total numbers of confirmed and suspected binding sites discovered in

Table 3.10 Known Transcription Factor Binding Sites Revealed from Discovered Clusters

	Sequence Revealed	Binding Site Name
Data Set 1		
C1	ACGCGT	MCB
C2	CGCGAA	SCB
C3	CCAGCA	Swi5; Ace2
C4	CCCAAA	Mcm1
C5	CTGTGG	Met31; Met32
C6	AAACAA	SFF
Data Set 2		
C1	TAAACA	Mcm1
C2	CTGTCC (potential variant of CTGTGG)	Met31; Met32
C3	CCAGCA	Swi5; Ace2
C4	AAGAAA	SCB
C5	ACGCGT	MCB

Table 3.11 Putative Transcription Factor Binding Sites Revealed from Discovered Clusters

Data Set 1	Sequence Revealed	Data Set 2	Sequence Revealed
C1	AACTCG	C1	GATGCC
	ACCTGG		CTCGAC
	ACGCGA		AGAAAC
	GCGTTT		GGTTGA
C2	GCAATG	C2	TGGACA
	GACGCG		GGTGAT
	TCATGG		TGTCCA
	ATCGTC		GGTGAC
C3	TGATCG	C3	CCAGCC
	GACTGT		AGATCG
	GAGCCA		AGGTGA
	TGGTTT		CCTTGC
C4	GGCTGG	C4	AGGAAA
	GAAATT		AGACCA
	GGTCAA		CTCTAA
	TTGGGT		TAGCAC
C5	TAGGAA	C5	GTCGCG
	GGCCCA		CGCGTT
	TGGATG		CGACGC
	TCCAAG		CGACGC
	GATTGA		CGTTGC
	ACTGAT		GAAGTT
C6	GCTAGA		
	CCACAG		
	GTGTGC		
	CCATGA		
	AACCGT		

the clusters found by the various clustering algorithms are also given in Table 3.13. In both data sets, EvoCluster is able to find many more such binding sites.

3.5 CONCLUSIONS

With the advent of microarray technology, we are now able to monitor simultaneously the expression levels of thousands of genes during important biological processes. Due to the large number of data collected everyday and due to the very noisy nature in the data collection process, interpreting and comprehending the experimental results has become a big challenge. To discover hidden patterns in gene expression microarray data, we presented a novel clustering algorithm, EvoCluster.

Table 3.12 Discovery of Known Transcription Factor Binding Sites in Each Cluster by Different Clustering Algorithms (site name and the no. of occurrences)

	EvoCluster	EA	*k* Means	SOM	Hierachical
Data Set 1					
C1	MCB (139)	MCB (95)	MCB (79)	—	MCB (61)
C2	SCB (86)	—	—	—	SCB (31)
C3	Swi5; Ace2 (44)	—	—	—	—
C4	Mcm 1 (62)	Mcm1 (55)	Mcm1 (40)	Mcm1 (36)	—
C5	Met31; Met32 (102)	—	—	—	—
C6	SFF (207)	SFF (116)	SFF (134)	SFF (89)	SFF (107)
Data Set 2					
C1	Mcm1 (64)	Mcm1 (32)	Mcm1 (23)	—	Mcm1 (26)
C2	Met31; Met32 (99)	Met31; Met32 (58)	—	—	—
C3	Swi5; Ace2 (32)	—	Swi5; Ace2 (21)	Swi5; Ace2 (25)	—
C4	SCB (138)	SCB (104)	SCB (119)	SCB (114)	SCB (96)
C5	MCB (101)	MCB (77)	MCB (85)	MCB (65)	MCB (57)

Table 3.13 Total Number of Known and Putative Transcription Factor Binding Sites Discovered in Each Cluster by Different Clustering Algorithms

	EvoCluster	EA	*k* Means	SOM	Hierachical
Data Set 1					
C1	12	5	8	4	5
C2	21	9	7	5	12
C3	9	6	3	—	—
C4	4	4	3	3	—
C5	24	8	13	7	4
C6	15	11	10	6	6
Data Set 2					
1	14	9	6	4	4
2	11	7	2	3	—
3	8	5	5	3	2
4	17	8	12	9	3
5	19	15	9	13	10

EvoCluster encodes an entire cluster grouping in a *chromosome* so that each *gene* encodes one cluster. Based on such a structure, it makes use of a set of reproduction operators to facilitate the exchange of grouping information between *chromosomes*. The fitness function it adopts is able to differentiate between how relevant a feature value is in determining a particular cluster grouping. As such, instead of just local pairwise distances, it also takes into consideration how clusters are arranged globally. EvoCluster does not require the number of clusters to be decided in advance. Patterns hidden in each cluster can be explicitly revealed and presented for easy interpretation. We have tested EvoCluster using both simulated and real data. Experimental results show that EvoCluster is very robust in the presence of noise. It is able to search for near-optimal solutions effectively and discover interesting association patterns in the noisy data for meaningful groupings. The results also show that, under some common performance measures, EvoCluster is better than other algorithms used in gene expression data analysis, and the discovered clusters contain more biologically meaningful patterns. In particular, we could correlate the clusters of coexpressed genes discovered by EvoCluster to their DNA sequences and found that we were able to uncover significant well-known and new biological binding sites in each cluster of sequences. Besides gene expression data, EvoCluster can be used to cluster other biological data such as DNA and protein sequence data, and most importantly, it can also be used for solving different kinds of clustering problems in other application areas. Moreover, in order to cope with very large gene expression data sets, the inherently parallel nature of the problem-solving process of EvoCluster can be exploited. In addition, the effect of varying the parameters used by EvoCluster has on its performance can also be investigated. For further work, we intend to look into how specifically these can be done and compare EvoCluster with other EA-based clustering methods that have emerged recently (Bleuler et al., 2004; Handl and Knowles, 2007; Speer et al., 2004).

REFERENCES

ALON, U., N. BARKAI, D. A. NOTTERMAN, K. GISH, S. YBARRA, D. MACK, and A. J. LEVINE (1999). "Broad patterns of gene expression revealed by clustering analysis of tumor and normal colon tissues probed by oligonucleotide arrays," *Proc. Natl. Acad. Sci., USA*, Vol. 96, pp. 6745–6750.

ASHLOCK, D. (2006). *Evolutionary Computation for Modeling and Optimization*. Springer, New York.

AU, W. H., K. C. C. CHAN, and X. YAO (2003) "A novel evolutionary data mining algorithm with applications to churn modeling," *IEEE Trans. Evolut. Computat., Special Issue on Data Mining and Knowledge Discovery with Evolutionary Algorithms*, Vol. 7, pp. 532–545.

BABU, G. P. and M. N. MURTY (1994). "Clustering with evolution strategies," *Pattern Recog.*, Vol. 27, pp. 321–329.

BAECK, T., D. FOGEL, and Z. MICHALEWICZ (2000a). *Evolutionary Computation 1: Basic Algorithms and Operators*. Institute of Physics, Bristol, United Kingdom.

BAECK, T., D. FOGEL, and Z. MICHALEWICZ (2000b). *Evolutionary Computation 2: Advanced Algorithms and Operators*. Institute of Physics, Bristol, United Kingdom.

BALL, C. A., H. JIN, G. SHERLOCK, S. WENG, J. C. MATESE, R. ANDRADA, G. BINKLEY, K. DOLINSKI, S. S. DWIGHT, M. A. HARRIS, L. ISSEL-TARVER, M. SCHROEDER, D. BOTSTEIN, and J. M. CHERRY (2001). "*Saccharomyces* genome database provides tools to survey gene expression and functional analysis data," *Nucleic Acids Res.*, Vol. 29, pp. 80–81.

BEN-DOR, A., R. SHAMIR, and Z. YAKHINI (1999). "Clustering gene expression patterns," *J. Comp. Biol.*, Vol. 6, pp. 281–297.

BERRAR, D. P., W. DUBITZKY, and M. GRANZOW (2003). *A Practical Approach to Microarray Data Analysis*. Kluwer Academic, Boston.

BLEULER, S., A. PRELIC, and E. ZITZLER (2004). "An EA framework for biclustering of gene expression data," *Proc. of Cong. Evolution. Computat.*, pp. 166–173.

CANO, J. R., F. HERRERA, and M. LOZANO (2003). "Using evolutionary algorithms as instance selection for data reduction in KDD: An experimental study," *IEEE Trans. Evolution. Computat.*, Vol. 7, pp. 561–575.

CHAN K. C. C. and L. L. H. CHUNG (1999). "Discovering clusters in databases containing mixed continuous and discrete-valued attributes," Proc. of SPIE AeroSense 99 Data Mining and Knowledge Discovery: Theory, Tools, and Technology, pp. 22–31.

CHAN, K. C. C. and A. K. C. WONG (1991). "A Statistical Technique for Extracting Classificatory Knowledge from Databases," in *Knowledge Discovery in Databases* (G. Piatesky-Shapiro and W.J. Frawley, Eds.). AAAI/MIT Press, Cambridge, MA, pp. 107–123.

CHAN, K. C. C., A. K. C. WONG, and D. K. Y. CHIU (1994). "Learning sequential patterns for probabilistic inductive prediction," *IEEE Trans. Systems, Man and Cybernetics*, Vol. 24, pp. 1532–1547.

CHI, Z., X. WEIMIN, T. M. TIRPAK, and P. C. NELSON (2003). "Evolving accurate and compact classification rules with gene expression programming," *IEEE Trans. Evolution. Computat.*, Vol. 7, pp. 519–531.

CHO, R. J., M. J. CAMPBELL, E. A. WINZELER, L. STEINMETZ, A. CONWAY, L. WODICKA, T. G. WOLFSBERG, A. E. GABRIELIAN, D. LANDSMAN, D. J. LOCKHART, and R. W. DAVIS (1998). "A genome-wide transcriptional analysis of the mitotic cell cycle," *Mol. Cell*, Vol. 2, pp. 65–73.

CHO, R. J., M. HUANG, M. J. CAMPBELL, H. DONG, L. STEINMETZ, L. SAPINOSO, G. HAMPTON, S. J. ELLEDGE, R. W. DAVIS, and D. J. LOCKHART (2001). "Transcriptional regulation and function during the human cell cycle," *Nat. Genet.*, Vol. 27, pp. 48–54.

COX, T. M. and J. SINCLAIR (1997). *Molecular Biology in Medicine*, Blackwell Science, Oxford.

DAVIES, D. L. and D. W. BOULDIN (1979). "A cluster separation measure," *IEEE Trans. Pattern Anal. Machine Intell.*, Vol. 1, pp. 224–227.

DEGROOT, M. H. and M. J. SCHERVISH (2002). *Probability and Statistics*, Addison-Wesley, Boston.

DE JONG, K. A. (2006). *Evolutionary Computation: A Unified Approach*. MIT Press, Cambridge, MA.

DERISI, J. L., V. R. IYER, and P. O. BROWN (1997). "Exploring the metabolic and genetic control of gene expression on a genomic scale," *Science*, Vol. 278, pp. 680–686.

DING, C. (2002). "Analysis of gene expression profiles: class discovery and leaf ordering," Proc. of the 6th International Conference of Research in Computational Molecular Biology (RECOMB), pp. 127–136.

EISEN, M. B., P. T. SPELLMAN, P. O. BROWN, and D. BOTSTEIN (1998). "Cluster analysis and display of genome-wide expression patterns," *Proc. Natl Acad. Sci. USA*, Vol. 95, pp. 14863–14868.

FALKENAUER, E. (1998). *Genetic Algorithms and Grouping Problems*. Wiley, New York.

FERNANDES, P. M., T. DOMITROVIC, C. M. KAO, and E. KURTENBACH (2004). "Genomic expression pattern in *Saccharomyces cerevisiae* cells in response to high hydrostatic pressure," *FEBS Lett.*, Vol. 556, pp. 153–160.

FRANTI, P., J. KIVIJARVI, T. KAUKORANTA, and O. NEVALAINEN (1997). "Genetic algorithms for large-scale clustering problems," *Computer J.*, Vol. 40, pp. 547–554.

GOLDBERG, D. E. (1989). *Genetic Algorithms in Search, Optimization and Machine Learning*. Addison-Wesley, New York.

GOLUB, T. R. (1999). "Molecular classification of cancer: Class discovery and class prediction by gene expression monitoring," *Science*, Vol. 286, pp. 531–537.

HALL, L. O., I. B. OZYURT, and J. C. BEZDEK (1999). "Clustering with a genetically optimized approach," *IEEE Trans. Evolut. Computat.*, Vol. 3, pp. 103–112.

HANDL J. and J. KNOWLES (2007) "An Evolutionary Approach to Multiobjective Clustering," *IEEE Trans. Evolut. Computat.*, Vol. 11, pp. 56–76.

HARTL, D. L. and E. W. JONES (2001). *Genetics: Analysis of Genes and Genomes*. Jones & Bartlett, Sudbury, MA.

Helden, J. V., B. Andre, and J. Collado-Vides (1998). "Extracting regulatory sites from the upstream region of yeast genes by computational analysis of oligonucleotide frequencies," *J. Mol. Biol.*, Vol. 281, pp. 827–842.

Helden, J. V., A. F. Rios, and J. Collado-Vides (2000). "Discovering regulatory elements in non-coding sequences by analysis of spaced dyads," *Nucleic Acids Res.*, Vol. 28, pp. 1808–1818.

Herrero, J., A. Valencia, and J. Dopazo (2001). "A hierarchical unsupervised growing neural network for clustering gene expression patterns," *Bioinformatics*, Vol. 17, pp. 126–136.

Jaeger, J., R. Sengupta, and W. L. Ruzzo (2003). "Improved gene selection for classification of microarrays," *Pacific Symp. on Biocomput.*, Vol. 8, pp. 53–64.

Jain, A. K. and R. C. Dubes (1998). *Algorithms for Clustering Data*. Prentice Hall, Englewood Cliffs, NJ.

Kohonen, T. (1989). *Self-Organization and Associative Memory*. Springer, New York.

Kowalski, G. (1997). *Information Retrieval Systems—Theory and Implementation*, Kluwer Academic, Boston.

Krishna, K. and M. N. Murty (1999). "Genetic *k*-means algorithm," *IEEE Trans. Syst., Man Cybern.–Part B*, Vol. 29, pp. 433–439.

Lapointe, J., C. Li, J. P. Higgins, M. Van De Rijn, E. Bair, K. Montgomery, M. Ferrari, L. Egevad, W. Rayford, U. Bergerheim, P. Ekman, A. M. DeMarzo, R. Tibshirani, D. Botstein, P. O. Brown, J. D. Brooks, and J. R. Pollack (2004). "Gene expression profiling identifies clinically relevant subtypes of prostate cancer," *Proc. Natl. Acad. Sci. USA*, Vol. 101, pp. 811–816.

Larsen, B. and C. Aone (1999). "Fast and effective text mining using linear-time document clustering," Proc. KDD-99 Workshop, San Diego, CA.

Lashkari, D. A., J. L. DeRisi, J. H. McCusker, A. F. Namath, C. Gentile, S. Y. Hwang, P. O. Brown, and R. W. Davis (1997). "Yeast microarrays for genome wide parallel genetic and gene expression analysis," *Proc. Natl. Acad. Sci. USA*, Vol. 94, pp. 13057–13062.

Li, W. and Y. Yang (2000). "How many genes are needed for a discriminant microarray data analysis," Critical Assessment of Techniques for Microarray Data Mining Workshop, pp. 137–150.

Lu Y. and J. Han (2003). "Cancer classification using gene expression data," *Info. Syst.*, Vol. 28, pp. 243–268.

Ma, P. C. H., K. C. C. Chan, X. Yao, and D. K. Y. Chiu (2006). "An evolutionary clustering algorithm for gene expression microarray data analysis," *IEEE Trans. Evolut. Comput.*, Vol. 10, pp. 296–314.

MacQueen, J. (1967). "Some methods for classification and analysis of multivariate observation," *Proc. Symp. Math. Stat. Prob. Berkeley*, Vol. 1, pp. 281–297.

Maulik, U. and S. Bandyopadhyay (2000). "Genetic algorithm-based clustering technique," *Pattern Recog.*, Vol. 33, pp. 1455–1465.

Muni, D. P., N. R. Pal, and J. Das (2004). "A novel approach to design classifiers using genetic programming," *IEEE Trans. Evolut. Comput.*, Vol. 8, pp. 183–196.

Murthy, C. A. and N. Chowdhury (1996). "In search of optimal clusters using genetic algorithms," *Pattern Recog. Lett.*, Vol. 17, pp. 825–832.

Park Y. and M. Song (1998). "A genetic algorithm for clustering problems," Genetic Programming 1998: Proc. 3rd. Annual Conf., pp. 568–575.

Parpinelli, R. S., H. S. Lopes, and A. A. Freitas (2002). "Data mining with an ant colony optimization algorithm," *IEEE Trans. Evolut. Comput.*, Vol. 6, pp. 321–332.

Quackenbush, J. (2001). "Computational analysis of microarray data," *Nat. Rev. Genet.*, Vol. 2, pp. 418–427.

Quinlan, J. R. (1993). *C4.5: Programs for Machine Learning*. Morgan Kaufmann, San Francisco.

Sherlock, G. (2000). "Analysis of large-scale gene expression data," *Curr. Opin. Immunol.*, Vol. 21, pp. 201–205.

Speer, N., C. Spieth, and A. Zell (2004). "A memetic clustering algorithm for the functional partition of genes based on the gene ontology," Proc. of the IEEE Symposium on Computational Intelligence in Bioinformatics and Computational Biology, pp. 252–259.

Spellman, P. T., G. Sherlock, M. Q. Zhang, V. R. Iyer, K. Anders, M. B. Eisen, P. O. Brown, D. Botstein, and B. Futcher (1998). "Comprehensive identification of cell cycle-regulated genes of the yeast *Saccharomyces cerevisiae* by microarray hybridization," *Mol. Biol. Cell*, Vol. 9, pp. 3273–3297.

SU, Y., T. M. MURALI, V. PAVLOVIC, M. SCHAFFER, and S. KASIF (2003). "RankGene: Identification of diagnostic genes based on expression data," *Bioinformatics*, Vol. 19, pp. 1578–1579.

SUZUKI, M., S. E. BRENNER, M. GERSTEIN, and N. YAGI (1995). "DNA recognition code of transcription factors," *Protein Eng.*, Vol. 8, pp. 319–328.

TAMAYO, P., D. SLONIM, J. MESIROV, Q. ZHU, S. KITAREEWAN, E. DMITROVSKY, E. S. LANDER, and T. R. GOLUB (1999). "Interpreting patterns of gene expression with self-organizing maps: methods and application to hematopoietic differentiation," *Proc. Natl. Acad. Sci. USA*, Vol. 96, pp. 2907–2912.

TAVAZOIE, S., J. D. HUGHES, M. J. CAMPBELL, R. J. CHO, and G. M. CHURCH (1999). "Systematic determination of genetic network architecture," *Nat. Genet.*, Vol. 22, pp. 281–285, 1999.

WANG, Y. and A. K. C. WONG (2003). "From association to classification: Inference using weight of evidence," *IEEE Trans. Knowledge Data Eng.*, Vol. 15, pp. 764–767.

WARD, J. H. (1963). "Hierarchical grouping to optimize an objective function," *J. Am. Stat. Assoc.*, Vol. 58, pp. 236–244.

XING, E. P. and R. M. KARP (2001). "CLIFF: Clustering of high-dimensional microarray data via iterative feature filtering using normalized cuts," *Bioinformatics*, Vol. 17, pp. S306–S315.

XIONG, M., X. FANG, and J. ZHAO (2001). "Biomarker identification by feature wrappers," *Genome Res.*, Vol. 11, pp. 1878–1887.

ZHENG, J., J. WU, and Z. SUN (2003). "An approach to identify over-represented cis-elements in related sequences," *Nucleic Acids Res.*, Vol. 31, pp. 1995–2005.

Chapter 4

Gene Networks and Evolutionary Computation

Jennifer Hallinan

4.1 INTRODUCTION

4.1.1 Gene Networks

There are approximately 23,000 genes[1] in the human genome, which produce thousands of proteins. The true size of the human proteome—the full complement of proteins—is still unknown, but Harrison et al. (2002) estimate, on the basis of a series of more-or-less reasonable assumptions, that "the human proteome size is likely to be significantly larger than approximately 90,000." These proteins interact with each and with myriad other biomolecules, such as noncoding RNAs (ribonucleic acids) (e.g., Mattick, 2001; Ambros, 2004), nutrients, and metabolites to build the complex metabolic machinery of the cell. Gene expression, and hence cellular phenotype, is controlled by a complex, dynamic network of intracellular interactions.

Gene networks are not only complex, they tend to be very large. Even the organism with the smallest known genome,[2] *Mycoplasma genitalium*, has 470 protein-coding genes (Fraser et al., 1995). This is not a large number of genes, but even if we assume that each gene only produces one product (an assumption that is certainly false in most eukaryotes), this genome could potentially produce a network of over 200,000 interactions. Understanding gene networks is essential to understanding the workings of cells, but their size and complexity make this a challenging undertaking.

[1] The Ensemble database (http://www.ensembl.org/Homo_sapiens/index.html) lists 22,205 known protein-coding genes as of August, 2006, and more probably remain to be identified.

[2] Counting only free-living organisms. Parasites, particularly intracellular parasites, can have even smaller genomes because they can utilize the products of host genes, allowing them to eliminate these genes from their own genomes.

Computational Intelligence in Bioinformatics. Edited by G. B. Fogel, D. Corne, and Y. Pan

Over the last decade advances in high-throughput technology have led to the generation of large amounts of data about various types of genetic interactions, such as protein–protein interactions (e.g., Xenarios et al., 2002), transcription factor binding (Wingender et al., 1996), gene expression (Sherlock et al., 2001), and many other aspects of gene regulation. The traditional approach of molecular biology (if a field that only extends back 50 years can be said to have traditions) is to concentrate on a single gene at a time, teasing out the details of its functions and interactions with exquisite precision. The (slightly ironic) rule of thumb in molecular biology laboratories in the twentieth century was "one gene–one Ph.D." However, the recent abundance of data has triggered an upsurge of interest in gene networks: their topology, their dynamics, and the relationship between the two. Systems biology is a rapidly growing field, and much of systems biology centers on gene networks.

Despite the fact that huge amounts of biological data are currently being produced, for many problems there is still not enough detailed information. Reconstructing gene networks in detail requires data about many different kinds of interactions, the circumstances under which they occur, and the kinetic parameters of each interaction. Biological data sets are noisy and incomplete in ways that are hard to quantify, and the systematic collection of essential elements such as kinetic parameters is labor intensive and hard, or even impossible, at present.

An alternative to the construction of detailed network models based on real data is the building and analysis of computational models. Computational models of gene networks yield different insights from completely data-based models. The primary difference is that computational models are built at a higher level of abstraction and generate information about global, emergent properties of networks. At our current level of biological understanding such insights are uniquely valuable. These models also serve to lay the foundation for future, more detailed, modeling. Computationally generated models can be of a size constrained only by the available computational facilities and are not dependent upon the time-consuming and expensive generation of data. They also have a major advantage over strictly biological networks in that multiple instances of a particular class of model can be generated and studied, permitting statistical analysis of their structure and behavior. Significant, generally applicable characteristics of networks can thus be distinguished from those arising by chance in a single network.

In this review we will consider the role of evolutionary computation in both optimization and the modeling of gene networks. We will then examine the application of evolutionary computing techniques to the understanding of network topology and end with speculation and suggestions as to the future of this exciting field of research.

4.1.2 Representing Gene Networks

Conceptually, a gene network can be thought of as a graph $G = (V, E)$, where V is a set of vertices and E is a set of edges between vertices. Edges may be directed, in which case they are known as arcs, or undirected, depending upon the nature of the interactions they represent. A genetic regulatory interaction, for example, is generally directed; the fact that the product of gene *A* activates gene *B* does not necessarily mean that the reverse is true. In contrast, a protein–protein interaction is generally symmetrical—if A binds to B, B necessarily binds to A—and would be represented by an undirected edge. Edges may also be weighted, with weights representing factors important to the interaction, such as rate constants or binding affinities.

Graph theory is a field of mathematics with a long history (see, e.g., Erdös and Rényi, 1959), and graph analysis has been applied to networks in fields as diverse as sociology, economics, computer science, physics, and, of course, biology. There is a large body of literature on computational modeling of genetic networks, of which this review covers only that concerned with the use of evolutionary algorithms. For reviews of the broader literature, see Schilling et al. (1999), Reil (2000), Endy and Brent (2001), De Jong (2002), Gilman and Larkin (2002), and Bolouri and Davidson (2002).

Although the graph representation of a network appears straightforward, when gene nets are constructed on the basis of real biological data the interpretation of an edge may not be as obvious as it appears. The information that *A* activates *B* may well subsume a multitude of biological interactions that have not been directly measured. Epigenetic modifications such as chromatin remodeling and DNA (deoxyribonucleic acid) methylation, the formation of complexes with other proteins and RNAs, and alterations to kinetic parameters such as biomolecular synthesis and degradation rates, as well as many other factors, can all affect genetic regulation. However, when measured using, for example, a microarray, all of these interactions would simply appear as "an increase in activity of gene *A* leads to increased transcription of gene *B*." Gene network representations must be interpreted with caution.

Computationally, graphs are often stored as adjacency matrices. An adjacency matrix is an *n* by *n* matrix of numbers, where *n* is the number of vertices in the network. In the simplest, unweighted, case a matrix entry of 0 indicates no interaction between the vertices in question, while an entry of 1 means that an edge exists. If the edge is weighted, the matrix entry can be a floating-point number representing the weight on that edge. Since biological networks are usually sparse, a more storage-efficient data structure, such as a linked list of adjacent vertices, is often used. Other computational representations are, of course, possible, but are less frequently used. With the advent of more complex integrated functional networks (see Section 4.4), in which a single edge can represent an amalgam of different types of interaction, network data is increasingly being stored in dedicated databases that can be queried in different ways as needed (e.g., Baitaluk et al., 2006).

The topological analysis of a network can be done directly from the graph representation. There are several excellent network analysis packages available, implementing a variety of published algorithms [e.g., Pajek (Batagelj and Mrvar, 1998), Osprey (Breitkreutz et al., 2003), Cytoscape (Shannon et al., 2003), ONDEX (Kohler et al., 2006), and many others].

The other major type of network analysis involves examination of the network's dynamic behavior; the patterns of node activation (usually interpreted as gene expression) over time. If the number of nodes is small enough, the entire state space of a network can be exhaustively analyzed, measured, and diagrammed [e.g., by using a tool such as DDLab (Wuensche, 2002)]. However, the number of states in a network's state space increases exponentially with the number of vertices; once a network contains more than around 30 to 50 vertices, exhaustive enumeration of the state space becomes computationally infeasible.

For large networks dynamics are usually assessed by "running" a network in a stochastic simulation mode. An appropriate start state is selected, and then the update rules of the network are applied, either synchronously (all vertices update their state simultaneously) or asynchronously (a subset of vertices is updated at any time step). The state of every vertex is tracked over time to generate a view of the activity of the entire network. Since different start states will lead into different trajectories through the network state space, this process must be iterated until an adequate picture of the total network dynamics is produced. This "statistical sampling" approach does have a tendency to underestimate the number of small basins of attraction in a network's state space.

Armed with an understanding of the practicalities of gene network analysis, we now turn to the question of how evolutionary computation can be used in a gene network analysis context.

4.2 EVOLUTIONARY OPTIMIZATION

4.2.1 Networks Based on Biological Data

Many researchers are interested in modeling small networks, closely tailored to experimental data. For example, a cancer researcher may be passionately interested in *p53*, a gene dubbed the "guardian of the genome," whose activity is vital to the smooth operation of the cell cycle. *p53* induces the production of a protein known as mdm2; mdm2 in turn inhibits the production of *p53*.[3] The details of this interaction—binding affinities, kinetic parameters, and the like—are known with considerable precision. As a consequence the *p53*/mdm2 loop, and its interaction with other pathways controlling the cell cycle, can be modeled accurately enough to yield useful predictions about issues such as the likely effects of anticancer drugs.

For small models like the *p53*/mdm2 network, the network topology and behavior can be strictly tailored to the data of interest, and the size of the network is limited by the amount of data available at the desired level of detail. Tailored models are particularly valuable as part of an experimental program, in which they can be used to generate hypotheses that can be tested in the laboratory, producing results that can be used, in turn, to refine the model. The process of building a tailored model is ideally a matter of iteration between experimental biology and computation.

Small tailored models tend to have a number of free parameters whose values must be set appropriately to produce the desired behavior in the network. There

[3] Actually, it increases the rate of degradation of *p53*, but the net effect is the same.

are numerous approaches to parameter estimation, both deterministic and stochastic (Mendes and Kell, 1998; Moles et al., 2003). The parameter sweep approach, in which a network is constructed and then the values of the parameters are systematically varied across a range deemed biologically plausible, can yield good results, but is computationally expensive, and may not stumble upon combinations of parameter values that are not intuitively obvious to the programmer. Simple guessing of sets of parameter values has been applied to good effect (e.g., von Dassow et al., 2000), but unless the topology of the network is robust to variations in parameter values, this approach is unlikely to be successful. An alternative approach is to establish the general framework in which the network is to operate and to use evolutionary computation to optimize it to the problem at hand. Stochastic approaches, such as evolutionary algorithms, have a number of advantages (Michalewicz and Schoenauer, 1996), one of which is potentially identifying multiple sets of parameter values, each of which produces the desired behavior. Critical comparison of the parameter values found by different runs of an algorithm can provide valuable information about the robustness of the underlying system and, particularly, which parameters are most important. Of the stochastic algorithms, evolutionary algorithms have been found to be particularly useful for parameter estimation in complex nonlinear systems such as models of biological pathways (e.g., Park et al., 1997; Moles et al., 2003).

Although evolutionary computation (EC) can be time consuming and may not produce the globally best solution to a problem, it is flexible, usually produces "good enough" solutions, and can find solutions that are completely nonintuitive to human minds, as does, all too often, biological evolution. Applied to a complex problem domain for which we usually have little or no idea what the "correct" solution is, EC can be a valuable addition to the modeler's toolbox.

This approach was taken by Mendoza and Alvarez-Buylla (1998) in their construction of a model of the genetic regulatory network controlling flower morphogenesis in the model organism beloved of biologists, the thale cress *Arabidopsis thaliana*. They constructed a network of 11 genes identified in the literature as being important in this process, with a connectivity pattern and interaction rules corresponding to a widely accepted model[4] of flower morphogenesis, the ABC model.

The major problem with modeling specific biological systems in detail is the lack of quantitative data. Essential parameters such as transcription and translation rates, decay rates of biomolecules, and reaction rates, are generally not known or exist only in qualitative form. Mendoza and Alvarez-Buylla (1998) used published data about the relative magnitudes of the interactions in their network, along with information about whether relationships are inhibitory or activatory, to constrain the weights on the edges. The actual values for the weights were found using a mutation-only genetic algorithm. The fitness function was based upon the presence in the state space of the networks of four attractors corresponding to stable gene expression

[4] The term *model* is used here as biologists use the term, to refer to a mental model that is clear enough to be depicted as a cartoon on a slide for a seminar but that is not necessarily as complete or internally consistent as a computational model.

patterns known to be exhibited in different *Arabidopsis* cell types. It must be emphasized that the values found for the weights on the network were not expected to be biologically plausible; they were selected to be the smallest possible integers that fitted the known constraints and produced the desired dynamic behavior. Only the relative magnitudes of the values are realistic. Because the *Arabidopsis* model was so small, the investigators were able to examine its entire state space, enumerating all of the basins of attraction. Six attractors were identified, four of which corresponded to the desired patterns of gene expression, one corresponding to cells that are not competent to flower, and one that has not been observed in wild plants but can be induced experimentally. The *Arabidopsis* flower morphogenesis model was strikingly successful in validating an existing model of gene regulation, producing a computational model that could be manipulated to generate further testable hypotheses.

Multiple runs of an evolutionary algorithm produce multiple sets of parameter values. If there exist several solutions that produce the same dynamic behavior, the actual parameter values may not matter, and the behavior of the model can be considered to be robust. This was the conclusion drawn by von Dassow et al. (2000) with regards to their model of the gene network controlling segment polarity during the development of the fruit fly *Drosophila*. They found that, in a network with 48 parameters, even a random choice of parameter value had a 90% chance of producing the desired behavior. Such robustness implies that the underlying network topology is more important than its parameterization and hence has more chance of accurately reflecting the network in the biological system under investigation.

Although the parameter values found using an optimization algorithm are unlikely to be biologically plausible, they do have the pragmatic advantage of producing a network that exhibits the desired behavior. The somewhat arbitrary nature of the numbers produced, however, means that the biological implications of the networks must be inferred with caution, and tested wherever possible by some other means.

4.2.2 Network Inference

Perhaps the most active area of research into biological gene networks in the past few years has been network inference from microarray data. DNA microarrays capture a snapshot of the gene expression levels of a cell at a given point in time. For example, microarrays have been used to track changes in mRNA (messenger RNA) level in yeast over the course of the cell cycle (Spellman et al., 1998) and to investigate altered transcription in disease conditions such as cancer (Ross et al., 2000). The technology is currently expensive, the data generated is statistically noisy (McLachlan et al., 2004), and a single experiment generates data applicable only to a particular cell type, under particular conditions, compared with a specific control condition, at a given point in time. With all these drawbacks, however, microarrays have one huge advantage: They can provide information on the expression levels of thousands of genes at once, making it possible, for the first time, to obtain a global view of gene expression. The complete yeast genome on a chip (~6200 genes) was published by DeRisi et al. (1997) and the technology continues to develop apace.

The advent of DNA microarray technology a decade ago (Schena et al., 1995) immediately suggested to bioinformaticians that it may be possible to use the data thus generated to infer the topology of the genetic regulatory networks underlying the observed gene expression patterns. The major drawback is that the problem is underdetermined. Data may be generated for literally tens of thousands of genes, but because of the investment of time and money required to run an experiment, most data sets consist of at most a few tens of time steps, usually far fewer. Data from different experiments can be combined to yield larger numbers of observations, but this is not a straightforward procedure. Hardware and software differences between microarray experiments, together with the noise inherent in the data, mean that simply combining data sets is not a valid approach; sophisticated statistical techniques are necessary and further noise is added to the data set. Further, combining data sets implicitly assumes that the coexpression of genes under different experimental conditions reflects coregulation. Several experiments have demonstrated that this is not necessarily the case (Allocco et al., 2004; Yeung et al., 2004).

A further problem with microarray data is the underlying assumption that mRNA levels more-or-less accurately reflect protein levels. Recent research indicates that this may not be so. Newman et al. (2006) examined the abundance of proteins tagged with green fluorescent protein (GFP) in single cells of the yeast *Saccharomyces cerevisiae* using a novel flow cytometry-based approach. They found considerable variability in protein abundance from cell to cell, a complicating factor in itself, since microarrays are by necessity carried out using populations of cells. When they compared protein levels with mRNA levels as determined by microarray analysis of the same cells, they found that while protein levels reflected mRNA levels for many genes, in some cases mRNA levels rose while protein levels were unaffected, and for some genes protein and mRNA levels changed in different directions. The generalizability of this result to other organisms and other condition is not yet known, and the work that has been done to date on gene network inference from microarray data has not taken this potential confounding factor into account.

Underdetermination of the gene network inference problem means that there are multiple possible gene network topologies consistent with the observed gene expression pattern. Determining which is the correct network is a nontrivial task and cannot be done completely computationally with current technology. However, intelligent computational approaches can be used to generate candidate network topologies whose details can then be confirmed or refuted in the laboratory.

A large amount of work has been done on the network inference problem, using an impressive variety of approaches [for reviews see van Someren et al. (2002) and Wessels et al. (2001)]. Of particular interest to this review are those approaches using evolutionary computation.

Evolutionary computation is a particularly useful approach when one or more good, but not necessarily optimal, solutions are needed to a problem that cannot be solved mathematically. It is therefore clearly applicable to the problem of gene network inference from microarray data, and several groups have so used it. The inferred networks may assume either Boolean or real-valued gene expression levels. There are two subproblems to the task of inferring gene networks from microarray data: identifying the network topology and estimating the kinetic parameters of the interactions between the genes. The former, although known to be NP-hard, is probably the "easier" task and is an essential prerequisite to the latter, and so most work has focused on it.

The simplest model of a gene network is as a system of differential equations of the form

$$\frac{dX_i}{dt} = f_i(X_1, X_2, \ldots, X_n), \tag{4.1}$$

where X_i is the expression level of gene i and f_i is a function of the expression levels of all the genes in the network. Sakamoto and Iba (2000) successfully used genetic programming (Koza, 1992) to fit parameter values to such a system.

A related formalism that has been widely used for modeling gene networks is S-systems (Savageau, 1976). S-systems are ordinary differential equations in which the component processes are characterized by power-law functions:

$$\frac{dX_i}{dt} - \alpha_i \prod_{j=1}^{n} x_{j=1}^{g_{ij}} - \beta_i \prod_{j=1}^{n} x_j^{h_{ij}} \qquad (i = 1, 2, \ldots, n), \tag{4.2}$$

where x_i is one of n state variables. The first term, g_{ij}, represents all the influences that increase x_i and the second term, h_{ij}, all those that decrease it. The parameters α_i and β_i are the relative inflow and outflow, respectively, of gene x_j.

S-systems have the advantage of being simple in representation. They are also nonlinear, which is desirable when modeling biological systems, and they incorporate terms that explicitly model different aspects of the network dynamics. They do, however, have several drawbacks. The major problem is the number of parameters that require optimization; this increases as $2n(n + 1)$ with increasing number of vertices, n. For large networks, solving the equations becomes computationally infeasible. Several workers have attempted to overcome this problem by using evolutionary computation techniques.

The major challenge when using EC to infer gene network architecture is to develop a useful fitness function. Since the topology of the target network is by definition unknown, the fitness function cannot involve direct topological comparison. The most widely used alternative is to compare time course gene activity data generated by the inferred network with that observed from the target network. If the target network is a real biological system, the training data can be time-course microarray data, while the activity of artificial networks can be simulated using an approach such as Runge–Kutta methods, a family of iterative methods for the approximation of solutions of ordinary differential equations (Forsythe et al., 1977). The fitness function then becomes the sum squared error of the gene expression values produced over time by the inferred network against those generated by the target network. The fitness of the final network thus reflects its accuracy in reproducing the gene expression patterns of the microarray training data. This approach has the advantage of being applicable to either real-valued or Boolean gene expression values. The major problem with it is the tacit assumption that a topology that produces the correct output must be the topology of the target network, an assumption that is almost certainly invalid since multiple networks can generate the same gene expression patterns over time. For example, Morishita et al. (2003) used an evolutionary algorithm to find parameters for an S-system representing a 5-node network. They found 207 different network structures that produced the target output. A further problem is that of selecting the "best" network from among the many that can be generated by a stochastic algorithm such as an EA. This is a topic rarely discussed by authors; the reader is left to assume that the highest fitness network provides the best fit to the data, but this is rarely stated explicitly. Pridgeon and Corne (2004) use the term "fidelity" to describe the fit of a reverse-engineered network to its biological counterpart.

These two, related questions—how reverse-engineered network accuracy and fidelity scale with network size—were addressed by Pridgeon and Corne (2004). They found that accurate networks were readily obtained for networks ranging from 100 to 6100 nodes. Network fidelity, however, declined sharply between $n = 100$ and $n = 1000$. They concluded that current data is probably inadequate to reverse engineer large gene networks with confidence.

Another, more practical problem with the use of S-systems is their requirement for large amounts of training data (in terms of number of time points and number of experiments), in general far surpassing the size of data sets that are actually produced. All of the algorithms developed to date appear to work reasonably well when applied to small, artificial data sets. None, however, has yet been applied in a convincing manner to large, real data sets. Small networks inherently have a smaller set of equivalencies, and it is not clear that methods that work well on a small data set will scale well to biological network sizes.

A common strategy used to reduce the number of nodes in the networks, and hence the number of parameters that need to be fitted, is to cluster genes based upon common expression profiles over time. The underlying assumption is that genes that are expressed in a similar manner are probably coregulated, or at least part of the same functional pathway, and hence can be considered as equivalent. This simplification is widely used in all aspects of microarray analysis but has been questioned (Allocco et al., 2004; Yeung et al., 2004).

A summary of the results of various evolutionary algorithms for gene network inference from microarray data is presented in Table 4.1. Results are presented for the largest artificial data set reported and for the largest real data set, if any. Because of the size of real data sets, genes are often clustered, with clusters based upon similarity of expression profile over time, and the number of clusters used is indicated in parentheses after the number of individual genes used. The assessment of the goodness of the results is my own.

The most striking observation arising from Table 4.1 is the small size of the data sets used. No matter what the algorithm, most researchers can obtain good

Table 4.1 Results of Evolutionary Network Inference on Artificial and Real Data[a]

Reference	Technique	Largest Artificial	Results	Largest Real (Clusters)	Results
Sakamoto and Iba (2000)	ODE	3	Good	N/A	N/A
Sakamoto and Iba (2001)	ODE/GP	3	Good	N/A	N/A
Koza et al. (2001)	GP	N/A	N/A	12	Good
Ando et al. (2002)	GP	8	Good	112 (5)	Good
Iba and Mimura (2002)					
Bongard and Lipson (2004)	Coevolutionary GA	10	Good	N/A	N/A
Morishita et al. (2003)	S-system	5	Good	N/A	N/A
Kimura et al. (2003)	S-system	30	Poor	N/A	N/A
Kikuchi et al. (2003)	S-system	5	Good	N/A	N/A
Kimura et al. (2005)	S-system	30	Poor	612 (24)	Poor
Maki et al. (2001)	S-system	30	Good	N/A	N/A
Shin and Iba (2003)	S-system	19	Good	14	Good
Spieth et al. (2004)	S-system	20	Fair	N/A	N/A
Kimura et al. (2004)	S-system	30	Poor	N/A	N/A
Tsai and Wang (2005)	S-system	8	Good	N/A	N/A

[a]If real data were clustered, the number of clusters is indicated in parentheses.

results on small data sets, whether real or artificial, but none of the articles scale to larger networks. Where somewhat larger real data sets are used, they are invariably clustered in order to reduce computational load. It appears that the inference of large networks from microarray data, while an active and important area of research, is still an unsolved problem.

4.3 COMPUTATIONAL NETWORK MODELING

Gene network models based on real biological data aim to answer questions about the topology and dynamics of specific biological systems. Another class of model, addressing a different set of questions, is that of abstract computational gene network models.

4.3.1 Random Boolean Networks

Gene network models were introduced in the 1960s (Kauffman, 1969) in the form of random Boolean networks (RBNs). RBNs are minimalist models that yet exhibit surprisingly orderly, and arguably biologically relevant, behavior.

In an RBN genes are represented as nodes in a graph, and regulatory interactions between genes are directed arrows (arcs) between nodes. Nodes have binary activation; at any point in time a node can be either on (1) or off (0). Each node receives inputs from exactly k other nodes, selected at random when the network is constructed. After construction the network topology remains constant. Each node updates its activation at each time step using a lookup table, the entries of which are, once again, assigned at random. For example, a node with two inputs, each of which may be either on or off, has a lookup table with four entries (Table 4.2).

The entries in the New State column in Table 4.2 would be set randomly at initialization and thereafter remain unchanged. Each node has an independent, randomly generated lookup table. Node update is synchronous, which means that at every time step all nodes simultaneously update themselves on the basis of the activation of their inputs. And that's it!

With so much randomness built into the system, it would not be surprising if the "gene expression" pattern of the network over time was also random. And indeed, under certain parameterizations, it is. If k is high, there is no detectable pattern in the gene expression, a state dubbed "chaotic" by RBN practitioners.[5] If k is low, most networks pass through a short transitional phase and then settle into a frozen pattern of ons and offs, a similarly biologically uninteresting state. However, under some parameterizations ($k \approx 2$ for the "plain vanilla" RBN) the picture is completely different (Fig. 4.1).

After a transitional phase a network tends to settle into a cyclical pattern of states that will reiterate endlessly, a phenomenon known as a limit cycle in the parlance of nonlinear dynamics. A single limit cycle can usually be reached from any of a number of starting states, which are collectively known as its basin of attraction. A limit cycle consisting of a single state is known as a point attractor. The state space of a network will typically be divided into several basins of attraction, each containing a single attractor. Kauffman suggests that the number of attractors in a network's state space increases as the square root of the number of nodes, just as the number of cell types in an organism increases as approximately the square root of the number of genes (Kauffman, 1995).

Table 4.2 Example Lookup Table for Node in RBN with $k = 2$

Input 1	Input 2	New State
0	0	1
1	0	0
0	1	0
1	1	1

[5] In practice, of course, a finite system like an RBN can never be truly chaotic. Eventually, a state previously observed is bound to recur, and then the whole set of states will be followed again, since the system is deterministic. But "a limit cycle the size of the entire state space" is a good working definition of chaos, in this system.

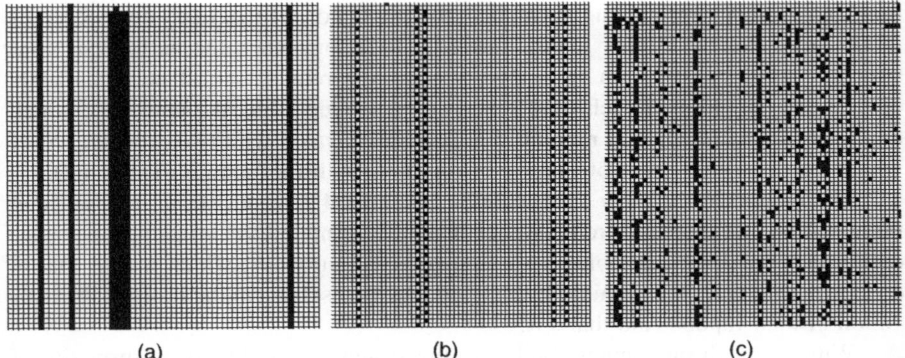

Figure 4.1 Dynamic behaviour of an RBN. Genes are along the x axis, and time goes down the page. A black square indicates a gene is on, while a white square indicates that it is off. **(a)** A point attractor, **(b)** a limit cycle of length 2, **(c)** chaotic behavior. Note the brief transitional period at the start of each run.

These observations led Kauffman to suggest that RBNs are useful models of some aspects of biological gene networks. According to this view, a limit cycle corresponds to a cell type in a multicellular organism, and the existence of basins of attraction explains the robustness of organisms to mutation or other minor damage to a cellular system. If an RBN is perturbed by randomly flipping the states of a small number of nodes, the resulting state is likely to still be within the basin of attraction of the original limit cycle, to which it will eventually return. Similarly, in a biological context, random mutation rarely changes one of our cells to a completely different cell type. Few biologists would quibble with the notion of a cell type as an attractor in gene expression space, although most would not use this terminology.

The analogy can be taken even further. In contrast to fully differentiated cells, totipotent stem cells exist in a state that can give rise to any type of cell under the appropriate conditions. This situation would correspond, in an RBN, to the network starting in a state close to the confluence of many basins of attraction, into any of which it could be nudged by the appropriate perturbation.

4.3.2 Evolving RBNs

Random Boolean networks are models so simplistic that many traditional biologists dismiss them out of hand. However, they have provided unexpected and significant insights into the global behavior of gene networks; before RBNs no one expected that highly ordered and broadly predictable global behavior could arise from randomly generated systems, and yet this is an important issue for the study of biological gene networks.

Much of the work on RBNs has focused on their intrinsic properties, but they also provide an excellent test bed for research into the interaction between inherent order and evolution. There is ongoing debate between evolutionary theorists about the relative importance to evolution of self-organization, selection, neutral drift, and chance [e.g., Dawkins (1996), Gould and Eldridge (1993), and Kauffman (1993); these workers espouse different views on this debate]. Evolutionary algorithms can be applied to RBNs to investigate such issues.

4.3.3 Robustness and Evolvability

The topic of network evolution can be approached in either of two complementary ways. The evolutionary system can be considered as a population of networks, with individual networks evolving in competition (or cooperation) with each other. Alternatively, a single network may be the system, and the agents that evolve are the nodes themselves. Both approaches have the potential to shed light upon the power of evolution to shape gene networks.

Considerable research has been carried out using RBNs to investigate the evolution of robustness. As already mentioned, appropriately parameterized RBNs are robust to a certain amount of perturbation of the system, as are cells and their underlying gene networks. But cells exhibit much more complexity than RBNs; during development most organisms can tolerate a surprisingly large amount of variability in both their structure and their environment without modifying their developmental trajectories. This phenomenon is known as canalization. Redundancy can occur in several ways. Some of the robustness arises via self-organization, but it appears that self-organization alone cannot account for all the developmental and metabolic robustness of organisms. An interesting question is the way in which self-organized robustness interacts with the evolutionary process.

Robustness in gene networks largely arises through redundancy. Within a network, redundancy in the paths between nodes leads to the ability to maintain functional pathways even if individual links are destroyed, for example, by mutation. There is also considerable redundancy in the mapping from genotype to phenotype, leading to developmental robustness.

Individual organisms must be robust to survive in the face of the slings and arrows of outrageous fortune. The species to which they belong, however, must also exhibit evolvability: the ability of a lineage to adapt to changing environmental conditions. Evolvability is a topic of considerable debate in evolutionary biology (Brookfield, 2001); even apart from the tension between the conflicting needs for robustness and adaptability, the suggestion that increased mutation rates are advantageous to organisms under some circumstances is controversial in itself. Since most mutations are disadvantageous, a population experiencing increased mutation rates would be expected to suffer poorer fitness, at least initially. Despite this theoretical caveat, increased mutation rates have been observed in several organisms under stressful conditions (e.g., Rudner et al., 1999).

Robust networks exhibiting limit cycle dynamics are poised, to use Kauffman's evocative phrase, on the "edge of chaos" (Kauffman and Johnsen, 1991); a point in the system's phase space where small changes to the parameterization of the network can alter the dynamics dramatically, either tipping it over into chaos or dropping it into a frozen point attractor. Such systems tend to resist evolutionary change, because most change is disastrous to them (Kauffman, 1990). Biological robustness, in contrast, is self-evidently accompanied by a high degree of evolvability, and yet there is some evidence that eukaryotic gene networks, as measured by their gene expression patterns, may still operate at the border between order and chaos (Shmulevich et al., 2005).

To investigate whether the link between robustness and evolvability is mediated by evolution, Bassler et al. (2004) used RBNs in which the update lookup table for each gene in an attractor evolved over time in competition with those of the other nodes in the same attractor. The fitness function penalized nodes that were in the same state as the majority of the other nodes in the attractor. The result was that the network evolved to a nonrandom critical steady state in which robustness is maintained along with the capacity for further evolution. The mechanism responsible for the evolution was a balance of two competing dynamical effects. Although this model is simplified to the point where it is not at all biologically plausible, it is an effective demonstration that evolutionary processes can produce biologically important phenomena such as robust evolvability.

4.3.4 Neutral Evolution

Not all evolution is adaptive. The definition of the term *evolution* is still, surprisingly, contentious, but perhaps the simplest definition is "evolution is a change in gene frequencies" (Mayr, 1980, p.12). Until the development of the technique of gel electrophoresis toward the end of the twentieth century, it was assumed that most genetic variability results in phenotypic variability and hence is immediately exposed to natural selection. Electrophoresis allows separation of proteins and nucleic acid on the basis of differences in their size and electric charge. When the technique was applied to a wide variety of proteins, it immediatly became apparent that there is a previously unexpected richness of variability at the genetic level, most of which is not apparent at the phenotypic level. It turns out that this variability arises via a number of mechanisms. For example, many substitutions in the third base of a codon do not alter the amino acid for which it codes. Further, different amino acid strings, arising from different DNA sequences, can fold into functionally identical proteins. It was suggested by some workers that much evolution occurs in the form of genetic drift that does not affect the phenotype (or at a minimum does not alter the fitness of the phenotype) and therefore is not subject to selection pressure The neutral evolution debate was born (Kimura, 1983; Ohta, 1973).

Computational gene networks provide an excellent test bed for the examination of the theoretical aspects of neutral evolution. The most widely used approach is the neutral networks model. A neutral network is defined by Newman and Engelhardt (1998) to be "a set of sequences that all possess the same fitness and that are connected together via . . . point mutations" (p. 1335). The sequences referred to here are usually regarded as the DNA sequences of the genes comprising a gene network, but they could also be interpreted as amino acids folding to form a ribosome, or even traits making up the phenotype of a complex organism. Evolutionary neutral networks can occur at any biological scale.

Much computational neutral network research builds upon Kauffman and Weinberger's (1989) *NK* landscape. This model consists of a genotype of *N* "genes," each of which is linked to *K* other genes. The fitness of each gene is dependent upon the value of the *K* other genes to which it is linked as well as to its own value. The fitness of the entire genotype is the average of the fitnesses of all the genes that it contains. Kauffman and Weinberger carried out an extensive analysis of the characteristics of the fitness landscapes of *NK* networks. They found that as the value of *K* increases the fitness landscape becomes gradually more rugged, moving from a single peak (which they dubbed a "Fujiyama" landscape) to a multipeaked ("badlands") landscape. Further, rugged landscapes have been shown to have the highest fitness peaks (Skellett et al., 2005). The evolutionary trajectory of an evolving system is profoundly affected by the characteristics of its fitness landscape, so the existence of a tuneable parameter for fitness landscape ruggedness makes the *NK* model well suited for investigations into network evolution.

Standard *NK* landscapes do not undergo neutral evolution, but several workers have extended Kauffman's model to incorporate a tuneable level of neutrality. Barnett (1998) introduced the *NKp* (probabilistic) model, in which a new parameter, p, is introduced to represent the probability that an arbitrary allelic combination makes no contribution to fitness. When $p = 0$ a normal *NK* landscape is generated, while $p = 1$ corresponds to a completely flat landscape. Intermediate values of p introduce areas of neutrality into the *NK* landscape. The proportion of the landscape that is neutral (flat) increases with increasing p. A similar extension to the basic *NK* model was proposed by Newman and Engelhardt (1998). They introduced a parameter, F, that controls the number of possible fitnesses of a network. As $F \rightarrow \infty$ the range of possible fitnesses becomes so large that no two networks will possess the same fitness, and the model reduces to the standard *NK* model. The smaller the value of F, the greater the degree of neutrality in the landscape.

The pattern of evolution evident in these models is broadly similar. Populations tend to go through long periods of neutral drift, followed by relatively abrupt transitions to a new phenotypic norm. This type of behavior, dubbed "punctuated equilibrium" by Eldredge and Gould (1972), has been observed in the fossil record, although as is so often the case in evolutionary biology, its significance and indeed, even its existence, has been hotly debated. As expected, the presence of neutrality in a fitness landscape appears to facilitate the diffusion of a population throughout the landscape, potentially bringing new phenotypes within its evolutionary reach. In addition, convergence to a local fitness optimum, a phenomenon that dogs many evolutionary algorithms, is avoided. This observation does not only apply to *NK*-based models; Shackleton and colleagues (Shackleton et al., 2000) discuss a range of genotype–phenotype mappings that produce redundancy of between 16 and 65,536 possible genotypes per phenotype. Evolution over these landscapes appears to be very similar to that on *NK*-based landscapes.

The different implementations of neutral networks do, however, produce search landscapes with different properties, some of which are more easily searchable than others (Geard et al., 2002). These findings suggest that there may be several different types of "neutral" evolution, and that "degree of neutrality" is not the only important characteristic of such networks. Knowles and Watson (2002) found that mutation-only EAs with appropriate choice of mutation rate evolved just as rapidly if they were directly encoded as when they were redundant. It appears that the amount and distribution of the neutrality present is determined by the way in which neutrality is implemented, whether this is by the addition of a neutrality parameter to a model system or via the existence of "wobble bases" in the biological genetic code. While computational simulations of gene networks support the theorized importance of neutral evolution to the evolutionary process, it is clear that more research is needed into the way in which neutrality affects biological fitness landscapes, and how biologically plausible neutral evolution can be modeled.

4.3.5. Extending the RBN Model

Classical RBNs are indisputably useful for modeling genetic networks, but they incorporate a number of simplifying assumptions that may affect their dynamics, and that are far from biologically plausible. The use of Boolean network activation, a constant connectivity pattern,[6] synchronous node updating, and random activation functions are all biologically implausible, and their effects have been investigated by a number of researchers.

A more subtle problem with the use of classical RBNs with evolutionary algorithms is that of representation. An RBN is generated as a network, and evolutionary operators applied to an RBN act, by necessity, at the level of nodes and links. They must therefore be tailored to the manipulation of networks. Operators such as node addition and edge rewiring perforce replace mutation and crossover. In an RBN there is no distinction between genotype and phenotype. In real organisms, in contrast, there is a considerable gap, both physical and temporal, between the genome, upon which mutations operate, and the phenotype in which their effects are eventually manifest. Along the long journey from DNA to body many other factors come into play—genetic context, environmental influences, and the effects of positive and negative feedback loops with and between cells, tissues, and the environment. The introduction of a distinction between genotype and phenotype in a computational gene network is a step toward understanding some of these processes. One family of models that takes this step is the artificial genome (AG) (Reil, 1999).

The AG starts with a randomly generated string of numbers from a given range [(0, . . . , 3) is widely used, reflecting the four bases A, C, G, and T in DNA, although any number of bases may be implemented]. A pattern of bases is selected to act as a marker for the beginning of a gene (e.g., 0101 mimicking the TATA box for promoter binding). The next n bases form the gene sequence. The stretch of

[6] Most biological networks have been shown to be scale free (e.g., Jeong et al., 2000).

chromosome between the end of one gene and the beginning of the TATA box marking the start of the next one becomes the *promoter region* for the second gene. Genes are *translated* into gene products by incrementing the value of the base at each position (0 → 1, 1 → 2, 2 → 3, 3 → 0). All of the promoter regions of the genome are then searched for an exact match with each gene product. If the product of gene *A* matches a sequence in the promoter region of gene *B, A* is taken to regulate *B* (Fig. 4.2).

Regulation may be either positive or negative, and the direction of regulation for a given interaction is determined by the value of the last base(s) of the gene product. For example, approximately 25% of interactions can be set to be inhibitory if a final value of 0 is taken to indicate inhibition, while 1, 2, or 3 indicates an activatory interaction.

Networks generated by the AG display the same categories of dynamics as RBNs (frozen/limit cycles/chaos) (Reil, 1999; Banzhaf, 2003), but they have a number of advantages over RBNs. For a start, topological features such as the number of nodes, number of edges, and proportion of inhibitory links do not have to be fixed but can be set approximately by choosing appropriate parameters, such as genome length, gene length, number of bases, and TATA box pattern. More importantly, from our point of view, the existence of a genome facilitates the use of standard evolutionary operators in an evolutionary algorithm and permits exploration of the relationship between genome and gene network, the first step on the path between genotype and phenotype.

Artificial genomes were used to explore the difference between mutation at the genome level and at the network level by Watson et al. (2004). They found that, although most mutations are likely to be neutral, with no impact upon the network topology, even small mutations to the genome could have dramatic effects upon network topology. A single mutation in, for example, a TATA box is sufficient to completely eliminate a gene, which in turn eliminates all edges to and from that gene. Mutations within a gene will affect all of its outgoing edges but none of its incoming edges. The result is that mappings between mutations at the genome level and those at the network level may not be as simple as they seem. Network level mutations usually take the form of something like a single rewiring event, in which an edge between gene *A* and gene *B* is moved to lie between gene *A* and gene *C*.

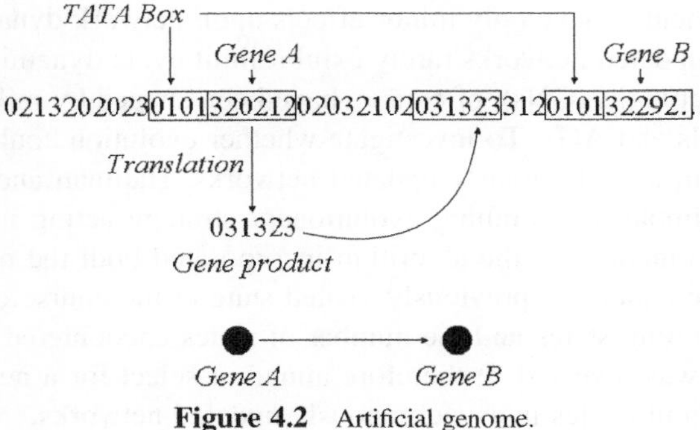

Figure 4.2 Artificial genome.

From a network point of view, this is a simple "mutation," but at the genome level this rewiring requires two relatively improbable mutations: one in the promoter region of gene *B*, to eliminate the binding site for gene *A*'s product, and one in the promoter region of gene *C*, to create a binding site. Apparently minor mutations at the network level may involve significant genetic rearrangement; possibly another observation supporting the importance of neutral evolution?

Mutations at the network level, while conceptually logical, may not be easy to achieve via mutations at the more realistic genome level. Evolutionary algorithms that operate at the network level are working in a very different fitness landscape from those operating at the genome level, which means that the course, and eventual outcome, of the evolutionary process may be different depending upon the model used. One of the major claims to fame of both RBNs and AGs as models of biological gene networks is the correspondence of their dynamics to gene expression dynamics in biological systems. However, this correspondence holds only under certain circumstances. As already discussed, the model parameters—connectivity, proportion of inhibitory edges, and so forth—must be set within relatively narrow limits. This is not a major objection to their use as models of gene networks; biological gene networks do, after all, have many characteristics in common, and their values tend to lie within relatively narrow ranges. A more significant problem is that limit cycle dynamics in Boolean networks, whether RBN or AG, only arise when the nodes are updated synchronously.

Asynchronously updated networks rapidly collapse to a frozen state and stay there. This behavior is clearly not biologically plausible (except insofar as death is a biological condition), but neither is synchronous gene update. Biological genes are turned on and off at different times as transcription factors bind and disengage, and considerable additional noise is contributed by variability in the rates of transcription, translation, and the transport and decay of biomolecules. Asynchronous updating can be implemented in a number of ways. Each node may be updated in turn, in the same order each time; or nodes may be selected at random to update. If nodes are selected at random, selection may be with replacement, so that different nodes may be updated a different numbers of times, or without replacement, so that all nodes in a network are updated an equal number of times. Asynchronous node update may even be probabilistic, with each node having a given probability of being updated at any time step (Hallinan and Wiles, 2004a). Variations in the way asynchrony is implemented have only minor effects upon network dynamics, however; asynchronously updated networks rarely exhibit limit cycle dynamics.

Real gene networks (GNs), of course, have been shaped by millennia of selection, unlike RBNs and AGs. To investigate whether evolution could produce limit cycle dynamics in asynchronously updated networks, Hallinan and Wiles (2004a, 2004b) used a simple hill-climbing evolutionary strategy acting upon a modified AG. The fitness function for the algorithm incorporated both the number of times each network encountered a previously visited state in the course of multiple runs from different starting states and the number of states encountered before a previously seen state was revisited. It therefore aimed to select for a network topology producing long limit cycles in asynchronously updated networks.

Asynchronous networks are nondeterministic since one state can be succeeded by many others (the exact number varies and depends on the update scheme used). Classical limit cycles, as seen in deterministic networks, therefore, cannot occur, but "fuzzy" limit cycles were observed, during which the networks repeatedly revisited a subset of the states in the state space, in roughly the same order. Fuzzy limit cycles are, in fact, more biologically plausible than the crisp alternative, given the amount of noise inherent in genetic regulation (Hasty and Collins, 2002; Ozbudak et al., 2002; Paulsson, 2004; Newman et al., 2006). It would appear that self-organization alone is not able to produce biologically plausible limit cycle dynamics in GNs, but self-organization in conjunction with selection can do so.

4.4 EXTENDING REACH OF GENE NETWORKS

The power and flexibility of gene networks need not, of course, be restricted to control of cellular processes. Several researchers have used them in the development of higher level control systems. One obvious direction in which to extend this research is toward the development of multicellular organisms. In order to model this process, the evolutionary algorithm must not act only upon the genome but must incorporate developmental processes such as cell division, differentiation, and death. There has been a considerable amount of work using evolutionary algorithms to evolve simulated organisms (e.g., Fleischer, 1996; Furasawa and Kaneko, 2000; Hogeweg, 2000; Kumar and Bentley, 2003), but most of it does not descend to the level of gene networks.

The first simulation experiment in this area was reported by Eggenberger (1997), who evolved an artificial genome to produce three-dimensional shapes of multicellular organisms. The simulations incorporated the developmental factors mentioned above, together with a *morphogen gradient* to provide positional information for the evolving cells. Using a fitness function that depended on the number of cells and their position along the x axis of the three-dimensional space in which they evolved, he was able to evolve a range of complex, bilaterally symmetrical forms made up of different "cell types" by using only a small number of genes.

Eggeneberger's work demonstrates that gene networks have the power to control the morphology of simple, abstract *multicellular organisms*. But what about real organisms?

Caenorhabditis elegans is a worm with a name considerably longer than its body, which is only 1 mm in length. This little nematode lives quietly in the soil in temperate areas, browsing on bacteria and reproducing as a self-fertilizing hermaphrodite. And that's all it did until Sydney Brenner selected it in the 1960s as the ideal organism with which to investigate the genetics of development. The worm body consists of 959 cells and follows a perfectly consistent pattern of development. Every cell division and death has been cataloged, making it the perfect subject upon which to explore the effects of mutations to its 20,000-odd genes. In addition, since they are self-fertilizing, it is easy to generate genetically identical inbred stocks. In the last half century thousands of worm biologists have raised and examined billions of worms in Petri dishes in labs around the world. *C. elegans* was the first animal to have its genome fully sequenced, a project which was completed in 1998.[7]

Because the complete developmental trajectory of *C. elegans* is well established, it is the perfect organism to use as a target for models of development. Geard and Wiles (2005) used an extension to the AG model, incorporating recurrent connections and thus dubbed the dynamic recurrent gene network (DRGN), to model the morphogenesis of *C. elegans*. Weight values for the links in the network were evolved using a simple evolutionary algorithm. They found that the DRGN could be evolved to reproduce aspects of the developmental sequence including cell diversification from an initial fertilized egg, cell differentiation, termination of cell lineages at appropriate points in time, using a minimum of external input (specifically, a signal to indicate whether a daughter cell was at the left or the right of the mother cell). The networks appear to be capable of producing moderately complex behavior from a small set of genes, and the addition of recurrent links add the ability to handle temporal patterns, after the manner of recurrent neural networks (Elman, 1990).

A different approach to the task of extending the reach of a gene network to the phenotypic level was taken by Watson et al. (2003). They used an AG to produce a gene network, the output of which was used to parameterize a Lindenmeyer system (L-system) model of development (Lindenmeyer, 1968). L-systems are parallel rewrite grammars and have been extensively used to model development, particularly that of plants. An L-system consists of a finite alphabet of values, an initial axiom, a set of rules for the replacement of string components, and a derivation process that involves the parallel rewriting of strings. Previous work on evolving L-systems has involved developing evolutionary algorithms that would operate upon the L-system itself to generate valid new L-systems, but the incorporation of an underlying gene network that is the focus of evolution is a novel approach. Watson and his colleagues used the characteristics of regulatory interactions within the network to specify three key parameters of their L-system (branching angle, growth rate, and delay before a new node was initiated). They then applied various mutation operators to the AG in order to investigate their effects upon the dynamics of the system. The project as described is a test bed for the examination of evolutionary dynamics, but it would be straightforward, and potentially very interesting, to extend this work to incorporate selection pressure upon the eventual phenotype.

Organic bodies are not the only ones to which simulated evolution based upon a gene network have been applied. Bongard (2002) used this approach to evolve robot bodies out of component morphological units, sensors, motors, neurons, and synapses. Once again, the primary external input required during development was an indicator of direction, in this case two different transcription factors, one in the anterior and one at the posterior pole of the unit. Gradients of morphogenic chemicals are widely used in biological systems to provide positional information to the

[7] Brenner received the Nobel Prize in 2002, along with H. Robert Horvitz and John Sulston "for their discoveries concerning 'genetic regulation of organ development and programmed cell death'."

cells of developing organisms (Gurdon and Bourillot, 2001); it would appear that computational evolution requires the same symmetry-breaking information in order to develop complex bodies.

A similar problem domain, and one to which gene-network-based controllers have also been applied, is that of robot control. Problems such as the development of autonomous control systems for robots are particularly difficult for two major reasons. First, the control system is embedded in an actual, physical object, the robot body ("embodiment"). No matter how realistic software simulations are, they can never fully capture the physics of the real world, let alone such individual-specific issues as a sticky wheel bearing or unreliable sensors. Even more important is the environment in which the robot is to act; highly controlled and predictable environments are relatively straightforward to deal with, as is apparent in the huge number of factory robots employed throughout the world. Controllers for robots performing repetitive tasks in a predictable environment can be programmed in a relatively straightforward manner. The real world, however, is a much messier place, and it is not possible to program appropriate responses to every possible eventuality into a robot controller. Learning algorithms, such as artificial neural networks, are frequently used for such tasks (Bin and Xiong, 2004), but that other great adapter, evolution, can also be employed. The two approaches are frequently combined (e. g., Beer and Gallagher, 1992; Floreano and Mondada, 1994).

A simple, simulated system is reported by Quick et al. (2003). They used an artificial genome to evolve a gene network to control positively phototactic behavior in a software simulation of the Khepera robot. Although their problem domain is very simple—a bounded flat physical surface with a light source suspended in the middle—and is simulated rather than embodied, they conclude that the approach is potentially valuable. This research is particularly interesting because it couples an evolved gene network to its environment via the amount and type of proteins produced. In contrast, Watson et al. (2003) used the topological characteristics of the evolved network to do the same task.

Taylor (2004) extends this approach to the evolution of controllers for a group of autonomous robots working together to perform high-level, coordinated tasks underwater. This group chose to use an evolutionary approach in order to develop a general controller that could be evolved to perform specific tasks in a collaborative manner, making it flexible enough for autonomous use underwater. Once again, he used simulation but planned to implement the controller in hardware eventually. Proteins produced by a gene network coded for by an artificial genome again act as the interface between the genome and the physical environment. The task required of the robots was modest: The robots were required merely to form a cluster, and this was successfully achieved. While this is preliminary work, the use of AG models to evolve controllers for a group of communicating robots is clearly valuable. The fact that each robot was evolving separately means that different robots ended up with slightly different, complementary controllers. This sort of coevolution promises to facilitate the production of robust, self-healing systems; if one or more robots was lost or added, the group could evolve to a new optimum set of controllers, taking into account group membership and physical environment.

Gene networks are powerful, flexible generators of apparently orderly nonlinear dynamics, and as such they can be used in a multitude of ways. In this section we have reviewed work in two main areas: (1) investigating the interaction of self-organization and evolution and the way in which evolution can shape the dynamics of networks, and (2) examining the evolved networks to control higher level processes such as building and operating bodies. The work in the former area is clearly at a more mature stage than that in the latter, an observation that is not unexpected given that researchers have been investigating the properties of large abstract networks for nearly 40 years. Although much has been learned in this time, it is clear that much remains to be learned.

The science of adaptive robotics is very much less developed, in part because the design and engineering of robot bodies is a complex and expensive prerequisite to the implementation of their controllers.[8] The work reviewed here is still primarily simulation based. However, it is clear that evolved gene networks have considerable promise in this area, and will undoubtedly take their place beside more widely used machine learning approaches such as neural networks.

4.5 NETWORK TOPOLOGY ANALYSIS

Yet another way in which evolutionary computation can be combined with gene networks is as a tool for the analysis of reconstructions of real interaction networks. Biological gene networks can be very large and complex, to the point where visualization of the network is of little value as an aid to analysis. Global metrics such as degree distribution, cluster coefficient, and diameter (Strogatz, 2001) provide limited information about networks as a whole, but they do not help to answer the questions of real interest to biologists, questions about individual genes and proteins and their relationships. Biological networks tend to be modular (Thieffry and Romero, 1999; Schuster et al., 2002; Snel et al., 2002; Rives and Galitski, 2003; Han et al., 2004), and many are hierarchically modular, with genes organized into small modules that are, in turn, organized into larger modules and so on (Ravasz et al., 2002; Ravasz and Barabasi, 2003; Hallinan, 2004). A module is defined as a subset of nodes the members of which are more tightly connected to each other than they are to the rest of the network.

Modules in a gene network are not necessarily isolated (Hartwell et al., 1999); a given component may belong to different modules at different times, and the function of a module can be affected by signals from other modules. Such cross talk between functional modules has been shown to be essential to the behavior of a variety of different biological systems (e.g., Amin, 2004; Natarajan et al., 2006). As with the identification of modules, the identification of the linkages constituting channels of cross talk in a network is not straightforward. If there are many edges between two clusters, they merge into a single cluster, but there is no obvious way of deter-

[8] Although there has been progress toward evolving robot bodies themselves (Lipson and Pollack, 2000).

mining the cutoff above which this merger should happen. Literally hundreds of clustering algorithms have been described (for an overview, see Hartigan, 1975), most of which can be modified to operate upon networks if a node distance metric can be specified. However, there are several drawbacks common to generic unsupervised clustering algorithms, particularly when applied to large, complex networks.

Many algorithms, such as *k* means and self-organizing maps (SOMs), need to know the number of clusters in advance and will partition data into the specified number of clusters whether or not that partitioning reflects real clustering in the network. Hierarchical clustering algorithms are widely used because they are fast, provide a useful overview of the cluster structure of the network, and reflect the generally hierarchically modular nature of biological networks, but for practical use a decision must be made as to where in the cluster tree to threshold. This is not a straightforward decision and often requires the use of further information about the network, which may not be available for all nodes of a large biological network.

Many clustering algorithms cannot use the information inherent in weightings on the edges in the graphs. While some biological networks, such as protein–protein interaction networks, are inherently unweighted, the edges in many networks represent interactions with which a metric can be associated. In metabolic networks, for example, kinetic parameters can be encoded as weights on edges between biochemical species. Clustering algorithms that do not incorporate weightings discard potentially valuable information about the network structure and function. Most algorithms, whether or not they use a predetermined number of clusters, cluster all of the data provided. In many problem domains this is not an issue, but gene networks inherently consist of structural and functional modules of varying sizes linked by nodes or short chains of nodes that lie, conceptually and topologically, outside the system of modules. Even more importantly, biological modules are essentially fuzzy in that a single node may belong to more than one module, and modules may overlap to a greater or lesser extent in different parts of the same network. Biological networks also have a temporal element, with different modules likely to be active at different times and in response to different external stimuli.

In order to usefully cluster a large biological network, then, an algorithm should have the following characteristics:

- Ability to identify overlapping clusters of varying sizes.
- Requires no foreknowledge of number of clusters to be found.
- Does not necessarily assign all nodes to clusters.
- Requires no information about the network except topological structure.
- Can utilize weights on edges if they are present.

An evolutionary algorithm is ideally suited for such a task. Hallinan and Wipat (2006) applied a simple genetic algorithm to the clustering of the nodes of a *Saccharomyces cerevisiae* gene network constructed by Lee et al. (2004) using a variety of data sources concerning genetic interactions. As well as direct genetic regulatory data, the network combined protein–protein interaction, coexpression, co-citation, and several other types of data using a Bayesian approach, to produce a large network with edges weighted to represent the computed probability of the existence of each interaction. Using as a fitness function a measure of modular coherence based upon the ratio of links internal to the module to those external, the algorithm identified several large clusters of genes, together with numerous smaller ones. According to the gene ontology annotation (Gene Ontology Consortium, 2000), the large clusters were primarily concerned with ribosomal biogenesis and protein biosynthesis, an observation in accordance with a clustering of the same network done by its original authors.

4.6 SUMMARY

In this review we have considered the application of evolutionary computation techniques to a variety of network models. We have identified a number of ways in which evolutionary algorithms and gene networks can interact:

- As an optimization technique, fitting parameters to an existing gene network model derived from the literature
- For the reverse engineering of gene networks from microarray data
- As a wrapper around an abstract network such as an RBN or an AG, to investigate the interaction between evolution and processes such as self-organization
- As a means of developing robust phenotypes, be they bodies or robot controllers
- For analyzing the topology of real biological gene networks

Evolutionary computation is clearly a powerful and flexible approach to optimization and modeling, particularly in complex nonlinear systems such as gene networks.

The optimization applications of EC to gene networks have yielded variable results. EC is an efficient and powerful way to fit parameters to complex models, and such models have successfully reproduced the behavior of small, detailed gene network systems. The application of EC to the reverse engineering of gene networks from microarray data has been less successful, as have all other approaches to this difficult and underdetermined problem. Evolution is, however, a powerful heuristic, and it is probably that variants on the algorithms currently being developed will scale more successfully to larger networks as the quality and quantity of microarray data available increases, as it inevitably will.

The combination of evolutionary computation with large abstract models of gene networks has already provided insights into the way in which the evolutionary process can interact with the inherent properties of large, complex networks (Kauffman's "order for free") to produce a system with the flexibility, evolvability, robustness, and coordination shown by the gene networks that control the development and metabolism of all living things. It is starting to be recognized that gene networks can be evolved to control larger scale systems such as bodies, whether real or simulated. There is still much to be done in this area; large networks have inherent emergent properties that may not be apparent in small systems, and abstract models currently provide the only feasible approach to the investigation of these properties.

The continued generation of high-throughput *omics*[9] data by biology labs around the world means that this situation is likely to change in the near future. There is already considerable interest in the subject of biological data integration, and networks combining data on gene interactions with that on protein–protein, transcriptional regulation, coexpression, and many other types of interaction are being constructed for organisms ranging from microbes to humans. The existence of these networks will provide fresh fodder for EC practitioners. Evolutionary approaches have already been used to map topological features of such networks to functional biological modules, but the possibilities for analysis of functional networks remain largely unexplored. Large functional networks will also be a valuable tool for testing some of the predictions generated by the abstract network research; the incorporation of new forms of interactions between genes will undoubtedly change the dynamics of large networks in ways that we currently have no understanding. The construction of dynamic models of such integrated systems will be both exciting and challenging.

As more biologists become aware of the potential of evolutionary computation approaches to answer questions of direct relevance to biology, particularly evolutionary biology, the generation and analysis of gene network models will become an iterative process, with models being used to generate hypotheses that are tested in the laboratory, leading to further modification of the models. Such as approach is already being used in some of the systems biology centers that are currently being established around the world. The combination of evolutionary computation and gene networks is a powerful one, and its potential is only beginning to be realized.

REFERENCES

Allocco, D. J., I. S. Kohane, and A. J. Butte (2004). "Quantifying the relationship between co-expression, co-regulation and gene function," *BMC Bioinform*, Vol. 5, No. 18.

Ambros, V. (2004). "The functions of animal microRNAs," *Nature*, Vol. 431, pp. 350–355.

Amin, A. (2004). "Genetic cross-talk during head development in Drosophila," *J. Biomed. Biotech.*, Vol. 1, pp. 16–23.

Ando, S., E. Sakamoto, and H. Iba (2002). "Evolutionary modelling and inference of gene networks," *Info. Sci.*, Vol 145; pp. 237–259.

Baitaluk, M., X. Qian, S. Godbole, A. Raval, A. Ray, and A. Gupta, (2006). "PathSys: Integrating molecular interaction graph for systems biology," *BMC Bioinformatics*, Vol. 7; p. 55.

Banzhaf, W. (2003). "On the dynamics of an artificial regulatory network," *Adv. Artifi. Life. Lecture Notes in Computer Sci.*, Vol. 2801, pp. 217–277.

[9] A catch-all term to include genomics, proteomics, metabolomics, and all the other high-throughput technologies of modern molecular biology (Palsson, 2002). Network analysis falls neatly into *interactomics*.

BARNETT, L. (1998). "Ruggedness and neutrality—the NKp family of fitness landscapes," in ALIFE VI, Proceedings of the Sixth International Conference on Artificial Life. C. Adami, R. K. Belew, H. Kitano and C. Taylor, Eds. MIT Press, Cambridge, MA, Vol. 18, p. 27.

BATAGELJ, V. and A. MRVAR (1998). "Pajek—Program for large network analysis," *Connections*, Vol. 21, pp. 47–57.

BASSLER, K. E., C. LEE, and Y. LEE (2004). "Evolution of developmental canalization in networks of competing Boolean nodes," *Phys. Rev. Lett.*, Vol. 93, No. 3, pp. 038101–038104.

BEER, R. D. and J. C. GALLAGHER (1992). "Evolving dynamic neural networks for adaptive behavior," *Adapt. Behav.*, Vol. 1, No. 1, pp. 91–122.

BIN, N. and C. XIONG (2004). "New approach of neural network for robot path planning," in Proceedings of the 2004 IEEE Conference on Systems, Man and Cybernetics, in N. Bin, ed., pp. 735–739.

BOLOURI, H. and E. H. DAVIDSON (2002). "Modelling transcriptional regulatory networks," *BioEssays*, Vol. 24, pp. 1118–1129.

BONGARD, J. (2002). "Evolving modular genetic regulatory networks," Proceedings of the 2002 IEEE Conference on Evolutionary Computation, pp. 1872–1877.

BONGARD, J. C. and H. LIPSON (2004). "Automating genetic network inference with minimal physical experimentation using coevolution," *Lecture Notes Comput. Sci.*, Vol. 3102, pp. 333–345.

BREITKREUTZ, B. J., C. STARK, and M. TYERS (2003). "Osprey: A network visualization system," *Genome Biol.*, Vol. 4, No. 3, pp. R22–R25.

BROOKFIELD, J. F. Y. (2001). "Evolution: The evolvability enigma," *Curr. Biol.*, Vol. 11, No. 3, pp. R106–R108.

DAWKINS, R. (1996). *Climbing Mount Improbable*. Penguin Books, London.

DE JONG, H. (2002). "Modeling and simulation of genetic regulatory systems: A literature review," *J. Comput. Biol.*, Vol. 9, No. 1, pp. 67–103.

DERISI, J. L., R. I. VISHWANATH, and P. O. BROWN (1997). "Exploring the metabolic and genetic control of gene expression on a genomic scale," *Science*, Vol. 278, pp. 680–686.

EGGENBERGER, P. (1997). "Evolving morphologies of simulated 3D organisms based on differential gene expression," in Proceedings of the 4th European Conference on Artificial Life (ECAL97), P. Husbands and I. Harvey, Eds. MIT Press, Cambridge, MA.

ELDREDGE, N. and S. J. GOULD (1972). "Punctuated equilibria: An alternative to phyletic gradualism," in *Models in Paleobiology*, T. J. M. Schope, Ed. Freeman, San Francisco, pp. 82–115.

ELMAN, J. L. (1990). "Finding patterns in time," *Cognitive Sci.*, Vol. 14, pp. 179–211.

ENDY, D. and R. BRENT (2001). "Modelling cellular behaviour," *Nature*, Vol. 409, pp. 391–395.

ERDÖS, P. and A. RÉNYI (1959). "On random graphs," *Pub. Math.*, Vol. 6, p. 290.

FLEISCHER, K. (1996). "Investigations with a multicellular developmental model," *Artif. Life,* Vol. 5, pp. 229–236.

FLOREANO, D. and F. MONDADA (1994). "Automatic creation of an autonomous agent: Genetic evolution of a neural-network driven robot," in Proceedings of the Third International Conference on Simulation of Adaptive Behavior: From Animals to Animats 3. MIT Press, Cambridge, MA, pp. 421–430.

FORSYTHE, G. E., M. A. MALCOLM, and C. B. MOLER (1977). *Computer Methods for Mathematical Computations*. Prentice-Hall, Englewood Cliffs, NJ.

FRASER, C. M., J. D. GOCAYNE, O. WHITE, M. D. ADAMS, C. A. CLAYTON, and R. D. FLEISCHMANN (1995). "The minimal gene complement of Mycoplasma genitalium," *Science*, Vol. 270, pp. 397–404.

FURASAWA, C. and K. KANEKO (2000). "Complex organization in multicellularity as a necessity in evolution," *Artif. Life*, Vol. 6, pp. 265–281.

GEARD, N., J. WILES, J. HALLINAN, B. TONKES, and B. SKELLETT (2002). "A comparison of neutral landscapes—NK, NKp and NKq," in Proceedings of the 2002 IEEE Congress on Evolutionary Computation (CEC2002). IEEE Press: Piskataway, pp. 205–210.

GENE ONTOLOGY CONSORTIUM (2000). "Gene Ontology: Tool for the unification of biology," *Nature Genet.*, Vol. 25, pp. 25–29.

GEARD, N. and J. WILES (2005). "A gene network model for developing cell lineages," *Artif. Life*, Vol. 11, No. 3, pp. 249–268.

GILMAN, A. and A. P. LARKIN (2002). "Genetic 'code': Representations and dynamical models of genetic components and networks," *Ann. Rev. Genom. Human Genet.* Vol. 3, pp. 341–369.

GOULD S. J. and ELDRIDGE N. (1993). "Punctuated equilibrium comes of age," *Nature*, Vol. 366, p. 223.

GURDON, J.B. and P.-Y. BOURILLOT (2001). "Morphogen gradient interpretation," *Nature*, Vol. 413, pp. 797–803.

HALLINAN, J. (2004). "Gene duplication and hierarchical modularity in intracellular interaction networks," *Biosystems*, Vol. 74, Nos. 1–3, pp. 51–62.

HALLINAN, J. and J. WILES (2004a). "Evolving genetic regulatory networks using an artificial genome," in 2nd Asia-Pacific Bioinformatics Conference (APBC2004), Dunedin, New Zealand. *Conferences in Research and Practice in Information Technology*, Vol. 29. Y.-P. Chen, Ed. Australian Computer Society, Vol. 29, pp. 291–296.

HALLINAN, J. and J. WILES (2004b). "Asynchronous dynamics of an artificial genetic regulatory network," Ninth International Conference on the Simulation and Synthesis of Living Systems (ALife9), Boston, September 12–15.

HALLINAN, J. and A. WIPAT (2006). "Clustering and crosstalk in a yeast functional interaction network," in *Proceedings of the 2006 IEEE Symposium on Computational Intelligence in Bioinformatics and Computational Biology*. IEEE Press, San Diego.

HAN, J.-D. J., N. BERTIN, T. HAO, D. T. GOLDBERG, G. F. BERRIZ, and L. V. ZHANG (2004). "Evidence for dynamically organized modularity in the yeast protein-protein interaction network." *Nature*, Vol. 430, pp. 88–93.

HARTWELL, L.H., J.J. HOPFIELD, S. LEIBLER, and A.W. MURRAY (1999). "From molecular to modular cell biology," *Nature*, Vol. 402, No. 6761 Supplement, pp. C47–C52.

HARRISON, P. M., A. KUMAR, N. LANG, M. SNYDER, and M. GERSTEIN (2002). "A question of size: The human proteome and the problems in defining it," *Nucleic Acids Res.*, Vol. 30, No. 5, pp. 1083–1090.

HARTIGAN, J. A. (1975). *Clustering Algorithms*. Wiley, New York.

HASTY, J. and J. J. COLLINS (2002). "Translating the noise," *Nature Genet.*, Vol. 31, pp. 13–14.

HOGEWEG, P. (2000). "Shapes in the shadow: Evolutionary dynamics of morphogenesis," *Artif. Life*, Vol. 6, pp. 856–101.

IBA, H. and A. MIMURA (2002). "Inference of a gene regulatory network by means of interactive evolutionary computing," *Info. Sci.*, Vol. 145, pp. 225–236.

JEONG, H., B. TOMBOR, R. ALBERT, Z. N. OLTVAI, and A.-L. BARABASI (2000). "The large-scale organization of metabolic networks." *Nature*, Vol. 407, pp. 651–654.

KAUFFMAN, S. (1969). "Metabolic stability and epigenesis in randomly constructed genetic nets." *J. Theor. Biol.* Vol. 22, pp. 437–467.

KAUFFMAN, S. (1995). *At Home in the Universe*. Penguin, London.

KAUFFMAN, S. A. (1990). "Requirements for evolvability in complex systems: Orderly dynamics and frozen components," *Physica D*, Vol. 42, Nos. 1–3, pp. 135–152.

KAUFFMAN, S. A. (1993). *The Origins of Order: Self-Organization and Selection in Evolution*. Oxford University Press, New York.

KAUFFMAN, S. A. and S. JOHNSEN (1991). "Coevolution to the edge of chaos: Coupled fitness landscapes, poised states and coevolutionary avalanches," *J. Theor. Biol.*, Vol. 149, No. 4, pp. 467–505.

KAUFFMAN, S. A. and E. D. WEINBERGER (1989). "The NK model of rugged fitness landscapes and its application to maturation of the immune response," *J. Theor. Biol.*, Vol. 141, No. 2, pp. 211–245.

KIKUCHI, S., D. TOMINAGA, M. ARITA, K. TAKAHASHI, and M. TOMITA (2003). "Dynamic modelling of genetic networks using genetic algorithm and S-system," *Bioinformatics*, Vol. 19, No. 5, pp. 643–650.

KIMURA, M. (1983). *The Neutral Theory of Molecular Evolution*. Cambridge University Press, Cambridge.

KIMURA, S., M. HATEKEYAMA, and A. KONAGAYA (2003). "Inference of S-system models of genetic networks using a genetic local search," Proceedings of the 2003 Congress on Evolutionary Computation (CEC 2003), pp. 631–638.

KIMURA, S., M. HATEKEYAMA, and A. KONAGAYA (2004). "Inference of S-system models of genetic networks from noisy time-series data," *Chem-Bio Inform. J.*, Vol. 4, pp. 1–14.

KIMURA, S., K. IDE, A. KASHIHARA, M. KANO, M. HATAKEYAMA, R. MASUI, N. NAKAGWA, S. YOKOYAMA, S. KURAMITSU, and A. KONAGAYA (2005). "Inference of S-system models of genetic networks using a cooperative coevolutionary algorithm," *Bioinformatics*, Vol. 21, No. 7, pp. 1154–1164.

KNOWLES, J. D. and R. A. WATSON (2002). "On the utility of redundant encodings in mutation-based evolutionary search," in *Parallel Problem Solving from Nature—PPSN VII: 7th International Conference. Lecture Notes in Computer Science* 2439. Springer, Berlin, pp. 88–98.

KOHLER, J., J. BAUMBACH, J. TAUBERT, M. SPECHT, A. SKUSA, A. RUEGG, C. RAWLINGS, P. VERRIER, and S. PHILIPPI (2006). "Graph-based analysis and visualization of experimental results with ONDEX," *Bioinformatics*, Vol. 22, No. 11, pp. 1383–1390.

KOZA, J. (1992). *Genetic Programming: On the Programming of Computers by Means of Natural Selection*. MIT Press, Cambridge, MA.

KOZA, J.R., W. MYDLOWEC, G. LAMZA, J. YU and M.A. KEANE (2001). "Reverse engineering of metabolic pathways from observed data using genetic programming," *Pacific Symp. Biocomp.*, pp. 434–445.

KUMAR, S. and P. BENTLEY (2003). "Biologically inspired evolutionary design," in Proceedings of the International Conference on Evolvable Systems 2003, A. M. Tyrell, P. C. Haddow and J. Torreson, Eds. Springer, Berlin, pp. 57–68.

LEE, I., S. V. DATE, A. T. ADAI, and E. M. MARCOTTE (2004). "A probabilistic functional network of yeast genes," *Science*, Vol. 306, No. 5701, pp. 1555–1558.

LINDENMEYER, A. (1968). "Mathematical models for cellular interactions in development, Parts I and II," *J. Theor. Biol.*, Vol. 18, pp. 280–315.

LIPSON, H. and J. POLLACK (2000). "Automatic design and manufacture of robotic life forms," *Nature*, Vol. 406, pp. 974–978.

MAKI, Y., D. TOMINAGA, M. OKAMOTO, S. WATANABE, and Y. EGUCHI (2001). "Development of a system for the inference of large scale genetic networks," Proceedings of the 2001 Pacific Symposium on Biocomputing, pp. 446–458.

MATTICK, J. S. (2001). "Non-coding RNAs: The architects of molecular complexity," *EMBO Repts.*, Vol. 2, No. 11, pp. 986–991.

MAYR, E. (1980). "Introduction," in *The Evolutionary Synthesis*, E. Mayr and W. Provine, Eds. Harvard University Press, Cambridge, MA.

MCLACHLAN, G. J., K.-A. DO, and C. AMBROISE (2004). *Analyzing Microarray Gene Expression Data*. Wiley, Hoboken, NJ.

MENDES, P. and D. KELL (1998). "Non-linear optimisation of biochemical pathways: Applications to metabolic engineering and parameter estimation," *Bioinformatics*, Vol. 14, pp. 869–883.

MENDOZA, L. and E. R. ALVAREZ-BUYLLA (1998). "Dynamics of the genetic regulatory network for Arabidopsis thaliana flower morphogenesis," *J. Theor. Biol.*, Vol. 193, pp. 307–319.

MICHALEWICZ, Z. and M. SCHOENAUER (1996). "Evolutionary algorithms for constrained parameter optimization problems," *Evolut. Comput.*, Vol. 4, No. 1, pp. 1–32.

MOLES, C. G., P. MENDES, and J. R. BANGA (2003). "Parameter estimation in biochemical pathways: A comparison of global optimization methods," *Genome Res.*, Vol. 13, pp. 2467–2474.

MORISHITA, R., H. IMADE, I. ONO, N. ONO, and M. OKAMOTO (2003). "Finding multiple solutions based on an evolutionary algorithm for inference of genetic networks by S-system," Proceedings of the 2003 Congress on Evolutionary Computation, Canberra, Australia, pp. 615–622.

NATARAJAN, M., K.-M. LIN, R. C. HSUEH, P. C. STERNWEIS, and R. RANGANATHAN (2006). "A global analysis of cross-talk in a mammalian cellular signalling network," *Nature Cell Biology*, Advance Online Publication DOI: 10.1038/ncb1418.

NEWMAN, J. R. S., S. GHAEMMAGHAMI, J. IHMELS, D. K. BRESLOW, M. NOBLE, J. L. DERISI, and J. S. WEISSMAN (2006). "Single-cell proteomic analysis of S. cerevisiae reveals the architecture of biological noise," *Nature*, Vol. 441, pp. 840–847.

NEWMAN, M. and R. ENGELHARDT (1998). "Effect of neutral selection on the evolution of molecular species," *Proc. Roy. Soc. London B*, Vol. 265, pp. 1333–1338.

OHTA, T. (1973). "Slightly deleterious mutant substitutions in evolution," *Nature*, Vol. 246, pp. 96–98.

Ozbudak, E. M., M. Thattai, I. Kurtser, A. D. Grossman, and A. van Oudenaarden (2002). "Regulation of noise in the expression of a single gene," *Nature Genet.*, Vol. 31, pp. 69–73.

Palsson, B. (2002). "In silico biology through 'omics'," *Nature Biotech.* Vol. 20, pp. 649–650.

Park, L. J., C. H. Park, C. Park, and T. Lee (1997). "Application of genetic algorithms to parameter estimation of bioprocesses," *Med. Biol. Eng. Comput.*, Vol. 35, No. 1, pp. 47–49.

Paulsson, J. (2004). "Summing up the noise in gene networks," *Nature*, Vol. 427, pp. 415–418.

Pridgeon, C. and D. Corne (2004). "Genetic network reverse-engineering and network size: Can we identify large GRNs?" Proceedings of the 2004 IEEE Symposium on Computational Intelligence in Bioinformatics and Computational Biology, San Diego, 7–8 October, pp. 32–36.

Quick, T., C. L. Nehaniv, K. Dautenhahn, and G. Roberts (2003). "Evolving embodied genetic regulatory network-driven control systems," in *European Conference on Artificial Life 2003 (ECAL 2003). Lecture Notes in Artificial Intelligence 2801*. Springer, Berlin, pp. 266–277.

Ravasz, E. and A.-L. Barabasi (2003). "Hierarchical organization in complex networks," *Phys. Rev.*, Vol. 67, pp. 026112-1–026112-7.

Ravasz, E., A. L. Somera, Z. N. Oltvai, and A.-L. Barabasi (2002). "Hierarchical organization of modularity in metabolic networks," *Science*, Vol. 297, pp. 1551–1555.

Reil, T. (1999). "Dynamics of gene expression in an artificial genome: Implications for biological and artificial ontogeny," in Proceedings of the 5th European Conference on Artificial Life, Floreano, D., F. Mondada, and J. D. Nicoud, Eds. Springer, New York, pp. 457–466.

Reil, T. (2000). "Models of gene regulation—A review," in *Artificial Life 7 Workshop Proceedings*, C. C. Maley and E. Boudreau, Eds. MIT Press, Cambridge, MA, pp. 107–113.

Rives, A. W. and T. Galitski (2003). "Modular organization of cellular networks," *Proc. Natl. Acad. Sci. USA*, Vol. 100, No. 3, pp. 1128–1133.

Ross, D. T., U. Scherf, M. B. Eisen, C. M. Perou, C. Rees, and P. Spellman (2000). "Systematic variation in gene expression patterns in human cancer cell lines," *Nature Genet.*, Vol. 24, pp. 227–235.

Rudner, R., A. Murray, and N. Huda (1999). "Is there a link between mutation rates and the stringent response in Bacillus subtilis?" *Ann. NY Acad. Sci.*, Vol. 870, No. 1, p. 418.

Sakamoto, E. and H. Iba (2000). "Identifying gene regulatory network as differential equation by genetic programming," *Genome Inform.*, Vol. 11, pp. 281–283.

Sakamoto, E. and H. Iba (2001). "Inferring a system of differential equations for a gene regulatory network by using genetic programming," Proceedings of the 2001 IEEE Congress on Evolutionary Computation, pp. 720–726.

Savageau, M. S. (1976). *Biochemical System Analysis: A Study of Function and Design in Molecular Biology*. Addison-Wesley, Reading, MA.

Schena, M., D. Shalon, R. W. Davis, and P. O. Brown (1995). "Quantitative monitoring of gene expression patterns with a complementary DNA microarray," *Science*, Vol. 270, No. 5235, pp. 467–470.

Schilling, C. H., S. Schuster, B. O. Palsson, and R. Heinrich (1999). "Metabolic pathway analysis: Basic concepts and scientific applications," *Biotech. Prog.*, Vol. 15, No. 3, pp. 296–303.

Schuster, S., T. Pfeiffer, F. Moldenhauer, I. Koch, and T. Dandekar (2002). "Exploring the pathway structure of metabolism: Decomposition into subnetworks and application to Mycoplasma pneumoniae," *Bioinformatics*, Vol. 18, pp. 351–351.

Shackleton, M., R. Shipman, and M. Ebner (2000). "An investigation of redundant genotype-phenotype mappings and their role in evolutionary search," in Proceedings of the 2000 Congress on Evolutionary Computation. IEEE Press, New York, pp. 493–500.

Shannon, P., A. Markiel, O. Ozier, N. S. Baglia, J. T. Wang, D. Ramage, N. Amin, B. Schwikowski, and T. Ideker (2003). "Cytoscape: A software environment for integrated models of biomolecular interaction networks," *Genome Res.*, Vol. 13, pp. 2498–2504.

Sherlock, G., T. Hernandez-Boussard, A. Kasarskis, G. Binkley, J. C. Matese, S. S. Dwight, M. Kaloper, S. Weng, H. Jin, C. A. Ball, M. B. Eisen, P. T. Spellman, P. O. Brown, D. Botstein, and J. M. Cherry (2001). "The Stanford microarray database," *Nucleic Acids Res.* Vol. 29, No. 1, pp. 152–155.

Shin, A. and H. Iba (2003). "Construction of genetic network using evolutionary algorithm and combined fitness function," *Genome Inform.*, Vol. 14, pp. 94–103.

Shmulevich, I., S. A. Kauffman, and M. Aldana (2005). "Eukaryotic cells are dynamically ordered or critical but not chaotic," *Proc. Natl. Acad. Sci. USA*, Vol. 102, No, 38, pp. 13439–13444.

Skellett, B., B. Cairns, N. Geard, B. Tonkes, and J. Wiles (2005). "Rugged NK landscapes contain the highest peaks," in Proceedings of the Genetic and Evolutionary Computation Conference, GECCO 2005. ACM Press, New York, pp. 579–584.

Snel, B., P. Bork, and M. A. Huynen (2002). "The identification of functional modules from the genomic association of genes," *Proc. Natl. Acad. Sci., USA*, Vol. 99, pp. 5890–5895.

Spellman, P. T., G. Sherlock, M. Zhang, K. Anders, M. B. Eisen, P. O. Brown, D. Botstein, and B. Futcher (1998). "Comprehensive identification of cell cycle-regulated genes of the yeast *Saccharomyces cerevisiae* by microarray hybridization," *Mol. Biol. Cell*, Vol. 9, pp. 3273–3297.

Spieth, C., F. Streichert, N. Speer, and A. Zell (2004). "Optimizing topology and parameters of gene regulatory network models from time-series experiments," *Lecture Notes Comput. Sci.*, Vol. 3102, pp. 461–470.

Strogatz, S. H. (2001). "Exploring complex networks," *Nature*, Vol. 410, pp. 268–276.

Taylor, T. (2004). "A genetic regulatory network inspired real-time controller for a group of underwater robots," Informatics research report EDI-INF-RR-0184, School of Informatics, University of Edinburgh, Scotland.

Thieffry, D. and D. Romero (1999). "The modularity of biological regulatory networks," *Biosystems*, Vol. 50, pp. 49–59.

Tsai, K.-Y. and F.-S. Wang (2005). "Evolutionary optimization with data collocation for the reverse engineering of biological networks," *Bioinformatics*, Vol. 21, No. 7, pp. 1180–1189.

van Someren, E. P., L. F. A. Wessels, E. Backer, and M. J. T. Reinders (2002). "Genetic network modelling," *Pharmacogenomics*, Vol. 3, No. 4, pp. 1–19.

von Dassow, G., E. Meir, E. M. Munro, and G. M. Odell (2000). "The segment polarity network is a robust developmental module," *Nature*, Vol. 406, pp. 188–192.

Watson, J., N. Geard, and J. Wiles (2004). "Towards more biological mutation operators in gene regulation studies," *Biosystems*, Vol. 76, Nos, 1–3, pp. 239–248.

Watson, J., J. Wiles, and J. Hanan (2003). "Towards more relevant evolutionary models: Integrating an artificial genome with a developmental phenotype," Proceedings of the Australian Conference on Artificial Life (ACAL 2003), pp. 288–298.

Wessels, L. F. A., E. P. van Someren, and M. J. T. Reinders (2001). "A comparison of genetic network models," in *Pacific Symposium on Biocomputing* 2001, pp. 508–519.

Wingender, E., P. Dietze, H. Karas, and R. Knuppel (1996). "TRANSFAC: A database on transcription factors and their DNA binding sites," *Nucleic Acids Res.*, Vol. 24, No. 1, pp. 238–241.

Wuensche, A. (2002). "Discrete Dynamics Lab: Tools for investigating cellular automata and discrete dynamical networks," *Kybernetes*, Vol. 32, Nos. 1/2.

Xenarios, I., L. Salwinsky, X. J. Duan, P. Higney, S.-M. Kim, and D. Eisenberg (2002). "DIP, the Database of Interacting Proteins: A research tool for studying cellular networks of protein interactions," *Nucleic Acids Res.*, Vol. 30, No. 1, pp. 303–305.

Yeung, K. Y., M. Medvedovic, and R. E. Bumgarner (2004). "From co-expression to co-regulation: How many microarray experiments do we need?" *Genome Biol.*, Vol. 5, p. R48.

Part Two

Sequence Analysis and Feature Detection

Chapter 5

Fuzzy-Granular Methods for Identifying Marker Genes from Microarray Expression Data

Yuanchen He, Yuchun Tang, Yan-Qing Zhang, and Rajshekhar Sunderraman

5.1 INTRODUCTION

It is well known that genes from cells in the same organism can be expressed in different combinations and/or quantities in different tissue types, development stages, and in different environmental conditions. Differential gene expression accounts for the huge variety of states and types of cells in the same organism (Shapiro and Tamayo, 2003). Recently, DNA (deoxyribonucleic acid) microarray technology, including cDNA (complementary DNA) microarrays and GeneChips, have been developed and represent powerful methods for the investigation of molecular genetics, by simultaneously measuring the mRNA (messenger ribonucleic acid) expression levels of thousands to tens of thousands of genes. A typical microarray expression experiment monitors the expression level of each gene, multiple times, under different conditions or in different tissue types. Such huge gene expression data sets give rise to the possibility of identifying marker genes that can be used to distinguish tissue types (Shapiro and Tamayo, 2003; Jiang et al., 2004; Bair and Tibshirani, 2003; Noble, 2004; Moler et al., 2000; Alon et al., 1999; Model et al., 2001).

From the viewpoint of predictive data mining (Hand et al., 2001), the goal with microarrays is to distinguish different tissue types to predict the unknown value of a variable (healthy or cancerous; if cancerous, which kind of cancer) of interest given known values of other variables (gene expression data). More specifically, this can

Computational Intelligence in Bioinformatics. Edited by G. B. Fogel, D. Corne, and Y. Pan

be approached as a classification problem. For example, one well-known problem utilizing microarray gene expression data is to distinguish between two variants of leukemia [e.g., acute myeloid leukemia (AML) and acute lymphoblastic leukemia (ALL)]. The AML/ALL problem can be modeled as a binary classification problem.

A typical gene expression data set is extremely sparse compared to a traditional classification data set. Microarray data typically uses only dozens of tissue samples but contains thousands or even tens of thousands of gene expression values. This extreme sparseness is believed to significantly deteriorate the performance of the resulting classifiers. As a result, the ability to extract a subset of informative genes while removing irrelevant or redundant genes is crucial for accurate classification. Gene selection can be modeled as a feature selection or dimensionality reduction problem. A good dimensionality reduction method should remove irrelevant or redundant gene features for classification. After removing these "noninformative" gene features, the inherent cancer-related data distribution is expected to be more easily recognized in the lower dimensional feature space formed by remaining informative or discriminative gene features. Consequently, a classifier modeled in the lower dimensional space can have better performance when making cancer diagnostic decisions of previously unseen samples.

For example, the AML/ALL leukemia data set has only 72 samples with 7129 gene expressions for each sample. This means, without gene selection, we have to discriminate and classify such a few samples in such a high-dimensional space. It is unnecessary or even harmful for classification because it is believed that no more than 10% of these 7129 genes are relevant to leukemia (Guyon et al., 2002). In summary there are two highly correlated and challenging tasks for bioinformaticians: (1) *gene selection*: given some tissues, extract cancer-related genes while removing irrelevant or redundant genes. A small set of genes is desirable for further biological study because it is very expensive or even impractical for biologists to pursue cancer study on a large number of genes; on the other hand, the prediction is unreliable if too few genes are selected. (2) *Cancer classification*: given a new tissue, predict if it is healthy or not or categorize it into correct classes.

The remaining sections of the chapter are organized as follows. In Section 5.2, previous work on cancer classification and gene selection is briefly reviewed. Following this, a new fuzzy-granular gene selection algorithm is proposed in Section 5.3. Section 5.4 evaluates the performance of this method on two microarray expression data sets. Finally, Section 5.5 concludes the chapter.

5.2 TRADITIONAL ALGORITHMS FOR GENE SELECTION

5.2.1 Support Vector Machines for Cancer Classification

Support vector machines (SVMs) are believed to be a superior model for high-dimensional classification problems including cancer classification on microarray expression data (Vapnik, 1998). SVMs are a new generation of learning system based on recent advances in statistical learning theory (Burges, 1998).

Due to the extreme sparseness of microarray data, the dimension of input space is already high enough so that the cancer classification becomes as simple as a linearly separable task (Vapnik, 1998). As a result, usually a SVM with a linear kernel [Eq. (5.1)] (Guyon et al., 2002; Schölkopf, et al. 2003) is adopted as the basic cancer classifier. We also adopt linear SVMs in this work.

$$K(x_i, x_j) = x_i x_j. \tag{5.1}$$

For a linear SVM, the margin width can be calculated by Eqs. (5.2) and (5.3):

$$w = \sum_{i=1}^{N_S} \alpha_i y_i x_i, \tag{5.2}$$

$$\text{Margin width} = \frac{2}{\|w\|}, \tag{5.3}$$

where N_s is the number of support vectors defined relative to the training samples with $0 < \alpha_i \leq C$. Note that C is a *regulation parameter* used to trade-off the training accuracy and the model complexity so that a capability for good generalization can be achieved. Interested readers may refer to the literature for additional information about SVMs (Vapnik, 1998; Burges, 1998; Schölkopf et al., 2003; Cristianini and Shawe-Taylor, 1999; Chin, 1999; Gunn, 1998).

The sparseness of microarray data is so extreme, however, that even an SVM classifier is unable to achieve reliable performance for cancer classification. A preprocessing step for gene selection is necessary for SVM modeling to achieve more reliable classification (Guyon et al., 2002).

5.2.2 Correlation-Based Feature Ranking Algorithms for Gene Selection

Correlation-based gene selection algorithms work by ranking genes individually in terms of a correlation metric, and then the top-ranked genes are selected as the most informative gene subset (Furey et al., 2000; Pavlidis et al., 2000; Duan and Rajapakse, 2004). Some commonly used ranking metrics are: signal-to-noise (S2N) (Furey et al., 2000), Fisher criterion (FC) (Pavlidis et al., 2000), and *t* statistics (TS) (Duan and Rajapakse, 2004); Eqs. (5.4) to (5.6), respectively:

$$w_i = \frac{|\mu_i(+) - \mu_i(-)|}{\sigma_i(+) + \sigma_i(-)}. \tag{5.4}$$

$$w_i = \frac{[\mu_i(+) - \mu_i(-)]^2}{\sigma_i(+)^2 + \sigma_i(-)^2}. \tag{5.5}$$

$$w_i = \frac{|\mu_i(+) - \mu_i(-)|}{\sqrt{\sigma_i(+)^2 / n(+) + \sigma_i(-)^2 / n(-)}}. \tag{5.6}$$

In these equations, $\mu_i(+)$ and $\mu_i(-)$ are the mean values of the ith gene's expression data over positive and negative samples in the training data set, respectively. σ_i (+) and σ_i (−) are the corresponding standard deviations of gene expression. n (+) and n (−) denote the numbers of positive and negative training samples, respectively. A larger w_i means that the ith gene is tending toward being informative for cancer classification.

Correlation-based algorithms are straightforward to understand and work efficiently. If there are d genes originally, the ranking process takes $O(d \log d)$ time. However, a common drawback of these approaches is that all of the genes are ranked together. Biologically, some genes may regulate cancers with a similar function and hence be similarly expressed. With correlation-based algorithms, these genes may be ranked closely together simply because they are coregulated. If they happen in the top of the ranking list, all of them may be selected as "informative" genes. As a result, the process of gene selection is biased to this single function, and genes with other functions are removed. Because multiple different gene groups may regulate cancers in different ways, the biological analysis on genes selected by traditional correlation-based algorithms may lose other cancer-related information.

Biologically, different groups of genes may regulate cancers with different functions on one hand, and one single gene may have more than one function to regulate cancers on the other hand. To select genes with more information from different function groups for reliable cancer classification and diagnosis, a novel fuzzy-granular-based algorithm is presented below. The algorithm is based on the principles of granular computing, which is described in the next section.

Finally, readers are referred to some other related work in the area of computational intelligence. Keedwell and Narayanan (2005) designed a novel method for determining gene interactions in temporal gene expression data using genetic algorithms combined with a neural network. Juliusdottir et al. (2005) identified genes that are significant regarding colon cancer and prostate cancer with a simple evolutionary algorithm/classifier combination. They also provided review material on evolutionary algorithms and k-Nearest Neighbor algorithm. There are also many related works described Fogel and Corne (2003).

5.3 NEW FUZZY-GRANULAR-BASED ALGORITHM FOR GENE SELECTION

5.3.1 Granular Computing

Granular computing (GrC) is a general computation theory for effectively using granules such as classes, clusters, subsets, groups, and intervals to build an efficient computational model for complex applications with huge amounts of data, information, and knowledge (Bargiela, 2002; Yao and Yao, 2002; Yao, 2001). The basic notions and principles of granular computing, though presented under different names previously, have appeared in many related fields (i.e., information hiding, granularity, divide and conquer, interval computing, cluster analysis, fuzzy and rough set theories, neutrosophic computing, quotient space theory, belief functions, machine learning, databases, etc.). In the just the past few years, there has been a renewed and growing interest in GrC. GrC has been applied not only in bioinformatics but also to e-business, security, machine learning, data mining, high-performance computing, and wireless mobile computing in terms of efficiency, effectiveness, robustness, and uncertainty. Some formal models using information granules include set theory and interval analysis, fuzzy sets, rough sets, probabilistic sets, decision trees, clusters, and association rules.

Lin (2000; 1999; 1998a,b) used GrC based on rough sets, fuzzy sets, and topology and used GrC theory in data mining applications. Pedrycz (2001a,b), Pedrycz and Vukovich (2002), and Bortolan and Pedrycz (1997) applied interval mathematics, fuzzy sets, rough sets, and random sets to GrC research and relevant applications such as pattern recognition. Interested readers may refer to http://www.cs.sjsu.edu/~grc for additional information on GrC.

Our algorithm utilizes a fuzzy *C*-means (FCM) clustering algorithm (Bezdek, 1981) to group genes into different function granules based on their expression patterns. The advantage of FCM clustering is that it can assign a sample (e.g., a gene) into multiple clusters wherein each gene has a possibly different membership value. Because a gene may regulate cancers with multiple functions, FCM matches the need to utilize this biological knowledge for granulation.

5.3.2 Relevance Index

A *relevance index* (RI) was used to measure the relevance of a feature to a cluster in Yip et al. (2004) to ease an unsupervised clustering process. We make use of this approach as a preprocessing step. The goal here was to filter irrelevant genes prior to gene selection and supervised classification. Given a binary cancer classification task, a gene may be negatively correlated or positively correlated. Eqs. (5.7) and (5.8) define the negative relevance index and the positive relevance index to measure the negative correlation and the positive correlation of a gene with the cancer being studied, respectively:

$$R_{i-} = 1 - \frac{\sigma_{i-}^2}{\sigma_i^2}, \tag{5.7}$$

$$R_{i+} = 1 - \frac{\sigma_{i+}^2}{\sigma_i^2}, \tag{5.8}$$

where σ_i^2, σ_{i-}^2, and σ_{i+}^2 are the variances of the projected values on the *i*th gene of the whole training samples, the negative training samples, and the positive training samples, respectively.

For example, gene *X1* in Figure 5.1 is positive related because the local variance among positive samples is much smaller than the global variance on the whole samples. Similarly, gene *X2* is negative related, gene *X3* is both negative related and

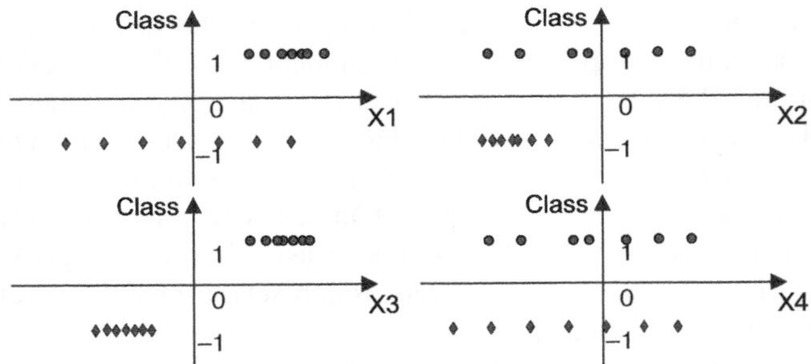

Figure 5.1 Positive-related gene, negative-related gene, both, neither.

positive related, and gene *X4* can be viewed as irrelevant in that it is neither negative related nor positive related.

To apply the RI metric for gene selection, a negative filtering threshold $\alpha_- \in [0,1)$ and a positive filtering threshold $\alpha_+ \in [0,1)$ needs to be determined. We define the ith gene to be negative related if $R_{i-} \geq \alpha_-$. Similarly, a gene is considered positive related if $R_{i+} \geq \alpha_+$. If $R_{i-} < \alpha_-$ and $R_{i+} < \alpha_+$, the gene is considered "irrelevant." A gene may be both negative related and positive related if both $R_{i-} \geq \alpha_-$ and $R_{i+} \geq \alpha_+$. These two filtering thresholds must be selected carefully. They cannot be too large else information loss may occur as some cancer-related genes are incorrectly eliminated. In addition the thresholds should be "balanced" otherwise genes in the minority side may be totally eliminated, leading to performance degradation, especially when negative-related genes and positive-related genes are significantly imbalanced in the original data set.

5.3.3 Fuzzy *C*-Means Clustering

The RI metric is commonly used to prune features. However, given that interactions among genes can be highly nonlinear, the success of this approach on its own is less than perfect. As a result, the thresholds α_- and α_+ should not be too large as to avoid removing informative genes.

The next step is to identify discriminative genes. Multiple genes may be coregulated. As a result, it is reasonable to select only a few representative genes while removing other "redundant" genes. By removing redundancy, genes with more regulation functions may be selected, assuming the number of genes is fixed.

Some genes may similarly regulate cancers and thus be similarly expressed. And hence these genes may play a similar role in cancer classification. As a result, if genes with similar expression patterns are grouped together into clusters, a few typical genes in a cluster may be selected, and other genes in the cluster may be safely eliminated without significant information loss. On the other hand, an informative gene may contribute to cancer classification with complex correlations with multiple different clusters. Therefore, after the prefiltering by RI metric, FCM is adopted to group genes into different function clusters.

Fuzzy C-means groups genes into K clusters with centers $c_1,\Lambda\ c_k,\Lambda\ c_K$ in the training sample space. (That is, each training sample is a dimension of the space.) FCM assigns a real-valued vector $U_i = \{\mu_{1i},\Lambda\ \mu_{ki},\Lambda,\ \mu_{Ki}\}$ to each gene; $\mu_{ki} \in [0,1]$ is the membership value of the ith gene in the kth cluster. The larger membership value indicates the stronger association of the gene to the cluster. Membership vector values μ_{ki} and cluster centers c_k can be obtained by minimizing:

$$J(K,m) = \sum_{k=1}^{K}\sum_{i=1}^{N}(\mu_{ki})^m d^2(x_i, c_k), \tag{5.9}$$

$$d^2(x_i, c_k) = (x_i - c_k)^{\mathrm{T}} A_k (x_i - c_k), \tag{5.10}$$

$$\sum_{k=1}^{K}\mu_{ki} = 1, \qquad 0 < \sum_{i=1}^{N}\mu_{ki} < N, \tag{5.11}$$

where $1 \le i \le N$ and $1 \le k \le K$ (Bezdek, 1981).

In Eq. (5.9), K and N are the number of clusters and the number of genes in the data set, respectively; $m > 1$ is a real-valued number that controls the "fuzziness" of the resulting clusters, μ_{ki} is the degree of membership of the ith gene in the kth cluster, and $d^2\ (x_i,\ c_k)$ is the square of distance from the ith gene to the center of the kth cluster. In Eq. (5.10), A_k is a symmetric and positive definite matrix. If A_k is the identity matrix, $d^2\ (x_i,\ c_k)$ corresponds to the square of the Euclidian distance. Equation (5.11) indicates that empty clusters are not allowed.

5.3.4 Fuzzy-Granular-Based Gene Selection

We categorize genes into three classes: (1) informative genes, which are essential for cancer classification and diagnosis; (2) redundant genes, which are also cancer-related but there are some other informative genes regulating cancers similarly but more significantly; and (3) irrelevant genes, which are not cancer related and do not affect cancer classification.

A desirable algorithm should extract genes of the first category while eliminating genes of the last two categories. However, it is difficult to perfectly implement this goal. First, inherent cancer-related factors are very possibly mixed with other non-cancer-related factors for classification. Second, some non-cancer-related factors may even have more significant effects on classifying the training data set. It is actually the notorious "overfitting" problem. It is even worse when the training data set is too small to embody the inherent real data distribution, which is common for microarray gene expression data analysis.

Correlation-based algorithms work by ranking genes in the same group. However, some truly informative genes can possibly be wrongly eliminated. For example, an informative gene is ranked the highest in a function group. However, the genes in this function group are all ranked below another group of genes. As a result, all of genes including the informative gene in this function group are possibly eliminated.

The fuzzy-granular-based algorithm is proposed in this work for more reliable gene selection. It works in two stages. Figure 5.2 sketches the algorithm.

At the first stage, RI metrics are used to coarsely group genes into two granules: a *relevant granule* and an *irrelevant granule*. The relevant granule consists of negative-related genes (with $R_{i-} \geq \alpha_-$) and positive-related genes (with $R_{i+} \geq \alpha_+$), while the irrelevant granule is comprised of irrelevant genes (with $R_{i-} < \alpha_-$ and $R_{i+} < \alpha_+$). Notice $\alpha_- \in [0,1)$ is the negative filtering threshold and $\alpha_+ \in [0,1)$ is the positive filtering threshold. Only genes in the relevant granule survive for the following stages. The assumption is that irrelevant genes are not so useful for cancer classification or even possible to correlate other genes in some unknown complex way to confuse FCM to get good clusters/granules or confuse SVMs to get good classification. This prefiltering process can dramatically decrease the number of candidate genes on which FCM works. Therefore, it can improve both the efficiency and the effectiveness of the following stages. Notice that the prefiltering step by RI metrics is targeted at minimizing information loss by eliminating most of the irrelevant genes.

Genes that survive this first stage are grouped by FCM into several *function granules*. In each function granule, a correlation-based metric is used to rank genes

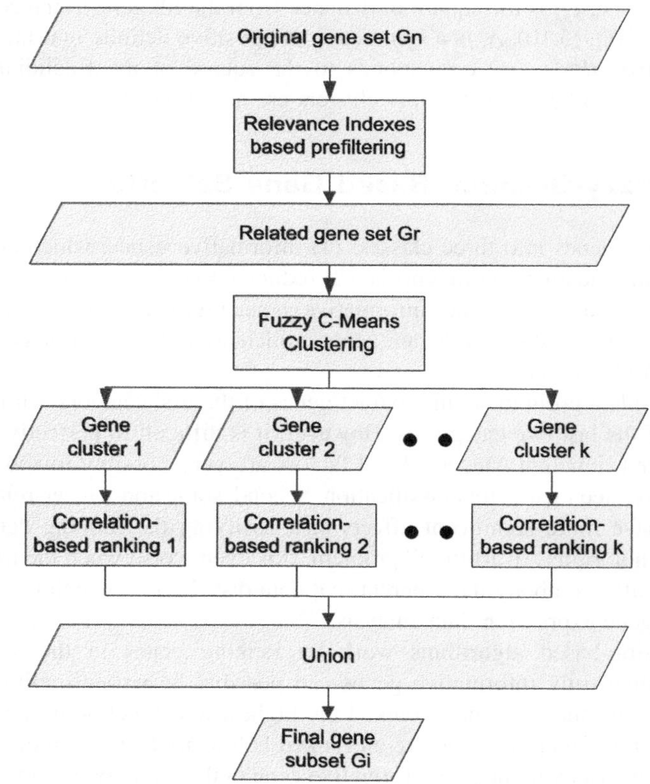

Figure 5.2 Fuzzy-granular gene selection.

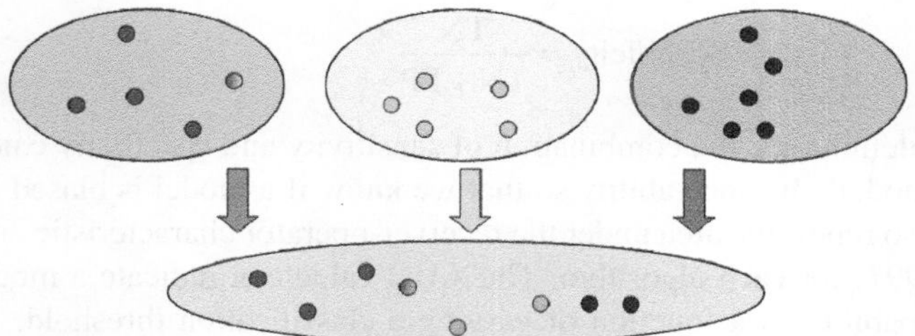

Figure 5.3 Clustering then aggregation (divide-and-conquer).

in descending order of correlation. The lower ranked genes are removed. Notice that the threshold to define "lower ranked" is data dependent and can be selected by human experts. The remaining genes in each function granule are combined disjunctively to form a final gene subset (Fig. 5.3). Through the use of FCM, our algorithm explicitly groups genes with similar expression patterns into clusters and then the lower ranked genes in each cluster can be safely removed while more significant genes with similar functions survive. Due to the complex correlation between genes, the similarity in expression is by no means a "crisp" concept. FCM deals with complex correlations between genes by assigning a gene into several clusters with different membership values. Therefore, a really informative gene achieves more than one opportunity to survive.

5.4 SIMULATION

Here we compare the new fuzzy-granular-based algorithm with three correlation-based algorithms, S2N, FC, and TS. The approaches were evaluated on a desktop PC (personal computer) with Pentium 4 processor operating at 2.8 MHz, with 256 MB of RAM (random-access memory). The software was based on the OSU SVM Classifier Matlab Toolbox (Ma et al., 2002), which implements a Matlab interface to LIBSVM (Chang and Lin, 2001).

5.4.1 Evaluation metrics

Three metrics (accuracy [Eq. (5.12)], sensitivity [Eq. (5.13)] and specificity [Eq. (5.14)]) were used to evaluate classification performance. Here, sensitivity is defined to be the fraction of the real positives that actually are correctly predicted as positives. Specificity is defined to be the fraction of the real negatives that actually are correctly predicted as negatives.

$$\text{Accuracy} = \frac{\text{TN} + \text{TP}}{\text{TN} + \text{FN} + \text{FP} + \text{TP}}. \tag{5.12}$$

$$\text{Sensitivity} = \frac{\text{TP}}{\text{TP} + \text{FN}}. \tag{5.13}$$

$$\text{Specificity} = \frac{\text{TN}}{\text{TN} + \text{FP}}. \tag{5.14}$$

By the definitions, the combination of sensitivity and specificity can be used to evaluate a model's balance ability so that we know if a model is biased to a special class. We also report the area under the receiver-operator characteristic curve (AUC) (Bradley, 1997) for each algorithm. The AUC value can indicate a model's generalization capability as a function of varying a classification threshold. An AUC = 1.0 represents a perfect classification, while an area of 0.5 represents random guessing for the classification problem.

5.4.2 Data Description

We evaluated this new classification approach using a prostate cancer data set for tumor versus normal classification (Li and Liu, 2002) consisting of a total of 136 prostate samples (77 tumor samples and 59 nontumor samples). A total of 12,600 normalized gene expression values resulted from microarray experiments. Here, negatives are defined to be the normal prostate samples (nontumor), while positives are tumor samples.

We also evaluated this approach using a colon cancer data set (Li and Liu, 2002) for comparison. This database consisted of 22 normal colon samples and 40 colon cancer samples. Gene expression information on more than 6500 genes were measured using microarrays, and 2000 with highest minimum intensity were extracted to form a matrix of 62 tissues × 2000 gene expression values.

5.4.3 Data Modeling

Similar to Golub et al. (1999), the original data set is simply normalized so that each gene vector has 0 for mean and 1 for standard deviation. For the jth gene vector $\mathbf{G}_j = <g_{1j}, g_{2j} \Lambda\, g_{nj}>$, where g_{ij} is the expression value of the jth gene on the ith tissue, $1 \le i \le n$, its mean is

$$m_j = \sum_{i=1}^{n} \frac{g_{ij}}{n}, \tag{5.15}$$

and standard deviation is

$$s_j = \sqrt{\frac{1}{n-1} \sum_{i=1}^{n} (g_{ij} - m_j)^2}. \tag{5.16}$$

The normalized vector should be norm $(\mathbf{G}_j) =< (g_{1j} - m_j)/s_j, (g_{2j} - m_j)/s_j \Lambda\, (g_{nj} - m_j)/s_j >$. To avoid overfitting, for leave-one-out or bootstrapping validation accuracy evaluation, validation samples were kept out from calculating these two values.

The regulation parameter was defined as $C \equiv 1$ for linear SVMs. For FCM, the "fuzziness degree" was defined as $m = 1.15$, the maximal iteration number set to 100, and the minimal improvement set to $\varepsilon = 10^{-5}$. For fuzzy-granular gene selection, genes were grouped into 10 clusters, in each of which S2N/FC/TS was used for gene ranking, and then 2 highest ranked genes in each of the 10 clusters were combined disjunctively to form the final gene set with a maximum of 20 genes. For comparison, top 20 ranked genes were also selected based on S2N, FC, and TS, respectively.

Notice that fuzzy membership values were defuzzified in such a way that a gene is always grouped into the cluster with the largest membership value and the cluster with the second largest membership value. The assumption is that different gene function groups are clustered based on their expression strengths. Some genes whose expression strengths are between two groups may be more suitable to be clustered into the two groups at the same time. This way, each gene achieves two opportunities to be selected.

The genes distribution in the prostate cancer data set is highly imbalanced between negative-related genes and positive-related genes. If $\alpha_+ = \alpha_- = 0.5$, 4761 positive-related genes and only 110 negative-related genes survive. To alleviate the imbalance, $\alpha_+ = 0.75$ and $\alpha_- = 0.5$ were used to select 721 positive-related genes and 110 negative-related genes. There was no overlap between positive-related genes and negative-related genes. Similarly, for the colon cancer data set, $\alpha_+ = 0.5$ and $\alpha_- = 0.1$.

Leave-one-out validation was used for performance evaluation: in each fold, one sample is left for validation and the other samples are used for training. Another evaluation heuristic adopted is balanced .632 bootstrapping (Braga-Neto and Dougherty, 2004): Random sampling with replacement is repeated for 100 times on each of the two data sets. Each tissue sample appeared exactly 100 times in the computation to reduce variance (Chernick, 1999).

5.4.4 Result Analysis

Table 5.1 and Figure 5.4 report the leave-one-out validation performance of six gene selection algorithms, named S2N, fuzzy-granular with S2N, FC, fuzzy-granular with

Table 5.1 Leave-One-Out Validation Performance on Prostate Cancer Data Set

Model	Accuracy	AUC	Sensitivity	Specificity
S2N [16]	0.8309	0.8388	0.7792	0.9091
FG+S2N	**0.9191**	**0.9226**	**0.8961**	**0.9583**
FC [17]	0.8824	0.8803	0.8961	0.8961
FG+FC	0.9118	0.9102	0.9221	0.9221
TS [18]	0.8603	0.8588	0.8701	0.8816
FG+TS	0.9191	0.9206	0.9091	0.9459

Note: The boldface row denotes the best method.

Table 5.2 .632 Bootstrapping Performance on Prostate Cancer Data Set

Model	Accuracy	AUC	Sensitivity	Specificity
S2N [16]	0.8323	0.8484	0.8047	0.9073
FG+S2N	**0.8684**	**0.9125**	**0.9045**	**0.9357**
FC [17]	0.8489	0.8688	0.8938	0.8829
FG+FC	0.8621	0.9054	0.9002	0.9280
TS [18]	0.8556	0.8734	0.8953	0.8880
FG+TS	0.8530	0.8864	0.8379	0.9436

Note: The boldface row denotes the best method.

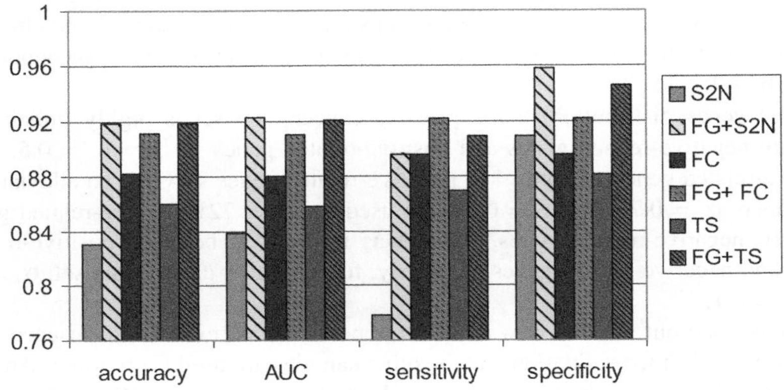

Figure 5.4 Leave-one-out validation performance on prostate cancer data set.

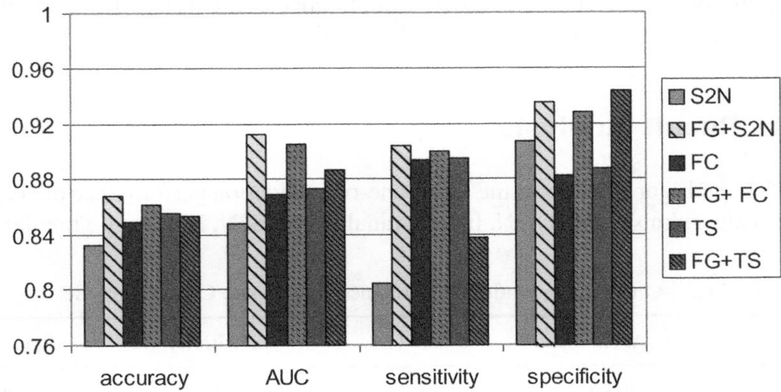

Figure 5.5 The .632 bootstrapping performance on prostate cancer data set.

FC, TS and fuzzy-granular with TS. Table 5.2 and Figure 5.5 report .632 bootstrapping performance. The results show that fuzzy-granular gene selection improves prediction performance compared to correlation-based algorithms both in terms of accuracy and AUC. Specifically, fuzzy-granular with S2N has the best performance under both leave-one-out validation and .632 bootstrapping validation. This suggests that the fuzzy-granular algorithm can select more informative genes for cancer classification.

There are two reasons for the improved performance of fuzzy-granular gene selection. First, RI-based prefiltering eliminates most of the irrelevant genes and hence decreases the correlation-induced noise. Second, FCM explicitly groups genes into different clusters with different expression patterns so that informative genes from different granules are selected in balance.

A similar gain in performance using fuzzy-granular gene selection was observed for the colon cancer dataset (Tables 5.3 and 5.4 and Figures 5.6 and 5.7).

Table 5.3 Leave-One-Out Validation Performance on Colon Cancer Data Set

Model	Accuracy	AUC	Sensitivity	Specificity
S2N [16]	0.8710	0.8591	0.9000	0.9000
FG+S2N	**0.9516**	**0.9523**	**0.9500**	**0.9744**
FC [17]	0.8710	0.8693	0.8750	0.9211
FG+FC	0.8710	0.8591	0.9000	0.9000
TS [18]	0.7903	0.7761	0.8250	0.8462
FG+TS	0.8710	0.8591	0.9000	0.9000

Note: The boldface row denotes the best method.

Table 5.4 .632 Bootstrapping Performance on Colon Cancer Data Set

Model	Accuracy	AUC	Sensitivity	Specificity
S2N [16]	0.8419	0.8701	0.9092	0.9116
FG+S2N	**0.8428**	**0.8881**	**0.9285**	**0.9212**
FC [17]	0.8314	0.8690	0.9087	0.9128
FG+FC	0.8323	0.8806	0.9236	0.9160
TS [18]	0.8150	0.8437	0.8741	0.9010
FG+TS	**0.8428**	**0.8881**	**0.9285**	**0.9212**

Note: The boldface rows denote the best method.

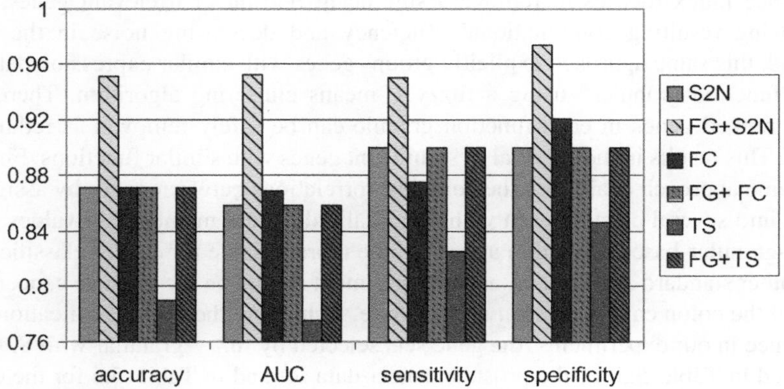

Figure 5.6 Leave-one-out validation performance on colon cancer data set.

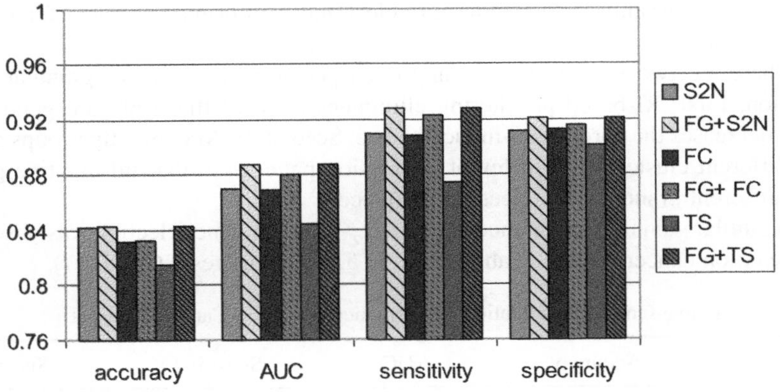

Figure 5.7 The .632 bootstrapping performance on colon cancer data set.

Table 5.5 Most Important Genes by FG+S2N on Prostate Cancer Data Set

Index	Probe ID	Index	Probe ID	Index	Probe ID
1,674	32312_at	7,167	40776_at	10,438	32542_at
2,791	36780_at	7,826	33825_at	10,833	1890_at
3,703	39608_at	8,058	34775_at	10,837	1894_f_at
4,173	41036_at	8,631	36666_at	11,050	1662_r_at
5,979	37193_at	8,986	37741_at		
6,643	39054_at	10,358	41812_s_at		

5.5 CONCLUSIONS

To select a more informative gene set for reliable cancer classification, a fuzzy-granular-based algorithm was described in this chapter. This approach utilizes relevance index metrics to remove a significant portion of irrelevant genes, thus improving resulting computational efficiency and decreasing noise in the data. Second, this same approach explicitly groups genes with similar expression patterns into "function granules" using a fuzzy *C*-means clustering algorithm. Therefore, lower ranked genes in each function granule can be safely removed as redundant genes. This results in the survival of significant genes with similar functions. Finally, this same approach deals with the complex correlations between genes by assigning genes into several clusters each with potentially different membership values. This fuzzy-granular-based algorithm appears to be more reliable for cancer classification than other standard approaches, as the experiment results on the prostate cancer data set and the colon cancer data set demonstrate. Achieving the best classification performance in our experiments, the gene sets selected by fuzzy-granular with S2N are reported in Table 5.5 for the prostate cancer data set and in Table 5.6 for the colon

Table 5.6 Most Important Genes by FG+S2N on Colon Cancer Data Set

Index	GAN	Index	GAN	Index	GAN
15	U14971	249	M63391	780	H40095
26	T95018	306	J00231	822	T92451
43	T57619	377	Z50753	878	M87789
62	T48804	493	R87126	1582	X63629
111	R78934	515	T56604	1772	H08393
138	M26697	765	M76378	1791	T48904

cancer data set. The gene sets are expected to be more helpful for biologists to uncover the inherent cancer-resulting mechanism. Because of the inherent advantage of eliminating irrelevant or redundant genes while selecting really informative genes, we expect that this superior performance can be generalized to other microarray data sets.

Acknowledgment

This work was supported in part by the National Institutes of Health under P20 GM065762.

REFERENCES

Alon, U., N. Barkai, D. A. Notterman, K. Gish, S. Ybarra, D. Mack, and A. J. Levine (1999). "Broad patterns of gene expression revealed by clustering analysis of tumor and normal colon tissues probed by oligonucleotide arrays," *Proc. Natl. Acad. Sci., USA*, Vol. 96, pp. 6745–6750.

Bair, E. and R. Tibshirani (2003). "Machine learning methods applied to DNA microarray data can improve the diagnosis of cancer," *SIGKDD Explorations*, Vol. 5, pp. 48–55.

Bargiela, A. (2002). *Granular Computing: An Introduction*, Kluwer Academic, Monte Carlo.

Bezdek, J. C. (1981). *Pattern Recognition with Fuzzy Objective Function Algorithms*. Plenum, New York.

Bortolan, G. and W. Pedrycz (1997). "Reconstruction problem and information granularity," *IEEE Trans. Fuzzy Systs.*, Vol. 2, pp. 234–248.

Bradley, A. P. (1997). "The use of the area under the ROC curve in the evaluation of machine learning algorithms," *Pattern Recogn.*, Vol. 30, pp. 1145–1159.

Braga-Neto, U. and E. R. Dougherty (2004). Is Cross-Validation Valid for Small-Sample Microarray Classification? *Bioinformatics*, Vol. 20, pp. 374–380.

Burges, C. J. C. (1998). "A tutorial on support vector machines for pattern recognition," *Data Mining Knowl. Disc.*, Vol. 2, pp. 121–167.

Chang, C.-C. and C.-J. Lin (2001). LIBSVM: A Library for Support Vector Machines. http://www.csie.ntu.edu.tw/~cjlin/libsvm.

Chernick, M. (1999). *Bootstrap Methods: A Practitioner's Guide*. Wiley, New York.

Chin, K. K. (1999). Support Vector Machines Applied to Speech Pattern Classification. Master's Thesis. Cambridge University.

Cristianini, N. and J. Shawe-Taylor (1999). *An Introduction to Support Vector Machines and Other Kernel-based Learning Methods*. Cambridge University Press, New York.

DUAN, K. and J. C. RAJAPAKSE (2004). "A variant of SVM-RFE for gene selection in cancer classification with expression data," in *Proc. of IEEE Symposium on Computational Intelligence in Computational Biology and Bioinformatics 2004*, IEEE Press, Piscataway, NJ, pp. 49–55.

FOGEL, G. B. and D. W. CORNE (2003). *Evolutionary Computation in Bioinformatics*. Morgan Kaufmann, San Francisco.

FUREY, T., N. CRISTIANINI, N. DUFFY, D. BEDNARSKI, M. SCHUMMER, and D. HAUSSLER (2000). "Support vector machine classification and validation of cancer tissue samples using microarray expression data," *Bioinformatics*, Vol. 16, pp. 906–914.

GOLUB, T. R., D. K. SLONIM, P. TAMAYO, C. HUARD, M. GAASENBEEK, J. P. MESIROV, H. COLLER, M. L. LOH, J. R. DOWNING, M. A. CALIGIURI, C. D. BLOOMFIELD, and E. S. LANDER (1999). "Molecular classification of cancer: Class discovery and class prediction by gene expression monitoring," *Science*, Vol. 286, pp. 531–537.

GUNN, S. (1998). "Support Vector Machines for Classification and Regression," ISIS Technical Report, Image Speech & Intelligent Systems Group, University of Southampton.

GUYON, I., J. WESTON, S. BARNHILL, and V. VAPNIK (2002). "Gene selection for cancer classification using support vector machines," *Machine Learn.*, Vol. 46, pp. 389–422.

HAND, D., H. MANNILA, and P. SMYTH (2001). *Principle of Data Mining*. MIT Press, Cambridge, MA.

JIANG, D., C. TANG, and A. ZHANG (2004). "Cluster analysis for gene expression data: A survey," *IEEE Trans. Knowledge Data Eng.*, Vol. 16, pp. 1370–1386.

JULIUSDOTTIR T., D. CORNE, E. KEEDWELL, and A. NARAYANAN (2005) "Two-Phase EA/k-NN for Feature Selection and Classification in Cancer Microarray Datasets," *Proc. of IEEE Symposium on Computational Intelligence in Bioinformatics and Computational Biology 2005*, pp. 1–8.

KEEDWELL, E. and A. NARAYANAN (2005). "Discovering gene regulatory networks with a neural-genetic hybrid," *IEEE/ACM Trans. Computat. Biol. Bioinform.*, Vol. 2, pp. 231–243.

LI, J. and H. LIU (2002). "Kent Ridge Bio-medical Data Set Repository," http://sdmc.lit.org.sg/GEDatasets/Datasets.html.

LIN, T. Y. (1998a). "Granular computing on binary relations I: Data mining and neighborhood systems," in *Rough Sets in Knowledge Discovery*, A. Skowron and L. Polkowski, Eds. Physica, Heidelberg, pp. 107–121.

LIN, T. Y. (1998b). "Granular computing on binary relations II: Rough set representations and belief functions," in *Rough Sets in Knowledge Discovery*, A. Skowron and L. Polkowski, Eds. Physica, Heidelberg, pp. 121–140.

LIN, T. Y. (1999). "Granular computing: Fuzzy logic and rough sets," in *Computing with Words in Information/Intelligent Systems*, L.A. Zadeh and J. Kacprzyk, Eds. Physica, Heidelberg, pp. 183–200.

LIN, T. Y. (2000). "Data mining and machine oriented modeling: A granular computing approach," *J. Appl. Intell.*, Vol. 13, pp. 113–124.

MA, J., Y. ZHAO, and S. AHALT (2002). OSU SVM Classifier Matlab Toolbox. http://www.ece.osu.edu/~maj/osu_svm/.

MODEL, F., P. ADORJAN, A. OLEK, and C. PIEPENBROCK (2001). "Feature selection for DNA methylation based cancer classification," *Bioinformatics*, Vol. 17, pp. S157–S164.

MOLER, E. J., M. L. CHOW, and I. S. MIAN (2000). "Analysis of molecular profile data using generative and discriminative methods," *Physiol. Genom.*, Vol. 4, pp. 109–126.

NOBLE, W. S. (2004). "Support vector machine applications in computational biology," in *Kernel Methods in Computational Biology*, B. Schoelkopf, K. Tsuda, and J.-P. Vert, Eds. MIT Press, Cambridge, HA, pp. 71–92.

PAVLIDIS, P., J. WESTON, J. CAI, and W. N. GRUNDY, (2000). "Gene functional analysis from heterogeneous data," *Proc. RECOMB*, ACM Press, New York, pp. 249–255.

PEDRYCZ, W. (2001a). *Granular Computing: An Emerging Paradigm*. Physica, Heidelberg.

PEDRYCZ, W. (2001b). "Granular computing in data mining," in *Data Mining & Computational Intelligence*, M. Last and A. Kandel, Eds., Physica, Heidelberg.

PEDRYCZ, W. and G. VUKOVICH (2002). "Granular computing in pattern recognition," in *Neuro-Fuzzy Pattern Recognition*, H. Bunke and A. Kandel, Eds. World Scientific, London.

SCHÖLKOPF, B., I. GUYON, and J. WESTON (2003). "Statistical learning and kernel methods in bioinformatics," in *Artificial Intelligence and Heuristic Methods in Bioinformatics 183*, P. Frasconi and R. Shamir, Eds. IOS Press, Amsterdam, pp. 1–21.

SHAPIRO, G. P. and P. TAMAYO (2003). "Microarray data mining: Facing the challenges," *SIGKDD Explorations*, Vol. 5, pp. 1–5.

VAPNIK, V. (1998). *Statistical Learning Theory*. Wiley, New York.

YAO, J. T. and Y. Y. YAO (2002). "A granular computing approach to machine learning," Proceedings of FSKD'02, Singapore, pp. 732–736.

YAO, Y. Y. (2001). "On modeling data mining with granular computing," Proceedings of COMPSAC 2001, pp. 638–643.

YIP, K. Y., D. W. CHEUNG, and M. K. NG (2004). "HARP: A practical projected clustering algorithm," *IEEE Trans. Knowledge Data Eng.*, Vol. 16, pp. 1387–1397.

Chapter 6

Evolutionary Feature Selection for Bioinformatics

Laetitia Jourdan, Clarisse Dhaenens, and El-Ghazali Talbi

6.1 INTRODUCTION

In the last decade, the rapid development of *omics* techniques[1] (i.e., genomics, proteomics, transcriptomics) has resulted in more and more data for biological analysis. This kind of data is known to include a large set of redundant, even perhaps irrelevant, features. Hence, feature selection is often considered as a required preprocessing step for data analysis. This task can indeed support dimensionality reduction and often results in better analysis (Yang and Honovar, 1998) by identifying and selecting a useful subset of features. Hence, if in some studies, feature selection is considered as an analysis tool, it is used jointly with learning algorithms that base themselves on the subset of features in order to proceed to a supervised or unsupervised classification.

Let us define the *feature selection problem*: Given a set of features $F = f_1, \ldots, f_i, \ldots, f_n$ the feature selection problem consists in finding a subset $F' \subseteq F$ that maximizes a scoring function $\Theta : \Gamma \rightarrow G$ (where Γ is the space of all possible feature subsets of F and G a subset of Γ) such that

$$F' = \arg\max_{G \subset \Gamma} \{\Theta(G)\}. \tag{6.1}$$

For example, in the context of microarrays, the number of studied genes is large. Hence, it is necessary to reduce this number of genes to better analyze experimental data. The entire set of genes will be F, whereas the set of selected genes will be F'. The scoring function Θ assigns to each possible subset of genes, G, a numerical value, that may indicate, for example, the classification rate, obtained using this

[1] The term *omics* refers to the comprehensive analysis of biological systems.

Computational Intelligence in Bioinformatics. Edited by G. B. Fogel, D. Corne, and Y. Pan

subset. F' must be the subset that optimizes the function Θ. The optimal feature selection problem has been shown to be nondeterministic polynomial-time hard (NP-hard) (Narendra and Fukunaga, 1977). This means that unless $P = \text{NP}$, no polynomial algorithm may solve this problem optimally. In this context, heuristic methods have been proposed to provide good solutions in a reasonable time.

Traditional feature subset selection (FSS) methods are sequential and based on greedy heuristics. For example, sequential forward selection (SFS) starts with an empty set and iteratively adds some features, whereas the sequential backward elimination (SBE, also referred to as "backward selection") starts with the full feature set and iteratively removes features (Whitney, 1971). These methods have a low complexity and may be applied to problems with large data sizes. Their main drawback is that they consider one feature at a time, which does not allow for the possibility of complex interactions between the features. Recently, more advanced methods have been used to explore the space of feature subsets. Among these methods, metaheuristics (general-purpose heuristics that can be used to solve different problems) such as evolutionary algorithms (Yang and Honoavar, 1998; Pei et al., 1997) have been proposed.

Two models of feature or attribute selection exist depending on whether the selection is coupled with the learning scheme or not. The first model, *a filter model*, treats feature subset selection and classification in two separated phases and uses a measure that is simple and fast to compute (see Fig. 6.1). Hence, a filter method is by definition, independent of the learning algorithm used after it.

The second model, *a wrapper method*, treats feature subset selection and classification in the same process and engages a learning algorithm to measure the classification accuracy (see Fig. 6.2). From a conceptual point of view, wrapper approaches are clearly advantageous since the features are selected by optimizing the discriminative power of the final induction algorithm. But using a specific algorithm to evaluate the quality of the subset selection may lead to a lack of robustness across different learning algorithms. We note that the vast majority of evolutionary approaches for feature selection use wrapper models.

In this chapter, we investigate the potential of evolutionary algorithms (metaheuristics) for feature selection in bioinformatics. First, in Section 6.2, generalities

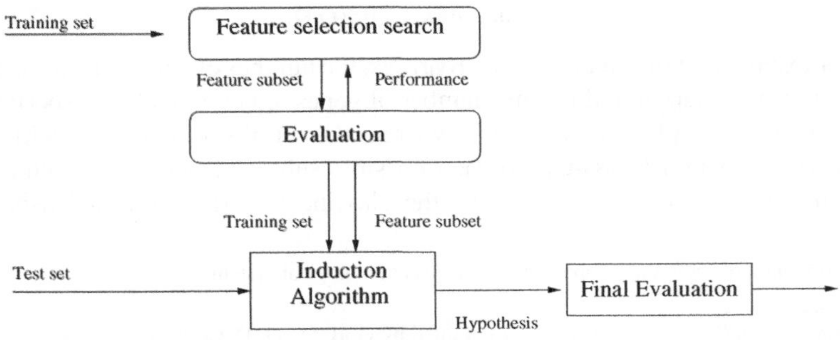

Figure 6.1 Filter approach for feature selection.

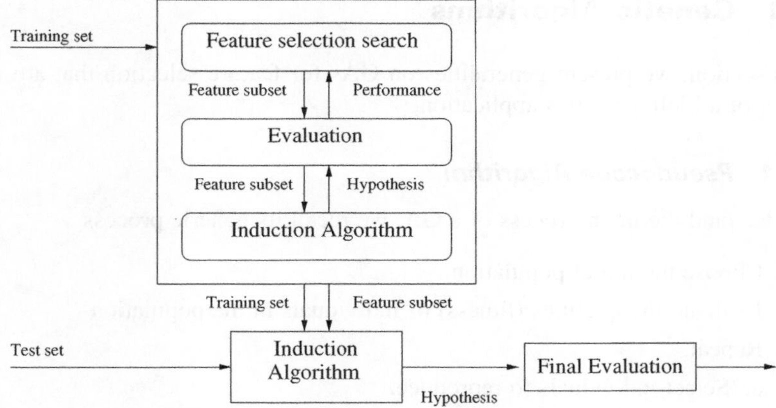

Figure 6.2 Wrapper approach for feature selection.

of evolutionary feature selection are presented and common characteristics of evolutionary algorithms for feature selection are shown, without considering the application. Section 6.3 is dedicated to feature selection in an unsupervised context. Section 6.4 stresses feature selection linked with supervised classification. Each of these two parts briefly presents the classical learning algorithms used in bioinformatics, the main objective functions, and some applications. Section 6.5 presents a framework and data sets used in this domain to help the reader to develop and test their own methods. Finally, Section 6.6 presents our conclusions and perspectives.

6.2 EVOLUTIONARY ALGORITHMS FOR FEATURE SELECTION

Evolutionary algorithms (Fraser, 1957; Bremermann, 1958; Fogel et al., 1966; Rechenberg, 1973; Holland, 1975; Koza, 1992; Smith, 1980; Cramer, 1985; Schwefel, 1981; Fogel, 1998; Davis et al., 2006) are based on the paradigm of evolution in nature. Three basic mechanisms are inspired from natural evolution: reproduction, mutation, and selection. The language used to describe the behavior of evolutionary algorithms is similar to biological notions (genotype, phenotype, crossover, mutation, etc.). There are several kinds of evolutionary algorithms (EAs): genetic algorithms (GAs), genetic programming (GP), evolutionary programming (EP), and evolution strategies (ES).

All evolutionary algorithms can be adopted for feature selection. For example, some methods use GP (Davis et al., 2006; Gray et al., 1998; Guo and Nandi, 2006; Langdon and Buxton, 2004), but the vast majority use GA. Therefore, we focus on this type of EA in this chapter.

6.2.1 Genetic Algorithms

In this section, we present generalities on GAs for feature selection that are independent of a bioinformatics application.

6.2.1.1 Pseudocode Algorithm

To understand the main process of a GA, we recall its generic process.

1. Choose the initial population.
2. Evaluate the qualities (fitness) of individuals in the population.
3. Repeat.
   a. Select individuals to reproduce.
   b. Breed new generation through crossover and mutation (genetic operations) and generate offspring.
   c. Evaluate offspring fitness.
   d. Replace part of population with offspring.
4. Until terminating condition.

6.2.1.2 Encoding

To describe a solution (an individual of the population), an encoding is used. In Holland's original work, a single binary "bit string" was used (Holland, 1975). This had computational advantages in terms of efficiency, but now, many different methods for encoding the genetic information are in common use today; tree encoding, real-valued arrays, permutations, gray encoding, and so on. In light of the no free lunch theorem (Wolpert and Macready, 1997), there is no optimal encoding over all problems. Thus, the encoding should be adapted to the problem at hand.

As the objective of feature selection is to find a subset of features that optimizes a quality function: A solution should be a subset of features. To achieve this, two simple encodings have been used. The first one uses a binary encoding. The length of the solution is equal to the number of features of the data set. A 1 at the nth position indicates that the nth feature is selected and a 0 indicates that the feature is not selected [see Fig. 6.3(a)]. This binary encoding allows using simple operators as one-bit mutation but presents serious scalability issues when databases are huge.

A second possible encoding is to indicate only the selected features by their indices. In the literature two possibilities are observed. (1) A fixed length encoding where zero values and indices of the features are combined (Cherkauer and Shavlik, 1996). This encoding allows multiple presences of features [see Fig. 6.3(b)]. (2) A variable length encoding wherein the length is bounded by the number of features in the data set [see Fig. 6.3(c)]. A length of zero indicates that no feature is selected. The advantage of this encoding is that it limits the size of the subset and to be adapted. This is useful when data sets become very large, as, for example, is the

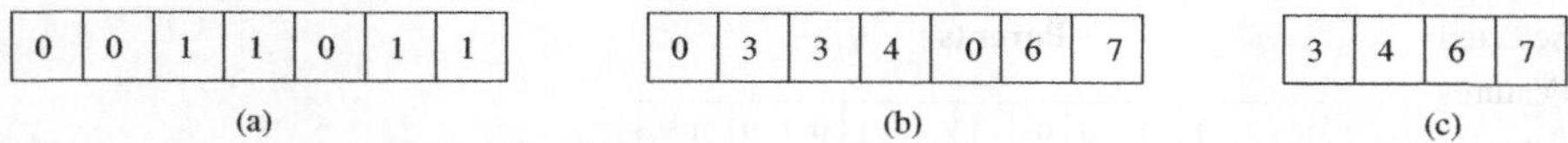

Figure 6.3 Different possibilities for encoding the selection of features 3, 4, 6, 7: (**a**) Binary encoding, (**b**) Cherkauer encoding, and (**c**) variable length encoding.

case in microarray experiments. The main disadvantage is that operators should be adapted to explore the search space.

Let us remark that in the case of unsupervised learning, the number of clusters is generally unknown and has to be determined. Thus, it may be useful, for some approaches, to integrate to the encoding the number of clusters [as realized in Handl and Knowles (2006a)].

6.2.1.3 Initialization

The initialization phase determines the initial population of solutions of the GA. The simplest initialization is to randomly select the initial features. Some authors (Ni and Liu, 2004) have controlled the size of the initial subsets of features by using a threshold parameter called size control probability since it is not useful to have too many selected features.

6.2.1.4 Operators

To evolve and to improve solutions, EAs use operators for variation. Operators are used to explore the search space and should be adapted to both the encoding and the problem.

Classical In many studies, authors have used classical operators: one-point crossover, two-point crossover, uniform crossover, and uniform mutation. Some authors introduce a bias in the mutation: As in feature selection the objective is to reduce the number of selected features. The mutation operator may favor the reduction by increasing the chance of mutating a gene to indicate that the feature is not selected (Jourdan et al., 2002; Juliusdottir et al., 2005).

Specific to Binary Encoding A specific crossover for feature selection is the *subset size-oriented common feature crossover operator* (SSOCF) (Emmanouilidis et al., 2000) that aims to keep useful "informative blocks" and produces offspring that have the same distribution of these structures as the parents (see Fig. 6.4). The shared features are kept by offspring and the nonshared features are inherited by the offspring corresponding to the ith parent with the probability $(n_i - n_c/n_u)$ where n_i is the number of selected features of the ith parent, n_c is the number of commonly selected features across both mating partners, and n_u is the number of nonshared selected features. We note that designing such specialized crossover operators is often found to be useful; however, the same is also true for the design of specialized

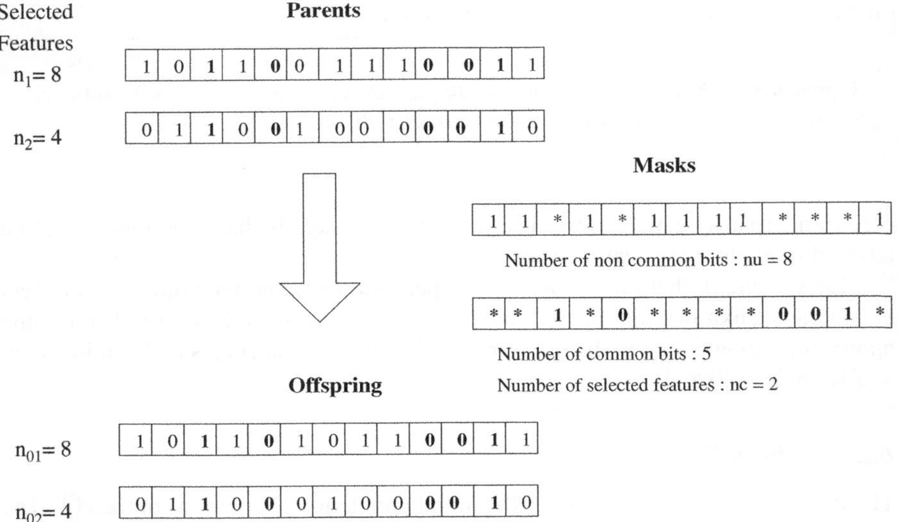

Figure 6.4 Subset size-oriented common feature crossover operator (SSOCF) (Emmanouilidis et al., 2000).

mutation operators. Theoretical research in evolutionary computation is currently unable to guide practitioners in such choices when it comes to dealing with realistic problems, so choices tend to be made based on the taste, intuition, and experience of the researcher.

6.3 FEATURE SELECTION FOR CLUSTERING IN BIOINFORMATICS

6.3.1 Motivation

Unsupervised learning (or clustering) is an important task for bioinformatics, where experiments generate a lot of data that have to be studied. As reported by Handl et al. (2005), unsupervised classification has many applications in postgenomics. It plays an important role in the analysis of gene expression data as well as the analysis of proteomics and metabolomics data. But it can also be applied directly to sequencing or in the context of protein comparison and structure prediction.

Unfortunately, before standard algorithms can be employed, the "curse of dimensionality" needs to be solved. Indeed, due to the large amount of information contained in data coming from most bioinformatics experiments, standard machine learning techniques cannot be directly applied. Instead, feature selection techniques are used to first reduce the dimensionality of the input space and thus enable the subsequent use of clustering algorithms. For example, high-resolution computed tomography images of the lungs have been studied with such an approach (Dy and Brodley, 2000).

When unsupervised learning (or clustering) is concerned, the feature selection task is difficult because to find the subset of features that maximizes a performance criterion, the clusters have to be defined. But unlike in supervised learning, there exists no class label for the data and so no obvious criteria to guide the search. The problem is made more difficult when the number of clusters is unknown in advance, which happens to be the case for most real-world situations.

In unsupervised learning, the filter/wrapper model distinction may also be used. Hence, *filter* methods use some intrinsic properties of the data to select features without using the clustering algorithm that will be ultimately applied. On the other hand, the *wrapper* approach applies the unsupervised learning algorithm to each feature subset in the search space and then evaluates the feature subset by a criterion function that uses the clustering result.

In this chapter, we focus on the use of EAs in the feature selection for clustering in bioinformatics. Therefore, we first give a brief presentation of the most common methods used for clustering in bioinformatics. This is followed by an overview of objective functions.

6.3.2 Some Clustering Algorithms

The predominant methods used in bioinformatics remain to be the classical ones, such as: hierarchical clustering, K-means, fuzzy c-means, finite mixture model, or self-organizing maps. We will not present, in detail, all of the existing clustering methods. Instead, the focus of this chapter is on methods that are often used in bioinformatics. To have a more detailed overview, the reader may refer to an online tutorial.[2] Some of these clustering approaches are considered classical approaches such that packages have been developed for the language R (language and environment for statistical computing and graphics). In particular, a specific package, Bioconductor (www.bioconductor.org), is an open source and open development software project for the analysis and comprehension of genomic data and offers an assortment of clustering algorithms.

6.3.2.1 Hierarchical Methods

In hierarchical clustering, the data is not partitioned into a particular cluster in a single step. Instead, a series of partitioning takes place. Hierarchical clustering is subdivided into agglomerative methods and dividing methods. Agglomerative techniques are more commonly used. They proceed by a series of fusions of the n objects into groups. Hence, they start with each data forming a singleton cluster. At each iteration, the two clusters with the highest similarity are merged to form one cluster. The process continues until no cluster can be merged (or the number of required clusters is attained). The differences between methods arise because of the different ways of defining distance (or similarity) between clusters. Hence, several classical

[2] http://www.elet.polimi.it/upload/matteucc/Clustering/tutorial_html/index.html.

ways to measure similarity have been proposed such as single linkage clustering, complete linkage clustering, average linkage clustering, and the like. A classical example of such an approach applied to microarray data can be found in Eisen et al. (1998). On the contrary, dividing methods start with a single cluster that groups every object and separates them successively into finer groupings.

6.3.2.2 K-Means

The K-means algorithm is one of the most often used nonhierarchical clustering methods. It starts with a random initial partition by defining centroids (center for each cluster). Then, it iteratively assigns each data point to the cluster whose centroid is nearest from the data point and recalculates the new centroids based on the new assignment until assignments are unchanged (McQueen, 1967).

If we assume that data (objects) may be represented as points of the space, and that a distance between objects has been defined, the K-means algorithm is as follows:

1. Select K points into the space. These points represent initial group centroids.
2. Assign each object to the group that has the closest centroid.
3. Once all objects have been assigned, recalculate the positions of the K centroids.
4. Repeat steps 2 and 3 while the centroids move.

6.3.2.3 Fuzzy c-Means

The fuzzy c-means is a method of clustering that allows one object to belong to two or more clusters (Dunn, 1973; Bezdek, 1981). This method is frequently used in pattern recognition but has also been used in bioinformatics. It is based on the minimization of an objective function that uses a degree of membership of an instance to a cluster. Fuzzy partitioning is carried out through an iterative optimization of the objective function.

6.3.2.4 Self-Organizing Maps

Self-organizing maps (SOMs) is a data visualization technique invented by Kohonen that reduces the dimension of data through the use of self-organizing neural networks (Kohonen, 1988, 1995). As remarked by Tamayo et al. (1999), SOMs have a number of characteristics that make them particularly well suited to clustering and analysis of gene expression patterns. They are ideally suited to exploratory data analysis, allowing one to impose partial structure on the clusters and facilitating easy visualization and interpretation (Tamayo et al., 1999).

6.3.3 Fitness Functions

As presented in Section 6.2, the development of an evolutionary approach for the feature selection process requires determining an appropriate objective function.

Table 6.1 Overview of Fitness Functions for Feature Selection in Clustering for Bioinformatics Applications

Criteria	Aim/Formulas	References
Feature cardinality	To reduce the number of selected features	Handl and Knowles (2006a)
Filter Approaches		
Entropy	Low if data has distinct clusters, high otherwise	Handl and Knowles (2005); Dash et al. (2002)
Wrapper Approaches		
F_{within}	Favor dense clusters	Kim et al. (2000)
F_{between}	Favor well-separated clusters	Kim et al. (2000)
F_{clusters}	Minimize number of clusters to be used with other metrics	Kim et al. (2000)
$F_{\text{complexity}}$	Minimize number of features: $F_{\text{complexity}} = 1 - \frac{\text{number of selected features} - 1}{\text{total number of features} - 1}$	Kim et al. (2000)
Silhouette	Reflect the confidence in the cluster assignment	Handl and Knowles (2005)
DB Index	Indicates both intracluster compactness and between-cluster separation	Davies and Bouldin (1979); Handl and Knowles (2005)

Table 6.1 presents some of these functions. A distinction has been made according to the approach (filter/wrapper) used.

6.3.3.1 Fitness Functions in Filter Approaches

Very few filter approaches may be found in the literature. However, when such approaches are used, the main question is to determine the selection criterion. For example, Dash et al. (2002) proposed an entropy measure that is low if the data has distinct clusters. This same entropy measure is high if the data have no distinct clusters. Dash et al. (2002) used this approach with success coupled with forward selection to select profiles for gene clustering. In their study, Handl et al. (2005) also proposed to use an entropy measure when filter approaches were developed.

6.3.3.2 Fitness Functions in Wrapper Approaches

When a wrapper approach is used, the objective function corresponds to classical objective functions of clustering. But unlike classification, which has a very quantifiable way of evaluating accuracy, there is no generally acceptable criterion to estimate the accuracy of clustering. As a result, a number of numerical measurements are available. Hence, Table 6.1 refers some objective functions used in bioinformatics applications.

6.3.3.3 Multiobjective Approaches

The feature selection problem is inherently a multiple objective problem. Some studies propose multiobjective modeling for bioinformatics problems. In this case, rather than combining several criteria, the goal is to produce solutions of best compromise between the objectives. Hence, the Pareto dominance is used and the Pareto front is to be obtained (Pareto, 1896).

For example, Kim et al. (2000, 2002) used an evolutionary local selection algorithm (ELSA) that was capable of maintaining a diverse population of solutions that approximated the Pareto front. In this wrapper approach, each evolved solution represented a feature subset and a number of clusters. Then a standard K-means algorithm was applied to form the given number of clusters on the selected features. The partitions obtained were evaluated according to four criteria: F_{within}, F_{between}, F_{clusters}, and $F_{\text{complexity}}$ (see Table 6.1).

Handl and Knowles (2006a) studied the particular problem of semisupervised feature selection in which some data were labeled and others were not. They addressed this problem via multiobjective optimization, which allowed them to use several objective functions coming from the unsupervised and the supervised class of methods (Handl and Knowles, 2006a). For this purpose, they adapted the Pareto envelope-based selection algorithm (PESA-II) (Corne et al., 2001).

6.3.4 Applications in Bioinformatics

Kim et al. (2000) tested their multiobjective approach ELSA on the Wisconsin Prognostic Breast Cancer (WPBC) data, which records 33 features for the 227 cases. Their algorithm managed to identify three well-separated clusters using only 7 selected features. Handl et al. (2005) used the data set developed by Golub et al. (1999) dealing with leukemia (http://www.broad.mit.edu/cgi-bin/cancer/datasets.cgi) to compare several objective functions.

Even if some authors are adopting an evolutionary feature selection to deal with clustering bioinformatics data, this is not a common approach for clustering in bioinformatics. This may be explained with different reasons. First, clustering algorithms such as the hierarchical approaches may deal with a large number of features, and the results they produce are quite easy to read. With such algorithms it is not necessarily essential to first select features. Second, when features are selected, this is not frequently done with an evolutionary approach. Maybe this is because the quality measures of clustering are not so widely adopted by the community, and people prefer selecting features regarding their own quality criteria. These two reasons may explain why, in bioinformatics, there are only a few references on evolutionary feature selection for clustering. Let us now better understand supervised classification where bioinformatics applications seem to be more frequent.

6.4 FEATURE SELECTION FOR CLASSIFICATION IN BIOINFORMATICS

6.4.1 Motivation

Classification is a major task of supervised machine learning and is widely used for cancer studies, protein identification, and other health issues. Unfortunately, in many bioinformatics problems, the number of features is significantly larger than the number of samples (high feature-to-sample ratio data sets). Such an observation implies that it could be useful to decrease the number of features to improve the classification or to help to identify interesting features in noisy environments. Feature selection is a good way to reduce the number of features. As feature selection is linked with the classification algorithms used, the most common algorithms are exposed below.

6.4.2 Overview of Classification Algorithms

Supervised classification methods are widely used in data mining. In feature selection literature all the types of classifiers are used. In Table 6.2 an overview of classification methods used in evolutionary feature selection for bioinformatics problems is realized. Applications can influence the choice of the classifier. We will present some of classification methods used in evolutionary feature selection for bioinformatics. In bioinformatics, as previously mentioned, most databases have few samples. To validate the classification, cross-validation techniques and bootstrapping are often used and are presented below.

6.4.2.1 Classification Algorithms

Classification algorithms used for feature selection in bioinformatics include *K*-nearest neighbor (KNN), maximum likelihood, naïve Bayes, support vector machine (SVMs), and decision trees. These methods are briefly explained below.

K-Nearest Neighbor A simple algorithm, KNN stores all available examples and classifies new instances based on a similarity measure (Fix and Hodges, 1951). The idea is to assign to the sample the same class than its *k* most similar examples (its neighbors). The algorithm is simple, fast, and has a good classification performance. Its main drawback is that the vector distance is not necessarily well suited for finding intuitively similar examples, especially if irrelevant attributes are present.

Maximum-Likelihood Classifier The maximum-likelihood classifier is a popular method, especially in image classification. Likelihood is defined as the posterior probability of a sample belonging to a class. The maximum-likelihood method has an advantage from the viewpoint of probability theory but cannot be applied when the distribution of the solution space does not follow a normal distribution.

Table 6.2 Overview of Induction Algorithms for Wrapper Methods in Classification for Bioinformatics Applications

Induction Algorithm	Application	References
GS classifier	Microarray	Liu et al. (2001); Liu and Iba (2002)
SVM	Microarray	Liu et al. (2005); Bonilla Huerta et al. (2006); Peng et al. (2003); Topon Kumar and Iba (2005); Borges and Nievola (2005); Ando and Iba (2004); Li et al. (2005)
MHLD	Microarray	Ooi and Tan (2003); Liu et al. (2005)
KNN	Microarray	Deutsch (2003); Jirapech-Umpai and Aitken (2005); Li et al. (2005); Ando and Iba (2004); Borges and Nievola (2005); Liu et al. (2005); Juliusdottir et al. (2005).
Naïve Bayes	Microarray/gene expression	Borges and Nievola (2005); Ando and Iba (2004); Topon Kumar and Iba (2004a,b)
Weighted voted	Microarray/gene expression	Ni and Liu (2004); Topon Kumar and Iba (2004a,b)
C4.5	Microarray/gene expression	Borges and Nievola (2005)
Bayesian network	Microarray/gene expression	Borges and Nievola (2005)
Decision table	Microarray/gene expression	Borges and Nievola (2005)
SVM	Protein fold recognition	Shi et al. (2004)
PLS/RMSEP	Spectroscopic data	Jarvis and Goodacre (2005)
KNN (Γ Test)	Nucleotide sequence	Chuzhanova et al. (1998)
Neural network	Microcalcification patterns	Zhang et al. (2005)
KNN	Chromatograms	Lavine et al. (2003)
SVM	MS proteomic data	Jong et al. (2004)
Decision tree	Gene/SNP selection	Shah and Kusiak (2004)

Naïve Bayes The naïve-Bayes classifier uses probabilistic approach based on applying Bayes' theorem with strong (naïve) independence assumptions to assign samples to classes. It computes the conditional probabilities of different classes given the values of the features and predicts the class with highest conditional probability. The naïve-Bayes method is simple and can be applied to multiclass classification problems, but it is based on independence between variables. In spite of its simplified assumptions, naïve-Bayes classifiers often work very well in many complex real-world situations.

Support Vector Machines The SVMs are derived from statistical learning theory, classifying points by assigning them to one of two disjoint half spaces (Cortes and Vapnik, 1995). These half spaces are either in the original input space of the problem for linear classifiers or in a higher dimensional feature space for nonlinear classifiers. The assignment is realized by a kernel function. A special property of SVMs is that they simultaneously minimize the empirical classification error and maximize the geometric margin. Hence it is also known as the maximum margin classifier. SVMs generally produce very good results in terms of accuracy.

Decision Trees Decision trees are also a common method used in data mining. Each node in the tree specifies a test on a given attribute of the instance, and each branch descending from that node corresponds to one of the possible values for this attribute. An instance is classified by starting at the root node of the decision tree, testing the attribute specified by this node, then moving down the tree branch corresponding to the value of the attribute. This process is then repeated at the node on this branch and so on until a leaf node is reached. A classical example of a decision tree algorithm is C4.5. To construct the decision tree, for each node and each feature, a value is computed and a test is made to determine the best feature. Decision trees are appreciated for their legibility. Their results are often good and they provide directly a kind of feature selection by choosing the good feature to create the node.

Huang et al. (2003) compared experimentally on several types of test beds naïve Bayes, C4.5, C4.4, and SVMs in terms of both accuracy and receiver-operator characteristic (ROC) area under the curve (AUC) to evaluate performances. They concluded that all of algorithms examined have a very similar predictive accuracy. In addition, naïve Bayes, C4.4, and SVM produce similar scores, and they all outperform C4.5 in AUC with statistical significance.

6.4.2.2 Resampling Methods

In many bioinformatics applications, the size of the available data sets is small and often, due to the experimental protocol, the data are not independent. To overcome this aspect, there are different techniques for assessing prediction error by implementing some form of partitioning or resampling of the available data set. When using these techniques, the data is typically divided into a *learning set* (also called a *training set*) and a *test set*. In situations where the amount of data samples is large enough, the data may be divided into three sets, where the extra set is called the *validation* set, or *holdout* set. This latter data set is used by different researchers in various ways, but typically with the goal of avoiding overfitting. For example, sampled performance on the validation set (which otherwise has no influence) will indicate when it is appropriate to stop the training process. In the following sections, however, we consider only the cases where we have a *learning* set and a *test* set, which is common when the amount of data is too small to wisely divide into three sets. We provide a brief review of cross validation, *k*-fold cross validation, leave-one-out cross validation, and bootstrapping.

Cross Validation Usually to validate a classification, a cross validation or at least a holdout validation is realized. Samples are chosen at random from the initial sample to form the validation data, and the remaining observations are retained as the training data. Normally, less than a third of the initial samples are used for validation data. But to better evaluate the classification, two other methods are used.

***k*-Fold Cross Validation** In k-fold cross validation, the data set is divided into k subsets, and the method is repeated k times. Each time, one of the k subsets is used as the test set and the other $k - 1$ subsets are put together to form a training set. Then the average error across all k trials is computed. The disadvantage of this method is that the training algorithm has to be rerun from scratch k times, which means that it takes k evaluation times.

Leave-One-Out Cross Validation Leave-one-out cross validation (LOOCV) is a k-fold cross validation taken to its logical extreme, with k equals to N, the number of samples in the set. That means that N separate times, the function approximator is trained on all the data except for one point and a prediction is made for that point. This method is useful when there are few samples in the data set, but it can be very expensive to compute.

6.4.2.3 Bootstrapping

There are numerous bootstrap estimators. Two are often used by the feature selection community: the e0 and the .632 bootstrap. For the e0 bootstrap estimator, a training group consists of n cases sampled with replacement from a sample of size n. Sampled with replacement means that the training samples are drawn from the data set and are placed back after they are used, so their repeated use is allowed. The error rate on the test group is the e0 estimator. For this technique, it turns out that the average or expected fraction of nonrepeated cases in the training group is .632, and the expected fraction of such cases in the test group is .368. The .632 bootstrap, .632B, is the simple linear combination of .368*app + .632*e0, where app is the apparent error rate on all the cases (both training and testing cases). It should be noted that e0 is approximated by repeated twofold cross validation (i.e., 50/50 splits of train-and-test cases).

In Molinaro et al. (2005), several sampling methods and in particular LOOCV, k-fold cross validation, and the .632+ bootstrap were compared. For small data sets, LOOCV, k-fold cross validation and the .632+ bootstrap had the smallest bias for diagonal discriminant analysis, nearest neighbor, and classification trees. LOOCV and k-fold cross validation had the smallest bias for linear discriminant analysis. Additionally, LOOCV, k-fold cross validation and the .632+ bootstrap had the lowest mean-square error. Taken over the space of all problems, these resampling methods give comparable results.

6.4.3 Overview of Fitness Functions

Generally for bioinformatics problems, EAs are often used as a wrapper approach and GP used as a filter approach. This observation is often due to the encoding, as GP encoding is more informative and does not necessarily need a classifier.

In Table 6.3, an overview of fitness functions used for feature selection in classification in the bioinformatics field is presented. We can distinguish different aims in the fitness functions. Some of them try to reduce the cardinality of the feature set by directly counting the number of features or by using a ratio of the cardinality of the subset over the cardinality of the global set of features.

For the wrapper model, the majority of fitness functions try to minimize the error rate of the used classifier. The error rate can be globally computed (Liu et al., 2001; Liu and Iba, 2002; Borges and Nievola, 2005; Zhang et al., 2005), computed on the testing set (Shi et al., 2004; Shah and Kusiak, 2004) or computed by using cross-validation accuracy (Juliusdottir et al., 2005; Ooi and Tan, 2003; Topon Kumar and Iba, 2004a,b; Shi et al., 2004). For specific applications, authors use also specific domain fitness functions. For example, in Jarvis and Goodacre, 2005, root-mean-squared error of prediction (RMSEP) was used to assess the predictive ability of multivariate models by comparing predictions with reference values for a test set in postgenomics spectroscopic techniques.

6.4.3.1 Toward Multiobjective Models

As mentioned previously, feature selection is inherently a multiple objective problem. Current research in bioinformatics proposes aggregations of criteria; however, some authors have proposed purely multiobjective methods.

Aggregation Techniques Two main objectives are often associated: minimizing the error rate and minimizing the number of selected features (Juliusdottir et al., 2005; Liu et al., 2001; Topon Kumar and Iba, 2004a,b). To aggregate the two objectives, researchers use parameters to control the trade-off between preferences for each objective.

Multiobjective Approaches An increasing number of authors utilize the strength of evolutionary computation for multiobjective optimization of feature selection for classification. As for aggregation work the objectives are often the maximization of the quality of the classifier for wrapper methods and the minimization of the number of selected features. In Shi et al. (2004), the authors proposed a multiobjective approach and resolution that maximized the cross-validation accuracy on the training set, the classification accuracy on the testing set, and minimized the cardinality of the feature subset. To solve the problem, the authors used the non-dominated sorting genetic algorithm (NSGA-II) approach.

Table 6.3 Overview of Fitness Function for Feature Selection in Classification for Bioinformatics Applications[a]

Criteria	Aim/Formula	References
Feature cardinality	Minimize number of selected features	Liu et al. (2001); Liu and Iba (2002); Topon Kumar and Iba (2004a,b); Shi et al. (2004); Juliusdottir et al. (2005)
Wrapper Model		
Global error rate	$\frac{\text{Number (missclassified samples)}}{n}$	Liu et al. (2001); Liu and Iba (2002); Borges and Nievola (2005); Zhang et al. (2005)
Difference in error rate among classes	Prevent from always predicting the same class $\frac{\sum_{j=i+1}^{m}(e_i - e_j)^2}{C_m^2}$	Liu and Iba (2002)
Prediction	Classification accuracy on the testing set	Shi et al. (2004); Shah and Kusiak (2004)
Cross-validation accuracy	Repeated classification accuracy	Shi et al. (2004); Topon Kumar and Iba (2004a,b); Ooi and Tan (2003); Juliusdottir et al. (2005); Jirapech-Umpai and Aitken (2005); Ni and Liu (2004); Shah and Kusiak (2004); Bonilla Huerta et al. (2006); Ando and Iba (2004)
Root-mean-square error of prediction (RMSEP)	Assess the predictive ability	Jarvis and Goodacre (2005)
Gamma test value	$\frac{1}{1+e^{\Gamma(F')/\kappa}}$	Chuzhanova et al. (1998)
Accuracy	Mean of the accuracy on the positive samples and the accuracy on negative samples	Langdon and Buxton (2004)
Filter Model		
Modified Fisher criterion	Maximized the ratio between within-class scatter and between-class scatter	Guo and Nandi (2006)

[a]Assume that there are m classes and n samples in the training set. Let F be the set of features, F' be the selected subset of features.

6.4.4 Applications

Feature selection for classification is mainly used for bioinformatics applications. Table 6.2 indicates applications and induction algorithms involved in some articles. Applications are varied, as well as the induction algorithms used. We can nevertheless remark that the majority of the articles use feature selection for microarray analysis (Ni and Liu, 2004; Juliusdottir et al., 2005; Liu et al., 2001; Liu and Iba, 2002; Liu et al., 2005; Peng et al., 2003; Topon Kumar and Iba, 2005; Borges and Nievola, 2005; Ando and Iba, 2004; Li et al., 2005; Ooi and Tan, 2003; Deutsch, 2003; Jirapech-Umpai and Aitken, 2005; Topon Kumar and Iba, 2004a,b). DNA (deoxyribonucleic acid) microarrays allow simultaneous analysis of thousands of genes and can give important insights about cell function since changes in the physiology of an organism are generally associated with changes in gene expression patterns. For example, microarrays are used in cancer studies. Feature selection for gene expression analysis in cancer prediction often uses wrapper classification methods to discriminate a type of tumor, to reduce the number of genes to investigate in case of a new patient, and/or to produce new treatment. As we can observe in Table 6.2, various classification algorithms can be used for wrapper methods (e.g., KNN, SVM). The results are very promising as researchers show a major reduction of the number of genes to consider and a great improvement of the classification accuracy. The other point is that feature selection with evolutionary algorithms has allowed to discover or to confirm genes directly involved in diseases.

Thus, in this context of classification, where new instances have to be assigned to classes, the feature selection phase has great importance. This will influence the quality of the classification algorithm. As in bioinformatics, classification represents an important aspect, many authors have studied the impact of feature selection, and some of them have used evolutionary approaches to deal with the combinatorics. In this context, some authors have used frameworks and data sets to develop and test their methods. These are identified next.

6.5 FRAMEWORKS AND DATA SETS

In this section we present some frameworks used in articles dealing with evolutionary feature selection in bioinformatics and previously cited in this chapter. The objective is not to provide an exhaustive list of all the existing frameworks but to recall frameworks that could be useful for developers. We focus on frameworks that either focus on the development of metaheuristics or utilize learning algorithms. By combining both, it is possible to easily develop an evolutionary approach for feature selection. In this part, some indications on data sets used in this domain are also given to give the reader the opportunity to test their methods.

6.5.1 Metaheuristics

Many frameworks are available to design metaheuristics (and in particular EAs). An extensive list of existing frameworks may be found at http://evonet.lri.fr/evoweb/

resources/software/. Here, only frameworks used in references dealing with feature selection in bioinformatics are presented.

6.5.1.1 GAlib

GAlib is a C++ Library of Genetic Algorithm Components (http://lancet.mit.edu/ga/). The library includes tools for using genetic algorithms to do optimization in any C++ program using any representation and genetic operators. The documentation includes an extensive overview of how to implement a GA as well as examples illustrating customization to the GAlib classes.

6.5.1.2 MATLAB

The Genetic Algorithm Toolbox is a module to be used with MATLAB that contains software routines for implementing genetic algorithms (GAs) and other evolutionary computing techniques (http://www.shef.ac.uk/acse/research/ecrg/gat.html).

6.5.1.3 ParadisEO

The PARAllel and DIStributed Evolving Objects (ParadisEO) is a C++ framework that enables the easy and fast development of the most common parallel and distributed models of hybrid metaheuristics (http://paradiseo.gforge.inria.fr).

6.5.2 Machine Learning

Many contributions in evolutionary feature selection for bioinformatics do not use homemade induction algorithms because there are many algorithms already suitable for this purpose; and for both classification and clustering, these are freely available. Here we identify available algorithms that use relevant articles. An extensive list may be found at http://www.kdnuggets.com/software/clustering.html for clustering algorithms and http://www.kdnuggets.com/software/classification.html for classification algorithms.

6.5.2.1 CLUTO

CLUTO is a family of computationally efficient and high-quality data clustering and cluster analysis programs and libraries that are well suited for low- and high-dimensional data sets, which can be found at http://glaros.dtc.umn.edu/gkhome/views/cluto. It consists of the following packages: CLUTO—Software for Clustering High-Dimensional Datasets, gCLUTO—Graphical Clustering Toolkit, and wCLUTO—Web-based Clustering of Microarray Data.

6.5.2.2 SVMs

There are numerous SVM algorithms and a large portion of them are listed at http://www.support-vector-machines.org/SVM_soft.html. Some of them are often used for evolutionary feature selection (SVM toolbox theoval.sys.uea.ac.uk/svm/toolbox/...), but many of them are designed for use with MATLAB.

6.5.3 Complete Packages

GALGO is an R package based on a GA variable selection strategy, primarily designed to develop statistical models from large-scale data sets (Trevino and Falciani, 2006). The feature selection approach uses a GA described in Li et al. (2001). Implemented classification methods are *k*-nearest-neighbors, discriminant functions, nearest centroid, SVMs, classification trees, and neural networks. The first three were implemented in C, whereas the others were adapted from original R packages.

6.5.4 Data Sets

Many articles use public databases. These databases are available directly from the Internet (references are given below) and are a good way to compare methods. Some of them have been used in the articles cited in the chapter.

6.5.4.1 Test Bed for Feature Selection

As reported by Choudhary et al. (2006) evaluating a feature selection algorithm may lead to several difficulties. Hence a common approach is to apply the method to data from a real sample and then using the training data to estimate the error of the feature set yielded by the algorithm. In light of this, Choudhary et al. (2006) proposed a genetic test bed for the evaluation of feature selection algorithms that is available at http://public.tgen.org/tgen-cb/support/testbed/. In this test bed, each project is based on a different microarray data set. Readers interested in trying their own feature selection algorithms are encouraged to review the data provided by Choudhary et al. (2006).

6.5.4.2 Microarray

Many articles present results on evolutionary feature selection using microarray data sets. Some frequently used data sets are:

- The Kent Ridge Bio-medical Data Set Repository contains several data sets issued of microarray experiments http://sdmc.lit.org.sg/GEDatasets/Datasets.html
- http://www.broad.mit.edu/tools/data.html including genetic variation and specific data relative to cancer
- http://llmpp.nci.nih.gov/lymphoma/
- http://microarray.princeton.edu/oncology/
- http://genome-www.stanford.edu/nci60/

6.6 CONCLUSION

Bioinformatics experiments generate significant amounts of data that need to be analyzed. Therefore, data mining techniques are used extensively. However, when huge data sets are addressed, feature selection should be carried out carefully before applying a classification or a clustering algorithm. Feature selection is useful in limiting the redundancy of features, promoting comprehensibility, and identifing structures hidden in high-dimensional data.

Hence in many applications in the field of bioinformatics, (un)supervised learning is very often coupled with a feature selection phase. This may be done either in an integrated manner (wrapper approach) or in a sequential approach (filter). In the first case, the learning algorithm is used to evaluate the feature subset. Despite the fact that this approach makes the feature selection phase dependent on the learning algorithm, the references presented in this chapter show that it is the most commonly used approach. This may be explained by the wish of authors to develop a complete approach for the feature selection and the learning phases. Hence, once the learning algorithm is chosen, the wrapper approach seems to give more interesting results.

Selecting features from a large set of potential features is clearly a combinatorial problem, and combinatorial optimization methods may be applied to this problem. The aim of this chapter was to present how EAs have been applied. We identified that many characteristics of such algorithms were similar from one application to the next, except for the objective function. The references contained in this chapter demonstrate that authors developing these evolutionary algorithms for feature selection have proposed different objective functions and do not necessarily agree on the use of a single one. Therefore, there is an increasing trend toward the use of multiobjective approaches. Such approaches afford the possibility of handling several objectives simultaneously. Multiobjective approaches have been proposed in particular by computer scientists, and many perspectives still exist within this field in order to develop adapted methods for bioinformatics problems.

Regarding frameworks that are available to develop such evolutionary approaches for the feature selection, some very general ones have been presented. These frameworks can be used to develop metaheuristics or can be used as a learning algorithm. However, with the exception of GALGO, few frameworks are adapted for this purpose in bioinformatics. Feature selection remains a central issue that requires powerful tools and remains an open problem for the combinatorial optimization community.

REFERENCES

Ando, S. and H. Iba (2004). "Classification of gene expression profile using combinatory method of evolutionary computation and machine learning," *Genet. Program. Evolvable Mach.*, Vol. 5, pp. 145–156.

Bezdek, J. C. (1981). *Pattern Recognition with Fuzzy Objective Function Algorithms*. Plenum, New York.

BONILLA HUERTA, E., B. DUVAL, and J. K. HAO (2006). "A hybrid GA/SVM approach for gene selection and classification of microarray data," in *EvoWorkshops, Lecture Notes in Computer Science*, Vol. 3907, Springer, Berlin, pp. 34–44.

BORGES, H. B. and J. C. NIEVOLA (2005). "Attribute selection methods comparison for classification of diffuse large B-cell lymphoma," *Trans. Eng. Comput. Tech.*, Vol. 8, pp. 193–197.

BREMERMANN, H. J. (1958). "The evolution of intelligence. The nervous system as a model of its environment," Technical Report, No.1, contract no. 477(17), Dept. Mathematics, Univ. of Washington, Seattle, July.

CHERKAUER, K. J., and J. W. SHAVLIK (1996). "Growing simpler decision trees to facilitate knowledge discovery," in Proceedings of the Second International Conference on Knowledge Discovery and Data Mining, pp. 315–318.

CHOUDHARY, A., M. BRUN, J. HUA, J. LOWEY, E. SUH, and E. R. DOUGHERTY (2006). "Genetic test bed for feature selection," *Bioinformatics*, Vol. 22, pp. 837–842.

CHUZHANOVA, N. A., A. J. JONES, and S. MARGETTS (1998). "Feature selection for genetic sequence classification," *Bioinformatics*, Vol. 14, pp. 139–143.

CORNE, D. W., N. R. JERRAM, J. D. KNOWLES, and M. J. OATES (2001). "PESA-II: Region based selection in evolutionary multiobjective optimization," in *Proceedings of the Genetic and Evolutionary Computation Conference (GECCO)*, L. Spector, E. Goodman, A. Wu, W. B. Langdon, H-M. Voigt, M. Gen, S. Sen, M. Dorigo, S. Pezeshk, M. Garzon, and E. Burke, Eds. Morgan Kaufmann, San Francisco, pp. 283–290.

CORTES, C. and V. VAPNIK (1995). "Support-vector networks," *Machine Learn.*, Vol. 20, pp. 273–297.

CRAMER, N. L. (1985). "A representation for the adaptive generation of simple sequential programs," International Conference on Genetic Algorithms and the Applications, Carnegie Mellon University, Pittsburgh.

DASH, M., K. CHOI, P. SCHEUERMANN, and H. LIU (2002). "Feature selection for clustering—A filter solution," IEEE International Conference on Data Mining (ICDM), pp. 115–122.

DAVIES, J. L. and D. W. BOULDIN (1979). "A cluster separation measure," *IEEE Trans. Pattern Anal. Machine Intell.*, Vol. 1, pp. 224–227.

DAVIS, R. A., A. J. CHARLTON, S. OEHLSCHLAGER, and J. C. WILSON (2006). "Novel feature selection method for genetic programming using metabolomic 1H NMR data," *Chemomet. Intell. Lab. Syst.*, Vol. 81, pp. 50–59.

DEUTSCH, J. M. (2003). "Evolutionary algorithms for finding optimal gene sets in microarray prediction," *Bioinformatics*, Vol. 19, pp. 45–52.

DUNN, J. C. (1973). "A fuzzy relative of the isodata process and its use in detecting compact well-separated clusters," *J. Cybernet.*, Vol. 3, pp. 32–57.

DY, J. G. and C. E. BRODLEY (2000). "Feature subset selection and order identification for unsupervised learning. *Proceedings of the 17th International Conf. on Machine Learning*. Morgan Kaufmann, San Francisco, pp. 247–254.

EISEN, M. B., P. T. SPELLMAN, P. PATRICK, O. BROWNDAGGER, and D. BOTSTEIN (1998). "Cluster analysis and display of genome-wide expression patterns," *Proc. Natl. Acad. Sci., USA* Vol. 95, pp. 14863–14868.

EMMANOUILIDIS, C., A. HUNTER, and J. MACINTYRE (2000). "A multiobjective evolutionary setting for feature selection and a commonality-based crossover operator." Proceedings of the 2000 IEEE Congress on Evolutionary Computation, pp. 309–316.

FOGEL, D. B. (1998). *Evolutionary Computation: The Fossil Record.* IEEE Press, New York.

FOGEL, L. J., A. J. OWENS and M. J. WALSH (1966). *Artificial Intelligence through Simulated Evolution*. Wiley, New York.

FRASER, A. S. (1957). "Simulation of genetic systems by automatic digital computers. I. introduction," *Austral. J. Biol. Sci.*, Vol. 10, pp. 484–491.

FIX, E. and J. L. HODGES (1951). *Discriminatory Analysis. Nonparametric Discrimination. Consistency Properties*. Technical Report 4, U.S. Air Force School of Aviation Medicine, Randolph Field, TX.

GOLUB, T. R., D. K. SLONIM, P. TAMAYO, C. HUARD, M. GAASENBEEK, J. P. MESIROV, H. COLLER, M. LOH, J. R. DOWNING, M. CALIGIURI, C. D. BLOOMFIELD, and E. S. Lander (1999). "Molecular classification of cancer: Class discovery and class prediction by gene," *ExpressionScience*, Vol. 286, pp. 531–537.

GRAY, H. F, R. J. MAXWELL, I. MARTINEZ-PEREZ, C. ARUS, and S. CERDAN (1998). "Genetic programming for classification and feature selection: Analysis of 1H nuclear magnetic resonance spectra from human brain tumour biopsies," *NMR Biomed.*, Vol. 11, pp. 217–224.

GUO, H. and A. K. NANDI (2006). "Breast cancer diagnosis using genetic programming generated feature," *Pattern Recog.*, Vol. 39, pp. 980–987.

HANDL, J. and J. KNOWLES (2006a). "Semi-supervised feature selection via multiobjective optimization," Proceedings of the International Joint Conference on Neural Networks (IJCNN), pp. 6351–6358.

HANDL, J. and J. KNOWLES (2006b). "Feature subset selection in unsupervised learning via multiobjective optimization," *Intl. J. Computat. Intell. Res.*, Vol. 2, pp. 217–238

HANDL, J., J. KNOWLES, and D. B. KELL (2005). "Computational cluster validation in post-genomic data analysis," *Bioinformatics*, Vol. 21, pp. 3201–3212.

HOLLAND, J. (1975). *Adaptation in Natural and Artificial Systems*. University of Michigan Press, Ann Arbor, MI.

HUANG, J., J. LU, and C. X. LING (2003). "Comparing naive Bayes, decision trees, and SVM with AUC and accuracy," Third IEEE International Conference on Data Mining, p. 553.

JARVIS, R. M. and R. GOODACRE (2005). "Genetic algorithm optimization for pre-processing and variable selection of spectroscopic data," *Bioinformatics*, Vol. 21, pp. 860–868.

JIRAPECH-UMPAI, T. and S. AITKEN (2005). "Feature selection and classification for microarray data analysis: Evolutionary methods for identifying predictive genes," *BMC Bioinform.*, Vol. 6, p. 148.

JONG, K., E. MARCHIORI, and A. VAN DER VAART (2004). "Analysis of proteomic pattern data for cancer detection," in EvoWorkshops, LNCS, pp. 41–51.

JOURDAN, L., C. DHAENENS, E. G. TALBI, and S. GALLINA (2002). "A data mining approach to discover genetic and environmental factors involved in multifactoral diseases," *Knowledge Based Syst.*, Vol. 15, pp. 235–242.

JULIUSDOTTIR T., D. CORNE, E. KEEDWELL, and A. NARAYANAN (2005). "Two-phase EA/K-NN for feature selection and classification in cancer microarray datasets," in *Proceedings of the 2005 IEEE Conference on Computational Intelligence in Bioinformatics and Computational Biology*. IEEE Press, New York, pp. 1–8.

KIM, Y. S, W. N. STREET, and F. MENCZER (2000). "Feature selection in unsupervised learning via evolutionary search," *Sixth ACM SIGKDD International Conference on Knowledge Discovery and Data Mining*, pp. 65–369.

KIM Y. S., W. N. STREET, and F. MENCZER (2002). "Evolutionary model selection in unsupervised learning. Intelligent data analysis," *Intell. Data Anal.*, Vol. 6, pp. 531–556.

KOHONEN, T. (1988). *Self-Organization and Associative Memory*. Springer, New York.

KOHONEN, T. (1995). *Self-Organizing Maps*. Springer, Berlin, Heidelberg.

KOZA, J. R. (1992). *Genetic Programming: On the Programming of Computers by Means of Natural Selection*. The MIT Press, Cambridge, MA.

LANGDON, W. B. and B. F. BUXTON (2004). "Genetic programming for mining DNA chip data from cancer patients," *Genet. Program. Evolvable Mach.*, Vol. 5, pp. 251–257.

LAVINE, B. K., C. DAVIDSON, S. LAHAV, R. K. VANDER MEER, V. SOROKER, and A. HEFETZ (2003). "Genetic algorithms for deciphering the complex chemosensory code of social insects," *Chemomet. Intell. Lab. Syst.*, Vol. 66, pp. 51–62.

LI, L., W. JIANG, X. LI, K. L. MOSER, Z. GUO, L. DU, Q. WANG, E. J. TOPOL, Q. WANG, and S. RAO (2005). "A robust hybrid between genetic algorithm and support vector machine for extracting an optimal feature gene subset," *Genomics*, Vol. 85, pp. 16–23.

LI, L., C. R. WEINBERG, T. A. DARDEN, and L. G. PEDERSEN (2001). "Gene selection for sample classification based on gene expression data: Study of sensitivity to choice of parameters of the GA/KNN method," *Bioinformatics*, Vol. 17, pp. 1131–1142.

LIU, J. and H. IBA (2002). "Selecting informative genes using a multiobjective evolutionary algorithm," in *Proceedings of the 2002 IEEE Congresse on Evolutionary Computation*. IEEE Press, New York, pp. 297–302.

LIU, J., H. IBA, and M. ISHIZUKA (2001). "Selecting informative genes with parallel genetic algorithms in tissue classification," *Genome Inform.*, Vol. 12, pp. 14–23.

Liu, J. J., G. Cutler, W. Li, Z. Pan, S. Peng, T. Hoey, L. Chen, and X. B. Ling (2005). "Multiclass cancer classification and biomarker discovery using GA-based algorithms," *Bioinformatics*, Vol. 21, pp. 2691–2697.

McQueen, J. (1967). "Some methods for classification and analysis of multivariate observations," Fifth Berkeley Symposium on Mathematical Statistics and Probability, pp. 281–297.

Molinaro, A. M., R. Simon, and R. M. Pfeiffer (2005). "Prediction error estimation: A comparison of resampling methods," *Bioinformatics*, Vol. 21, pp. 3301–3307.

Narendra, P. M. and K. Fukunaga (1977). "A branch and bound algorithm for feature subset selection," *IEEE Trans. Comput.*, Vol. C-26, pp. 917–922.

Ni, B. and J. Liu (2004)."A novel method of searching the microarray data for the best gene subsets by using a genetic algorithm," in *PPSN, Lecture Notes in Computer Science*, X. Yao, E. K. Burke, J. A. Lozano, J. Smith, J. J. Merelo Guervos, J. A. Bullinaria, J. E. Rowe, P. Tino, A. Kaban, and H. P. Schwefel, Eds. Vol. 3242, Springer Berlin, pp. 1153–1162.

Ooi, C. H. and P. Tan (2003). "Genetic algorithms applied to multi-class prediction for the analysis of gene expression data," *Bioinformatics*, Vol. 9, pp. 37–44.

Pareto V. (1896). *Cours d'économie Politique*. Rouge, Lausanne.

Pei, M., E. D. Goodman, and W. F. Punch (1997). *Feature Extraction Using Genetic Algorithms*. Technical Report, Michigan State University, East Lansing, MI.

Peng, S., Q. Xu, X. B. Ling, X. Peng, W. Du, and L. Chen (2003). "Molecular classification of cancer types from microarray data using the combination of genetic algorithms and support vector machines. *FEBS Lett.*, Vol. 555, pp. 358–362.

Rechenberg, I. (1973). *Evolutionsstrategie—Optimierung technischer Systeme nach Prinzipien der Biologischen Evolution*. Frommann-Holzboog, Stuttgart.

Schwefel H.-P. (1981). *Numerical Optimization of Computer Models*. Wiley, Chicester.

Shah, S. C., and A. Kusiak (2004). "Data mining and genetic algorithm based gene/SNP selection," *Artif. Intell. Med.*, Vol. 31, pp. 183–196.

Shi, S. Y. M., P. N. Suganthan, and K. Deb (2004). "Multiclass protein fold recognition using multi-objective evolutionary algorithms," Proceedings of the 2004 IEEE Symposium on Computational Intelligence in Bioinformatics and Computational Biology, pp. 61–66.

Smith, S. F. (1980). A Learning System Based on Genetic Adaptive Algorithms. Ph.D. Dissertation, University of Pittsburgh, Pittsburgh.

Tamayo, P., D. Slonim, J. Mesirov, Q. Zhu, S. Kitareewan, E. Dmitrovsky, E. S. Lander, and R. R. Golub (1999). "Interpreting patterns of gene expression with self-organizing maps: Methods and application to hematopoietic differentiation," *Proc. Natl. Acad. Sci., USA*, Vol. 96, pp. 2907–2912.

Topon Kumar, P. and H. Iba (2004a). "Identification of informative genes for molecular classification using probabilistic model building genetic algorithm," *Proc. Genet. Evolutionary Computation Conf.* (GECCO), Vol. 3102, pp. 414–425.

Topon Kumar, P. and H. Iba (2004b). "Selection of the most useful subset of genes for gene expression-based classification," Proceedings of the 2004 IEEE Congress on Evolutionary Computation, pp. 2076–2083.

Topon Kumar, P. and H. Iba (2005). "Extraction of informative genes from microarray data," Proceedings of the 2005 Genetic and Evolutionary Computation Conference (GECCO), pp. 453–460.

Trevino, V. and F. Falciani (2006). "GALGO: An R package for multivariate variable selection using genetic algorithms," *Bioinformatics*, Vol. 22, pp. 1154–1156.

Whitney, A. W. (1971). "A direct method of nonparametric measurement selection," *IEEE Trans. Comput.*, Vol. C-20, pp. 1100–1103.

Wolpert, D. and W. Macready (1997). "No free lunch theorems for optimization," *IEEE Trans. Evolutionary Computat.*, Vol. 1, pp. 67–82.

Yang, J. and V. Honoavar (1998). "Feature subset selection using a genetic algorithm," in *Feature Extraction, Construction and Selection: A Data Mining Perspective*, H. Liu and H. Motoda, Eds. Kluwer, Boston, pp. 117–136.

Zhang, P., B. Verma, and K. Kumar (2005). "Neural vs. statistical classifier in conjunction with genetic algorithm based feature selection," *Pattern Recogn. Lett.*, Vol. 26, pp. 909–919.

Chapter 7

Fuzzy Approaches for the Analysis CpG Island Methylation Patterns

Ozy Sjahputera, Mihail Popescu, James M. Keller, and Charles W. Caldwell

7.1 INTRODUCTION

Current knowledge indicates that there are two major types of mechanisms implicated in the development of cancer: genetic and epigenetic. Genetic causes include various types of changes in the actual DNA (deoxyribonucleic acid) sequence such as mutations, deletions, and translocations. These may be inherited and present at birth or arise later in life through exposure to known and unknown mutagenic agents. Cytogenetic screening can detect many of these genetic changes. In contrast, epigenetic causes produce no change in the DNA sequence but are thought to act through the biochemical changes of cytosine methylation and histone acetylation. Epigenetic modification is used in normal cells to silence gene expression and, consequently, the production of proteins inappropriate for that cell, for example, to prevent the manufacture of pancreatic enzymes by a brain neuron. If there is significant methylation in the promoter region of the gene, the transcription machinery is unable to produce the encoded mRNA (messenger ribonucleic acid). The most commonly utilized technique for analyzing levels of mRNA expression, which is a refeection of gene activity, is a DNA expression microarray.

Microarray technology is based on the principle of DNA hybridization (i.e., two single-stranded DNA fragments tend to bind together (anneal) at regions in which

Computational Intelligence in Bioinformatics. Edited by G. B. Fogel, D. Corne, and Y. Pan

the two molecules display sequence complementarity). A microarray typically has tens of thousands of spots, each containing different DNA fragments deposited on a solid surface such as glass or plastic. These DNA molecules comprise part of a gene sequence and are referred to as a *probe*. Fragments of DNA from the unknown sample, *targets*, are fluorescently labeled to permit detection and measurement. When the DNA fragments from the unknown sample are applied to the microarray, they will stick or hybridize only to the probes with identical or nearly identical complementary sequences. The amount of DNA bound to every probe is directly proportional to the intensity of the fluorescent emission, and this can be accurately measured by a laser scanner. In this fashion, a microarray can simultaneously monitor expression levels for all the genes in a genome in a single experiment.

In this chapter we study two types of microarrays: expression and methylation. A large number of different types of microarrays are available, differing in their basic chip technology, probe type, and experimental methodology. Expression microarrays quantify the level of gene activity, while methylation arrays measure amounts of gene methylation. Using both types of microarrays in the same tumor sample permits investigation of the molecular genetic mechanisms of cancer by detecting genes that are both silenced and inactivated by methylation.

The focus of our research has been on non-Hodgkin's lymphomas (NHL), which are malignancies of the lymphatic system and consist of cells involved in the body's immune defenses. These are distinguished from the other main form of lymphoid tumors that are classified as Hodgkin's disease. The latter tumor type is identified histologically by the presence of large, atypical cells known as Reed–Sternberg cells. NHL is a diverse group of diseases with multiple heterogeneous genetic and epigenetic alterations. In recent years, microarray-based expression profiling has shown its usefulness in classifying different tumor types and tumor subtypes and in guiding treatment choices such as in NHL patients (Glas et al., 2005). Similarly, microarray-based DNA methylation profiling has also been used in tumor classifications of NHL (Husson et al., 2002; Adorjan et al., 2002). These studies provide important insight into gene expression and alterations in NHLs but have mainly focused on up-regulated genes without necessarily exploring potential causes for these disturbances other than genetic lesions. One method to perform DNA methylation profiling is called *differential methylation hybridization* (DMH) (Huang et al., 1999). Based on DMH, a method called *expressed CpG island sequence tags* (ECISTs) was developed to detect both CpG hypermethylation and gene silencing in breast and ovarian cancer cell lines (Shi et al., 2002). ECISTs are present in the human genome, and their GC-rich fragments can be used to screen aberrantly methylated CpG sites in cancer cells and the exon-containing portions can be employed to measure levels of gene expression.

In this chapter we introduce two sets of fuzzy algorithms to analyze the relationships between DNA methylation and gene expression in two small B-cell lymphomas (SBCL): follicular lymphoma (FL) and chronic lymphocytic leukemia (CLL). There are various statistical approaches for processing microarray data (Lee, 2004).

Instead, we have adopted a fuzzy logic framework (Woolf and Wang, 2000) for our application because it is reportedly better at accounting for noise in the data, as the fuzzy algorithms tend to extract trends rather than precise values from the microarray data. The noise-handling requirement was imposed by the poor quality of our data set. Since our data set was rather unique (two microarray types for the same set of genes and patients), we could not find similar algorithms in the literature. Moreover, we had to try several approaches to make sure we obtain consistent results. The first set of algorithms employs a fuzzy rule approach with linguistic values defined for the normalized ratios of methylation and expression. This approach does not take into account the spot quality (hybridization level). In the second set of algorithms, we propose a more sophisticated approach that takes into account the spot quality. We use fuzzy sets (Zadeh, 1965) to represent the methylation–expression relationships and the relationships are analyzed using the possibilistic c-means (PCM) clustering proposed by Krishnapuram and Keller (1993, 1996) and a region cluster density similar to Gath and Geva's (1989) fuzzy cluster density. The relationships analyzed are {hypermethylated, down-regulated}, {hypomethylated, up-regulated}, {hypermethylated, up-regulated}, and {hypomethylated, down-regulated}. Genes that best fit these relationships can be selected to provide biologists with working hypotheses.

In this chapter we focus on the inverse relationships {hypermethylated, down-regulated} and {hypomethylated, up-regulated}. These relationships may explain the roles of DNA methylation in the physiological regulation of some genes during both normal and neoplastic B-cell differentiation. For example, DNA hypermethylation of some genes is a frequent occurrence in B-cell NHL (Li et al., 2002; Cameron et al., 1999; Katzenellenbogen et al., 1999). Aberrant hypermethylation of CpG islands may cause the silencing of tumor suppressor genes (Fig. 7.1) (i.e., hypermethylation, down-regulation). On the other hand, hypomethylation may produce overexpression of oncogenes (i.e., hypomethylation, up-regulation). Both processes can lead to tumorigenesis (Egger et al., 2004). Costello et al. (2000) reported that aberrant patterns of CpG island methylation are not random but tumor specific. Thus, hypermethylated CpG islands can serve as markers for novel genes that are subject to this epigenetic control. Understanding the effects of aberrant methylation on the expression level of tumor suppressor and oncogenes may open possible gene-specific treatment (e.g., the use of a DNA methyltransferase inhibitor to reverse promoter CpG methylation in silenced tumor suppressor genes that may also reverse gene silencing) (Nagasawa et al., 2006).

To demonstrate the biological significance of the fuzzy approach, we compared the methylation–expression relationships on some genes to the confirmation results reported in other NHL studies (Rahmatpanah et al., 2006; Shi et al., 2007; Taylor et al., 2006). Among these genes are *EFNA5*, *LHX2*, *LRP1B*, and *DDX51*. The methylation confirmation methods included combined bisulfite restriction analysis (COBRA) (Xiong and Laird, 1997), methylation-specific polymerase chain reaction (MSP-PCR) (Herman et al., 1996), and bisulfite sequencing (Xiong and Laird, 1997). Gene expression was confirmed using real-time PCR (RT-PCR).

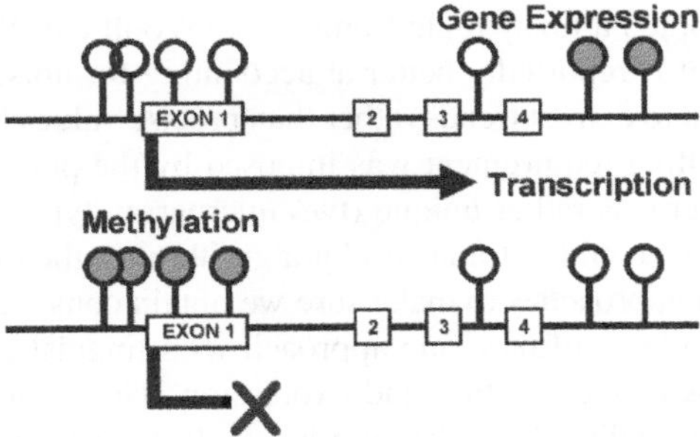

Figure 7.1 Two typical genes are shown: one methylated (lower figure) and one not (upper figure). The circles represent sites rich in guanine–cytosine (also called CpG islands); filled circles indicate methylated CpG islands where cytosines are methylated, while empty circles represent unmethylated CpG islands. The upper gene has the region around the first exon unmethylated. Consequently, the transcription machinery is able to attach to the DNA strand and produce the protein encoded by the gene. Some methylation may occur elsewhere in the DNA, mainly in the intronic regions, but it has no effect on transcription. The lower gene has the beginning of the gene (promoter and first exon) methylated. As a result, the gene is silenced since the transcription machinery is repelled.

7.2 METHODS

7.2.1 Microarray Processing

CpG island methylation and gene expression in each patient sample were assessed using DMH (Costello et al., 2000; Herman et al., 1996) and ECIST expression (Huang et al., 1999; Shi et al., 2002) microarray techniques. Both the DMH and ECIST expression arrays used the same clone library and array layout (Yan et al., 2001). Cancer and normal samples were dyed red (Cy5) and green (Cy3). Both specimens were co-hybridized on the array. After hybridization, the microarray slide was scanned using a GenePix 4200A scanner (Molecular Devices Corp., Sunnyvale, CA) and the image was processed using GenePix Pro software (Molecular Devices Corp., Sunnyvale, CA). Each spot was represented by the ratio *R/G*, where *R* and *G* were background-corrected Cy5 and Cy3 intensities. An *M–A* plot (Yang et al., 2002) was created for each array where $M = \log_2(R/G)$ and $A = \frac{1}{2}\log_2(RG)$ were the *y* and *x* axes. Pin-based robust loess normalization (Cleveland, 1979, Cleveland et al., 1988; Cleveland and Grosse, 1991) and the control-driven global normalization (Yang et al., 2002) were used. The nonlinear loess regression (available in the MATLAB Bioinformatics Toolbox Ver. 2.2, The MathWorks, Inc., http://www.mathworks.com/products/bioinfo/index.html?fp) handles the nonlinear aspects of the microarray data possibly due to nonadditive interaction between the background and feature components, signal saturation, or dye variation (Simon et al., 2004). For global normalization, spots whose DNA fragments did not contain cut sites targeted by the methylation-sensitive enzymes were used as controls in DMH arrays, while spots linked to known housekeeping genes (Eisenberg and Levanon, 2003) were used as control in ECIST arrays. Using a linear function, the loess and global normalization factors were combined into a composite normalization factor and was used to normalize the microarray data.

7.2.2 Fuzzy Rule Gene Selection (FRGS) Method

7.2.2.1 Fuzzy Models for Methylation and Expression Levels

Consider a medical condition with N patient samples. DMH and ECIST cDNA (complementary DNA) arrays are prepared using each sample. Each microarray (methylation or expression) has probes from M clones (gene fragments). In this chapter we denote a microarray spot as a clone or a gene interchangably. However, technically, "clone" is more precise since a gene may be represented by several clones in the biological sample library used to make the arrays. For clone i in sample n we denote the normalized log-ratio of Cy5 and Cy3 intensities for the expression as $\{e_{in}\}_{n=1,\ldots,N}$ and the methylation as $\{m_{in}\}_{n=1,\ldots,N}$. Biologists may want to investigate genes with certain methylation–expression relationships in different diseases, such as those that are {hypermethylated, down-regulated}. We define linguistic values *HIGH* and *LOW* for e_{in} and m_{in} using membership functions shown in Figure 7.2 with parameters ρ_{high} and ρ_{low}. The linguistic values *HIGH* and *LOW* are used to represent hypermethylation and hypomethylation in DMH methylation arrays, and up-regulation (overexpression) and down-regulation (silencing) in ECIST cDNA expression arrays. For now, we chose $\rho_{low} = 0$ and $\rho_{high} = 1$ for both array types. However, different values of ρ_{high} and ρ_{low} may be used for expression and methylation data.

7.2.2.2 Fuzzy Rule with Ordered Weighted Average (OWA) Operators

For each expression value of clone i in sample n, e_{in}, the individual *HIGH* and *LOW* memberships, $\mu_{in}^{H,e}$ and $\mu_{in}^{L,e}$, were calculated as follows:

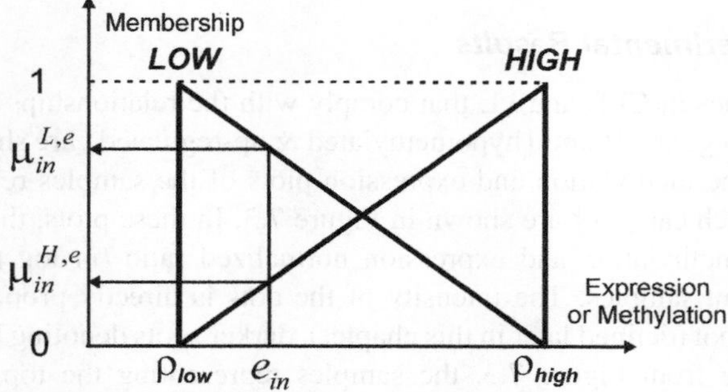

Figure 7.2 *HIGH* and *LOW* membership functions used to calculate the membership of each individual expression or methylation value. Here, $\mu_{in}^{H,e}$ and $\mu_{in}^{L,e}$ are memberships for *HIGH* and *LOW* for the expression value of locus i from sample n (e_{in}).

$$\mu_{in}^{H,e} = \frac{e_{in} - \rho_{\text{low}}}{\rho_{\text{high}} - \rho_{\text{low}}}, \qquad \mu_{in}^{L,e} = \frac{\rho_{\text{high}} - e_{in}}{\rho_{\text{high}} - \rho_{\text{low}}}. \tag{7.1}$$

Individual *HIGH* and *LOW* memberships for the corresponding methylation m_{in} were calculated similarly. To assess the overall methylation and expression for clone *i*, we aggregated *HIGH* and *LOW* memberships from all *N* samples. Using an OWA aggregation operator (Yager, 1993), the aggregate *HIGH* memberships for methylation ($\mu_i^{H,m}$) and expression ($\mu_i^{H,e}$) are calculated as follows:

$$\mu_i^{H,m} = \sum_{n=1}^{N} w_n \mu_{in}^{H,m}, \qquad \mu_i^{H,e} = \sum_{n=1}^{N} w_n \mu_{in}^{H,e}, \tag{7.2}$$

The aggregate *LOW* memberships for methylation ($\mu_i^{L,m}$) and expression ($\mu_i^{L,e}$) were calculated similarly. In our example we use the average operator, $w_n = 1/N$ in Eq. (7.2). Using these aggregate membership values, we can translate various methylation–expression relationships into fuzzy rules. For example, we can find genes following the relationship {hypermethylated, down-regulated} using the fuzzy rule "*IF* metylation is *HIGH* and expression is *LOW, THEN* gene *i* is *Hypemerthylated & DownRegulated*". The output membership of this rule ($\mu_i^{Hypermethylated\ \&\ DownRegulated}$) can be calculated by aggregating $\mu_i^{H,m}$ and $\mu_i^{L,e}$. Using the product aggregation, we define the membership of several methylation–expression relationships as follows:

$$\begin{aligned}
\mu_i^{Hypermethylated\ \&\ DownRegulated} &= \mu_i^{H,m}\mu_i^{L,e}, \\
\mu_i^{Hypomethylated\ \&\ UpRegulated} &= \mu_i^{L,m}\mu_i^{H,e}, \\
\mu_i^{Hypermethylated\ \&\ UpRegulated} &= \mu_i^{H,m}\mu_i^{H,e}, \\
\mu_i^{Hypomethylated\ \&\ DownRegulated} &= \mu_i^{L,m}\mu_i^{L,e}.
\end{aligned} \tag{7.3}$$

Genes are then ranked in decreasing order of one of the methylation–expression memberships given in Eq. (7.3). Biologists can use the ordered list to define a set of "interesting" genes in each methylation–expression relationship category for further biological analysis.

7.2.2.3 Experimental Results

The top 10 genes in CLL and FL that comply with the relationships {hypermethylated & down-regulated} and {hypomethylated & up-regulated} are shown in Tables 7.1 and 7.2. The methylation and expression plots of the samples representing the top genes in each category are shown in Figure 7.3. In these plots, the *x* and *y* axes represent the methylation and expression normalized ratio (in $\log_2$) and the dots represent patient samples. The intensity of the dots is directly proportional to the quality of the spot (defined later in this chapter), darker spots denoting higher quality. As we can see from Figure 7.3. the samples representing the top-ranked genes were not of high quality (except for *MCF2L2*). The four quadrants capture the four methylation–expression relationships, with quadrants 2 and 4 representing {hypermethylated & down-regulated} and {hypomethylated & up-regulated}. In the next

Table 7.1 Genes Found Hypermethylated and Down-Regulated for CLL and FL Using FRGS Approach

CLL		FL	
MCF2L2	Rho family guanine-nucleotide exchange factor	FLJ10408	Family with sequence similarity 90, member A1
MAST2	Microtubule associated serine/threonine kinase 2	UBE2S	Ubiquitin-conjugating enzyme E2S
EFNA5	EPH-related receptor tyrosine kinase ligand 7	GRIK1	Glutamate receptor, ionotropic, kainate 1
ZNF546	Zinc finger protein 546	FLJ37927	Cell division cycle 20 homolog B
HIRIP5	HIRA interacting protein 5	ZNF77	Zinc finger protein 77
DDX51	DEAD (Asp-Glu-Ala-Asp) box polypeptide 51	TNFRSF6	Fas-associated factor 1
H3F3A	H3 histone, family 3A	MUC4	Mucin 4, cell surface associated
MGC10561	Cell division cycle associated 6	FBLN5	Fibulin 5
BCL10	B-cell CLL/lymphoma 10	LRP2	Low-density lipoprotein-related protein 2
TNFRSF11A	Receptor activator of nuclear factor-kappa B	AURKA	Aurora kinase A

Table 7.2 Genes Found Hypomethylated and Up-Regulated for CLL and FL Using FRGS Approach

CLL		FL	
DPAGT1	GlcNAc-1-P transferase	HRB2	HIV-1 Rev binding protein 2
FLJ38564	Choroideremia (Rab escort protein 1)	VGCNL1	Voltage gated channel like 1
BET1	Blocked early in transport 1	ATP6V1G1	ATPase
MYOD1	Myoblast determination protein 1	FLJ32370	Transmembrane protein 68
FLJ10514	Aspartyl-tRNA synthetase 2	HNRPA2B1	Heterogeneous nuclear ribonucleoprotein A2/B1
CMAS	CMP-Neu5Ac synthetase	FIP1L1	Rearranged in hypereosinophilia
ANKRD13	Ankyrin repeat domain 13	COL1A2	Procollagen, type I, alpha 2
USP16	Ubiquitin-specific peptidase 16	DKFZP586C1924	Transmembrane protein 126A
M6PR	Mannose-6-phosphate receptor	BCL6	B-cell CLL/lymphoma 6
SYNJ2	Synaptojanin 2	FXR1	Fragile X mental retardation-related protein 1

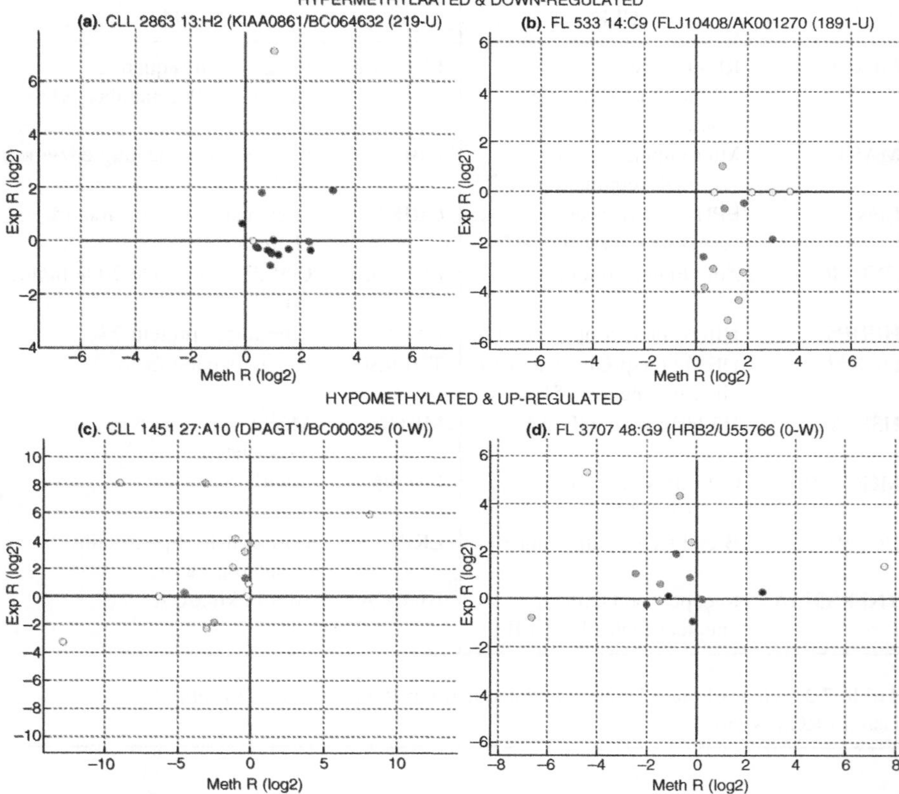

Figure 7.3 Spot quality for the top genes identified by the FRGS approach (Table 7.1). *A-MCF2L2*, *B-FLJ10408*, *C-DPAGT1*, and *D-HRB2*. Darker spots denote higher quality samples. Except for case (**a**), the samples representing the top genes have low quality.

approach, we exploit the quadrant representation of these relationships to build more sophisticated relational analysis involving clustering techniques. The comparison of results given by FRGS and those of the clustering approach will be discussed in detail later in this chapter.

7.2.3 Clustering Approach in Methylation–Expression Relationship Analysis

The previous gene selection method does not take into account the intensity of background-corrected Cy5 or Cy3 channels. Low background-corrected Cy5 or Cy3 can produce log ratios with higher variance and may lead to gene selection prone to failing in biological confirmation. Low background-corrected Cy5 or Cy3 may be due to low methylation and/or expression levels, but it may also be due to hybridization problems or high background intensity from noise or artifacts. In this section,

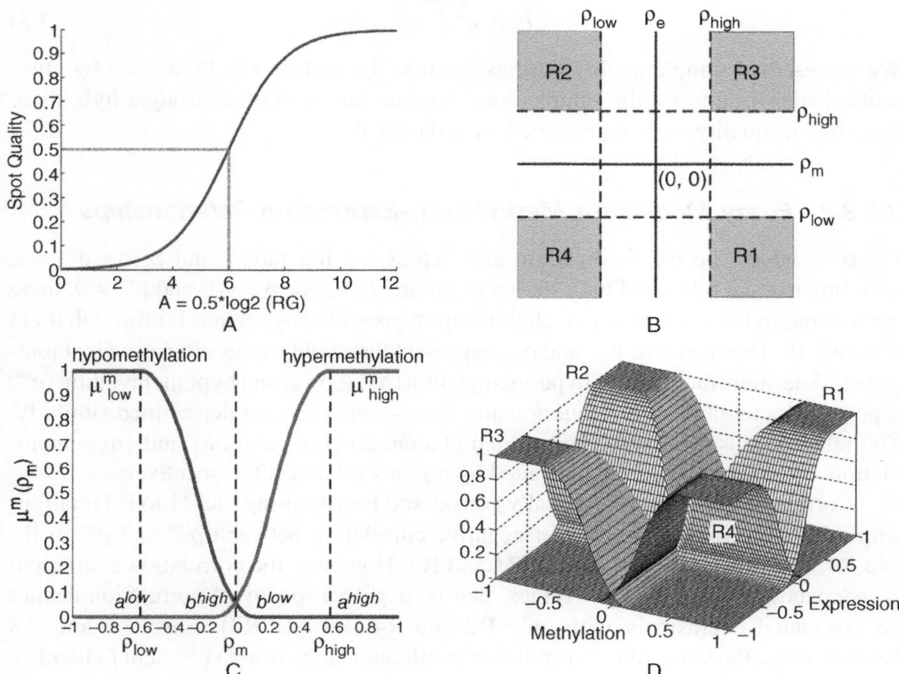

Figure 7.4 (a) Membership function for spot quality (τ) based on aggregate intensity *A*. (b) Relational map between gene methylation and expression ratio. (c) Fuzzy membership functions of hypermethylation and hypomethylation. (d) Three-dimensional fuzzy membership functions for methylation–expression relationships in all four quadrants.

we consider a quality measure for the locus hybridization levels in methylation and expression arrays. Unlike the approach in FRGS where fuzzy sets are used to capture the level of methylation and expression separately, here we used fuzzy sets to model the methylation–expression relationships directly. These relationships are assessed using various measures by taking into account individual clone quality. These techniques were proposed by Sjahputera et al. (2007).

7.2.3.1 Clone Quality

We used the sigmoid membership function (Fig. 7.4a) to assign the hybridization quality (τ) for each clone in an array. For clone *i* with aggregate intensity *A*, the clone quality $\tau_i(A)$ was defined as:

$$\tau_i(A) = (1 + e^{-a(A-b)})^{-1}, \tag{7.4}$$

where $\tau_i(A) \in [0, 1]$, *a* is the steepness coefficient, and *b* is the inflection point where $\tau_i(b) = 0.5$. We use $a = 1$ and $b = 6$ such that $\tau_i(0) = 0$ and $\tau_i(A \geq 12) = 1$ (*A* in $\log_2$ scale). Let $\tau_{in}^{(m)}$ and $\tau_{in}^{(e)}$ be the quality of clone *i* in patient *n* of the methylation (*m*) and expression (*e*) data set. The clone quality T_{in} is given as

$$T_{in} = \sqrt{\tau_{in}^{(m)}\tau_{in}^{(e)}}. \tag{7.5}$$

We represented sample quality in plots, such as those shown in Figure 7.3 by grayscale shades used to fill the sample dots. A clone sample with a stronger hybridization (higher quality) was represented by a darker dot.

7.2.3.2 Fuzzy Models for Methylation–Expression Relationships

Let $\rho^{(m)}$ and $\rho^{(e)}$ be the methylation and expression log ratios, and $\rho^{(m)} \times \rho^{(e)}$ is a two-dimensional relational map shown in Figure 7.4b. For $\rho^{(m)} \neq 0$ and $\rho^{(e)} \neq 0$, there are four methylation–expression relationships represented by regions R1 through R4 in Figure 7.4b. The constants ρ_{low} and ρ_{high} represent threshold values often used by biologists to determine the status of hypermethylation ($\rho^{(m)} \geq \rho_{\text{high}}$) and hypomethylation ($\rho^{(m)} \leq \rho_{\text{low}}$). The conditions *up-regulation* and *down-regulation* are determined similarly. The relationships of interest, {hypermethylation, down-regulation} and {hypomethylation, up-regulation}, are represented by regions R4 and R2, respectively.

Correlation analysis was initially considered for capturing the R1 to R4 relationships. Intuitively, one may expect negative correlation between $\rho^{(m)}$ and $\rho^{(e)}$ in R2 and R4, and positive correlation in R1 and R3. However, the correlation coefficient is independent of $\rho^{(m)}$ and $\rho^{(e)}$ values, that is, negative (positive) correlation cannot be associated exclusively with only R2 and R4 (R1 and R3) (Figs. 7.5 and 7.8 below). Thus, the value of the correlation coefficient alone is not sufficient to identify methylation–expression relationships as defined by biologists. The uncertainty from selecting ρ_{high} and ρ_{low} is not strictly probabilistic, but it can be modeled appropriately using fuzzy sets in Figures 7.4d and 7.4c.

The spline-based fuzzy membership functions $\mu_{\text{high}}^{(m)}$ ($\mu_{\text{high}}^{(e)}$) and $\mu_{\text{low}}^{(m)}$ ($\mu_{\text{low}}^{(e)}$) were used to model the uncertainty of thresholds ρ_{high} and ρ_{low} (Fig. 7.4c). In MATLAB Fuzzy Logic Toolbox (http://www.mathworks.com/products/fuzzylogic/?BB=1) these membership functions are implemented as the S-type ($\mu_{\text{high}}^{(m)}$ and $\mu_{\text{high}}^{(e)}$) and the Z-type ($\mu_{\text{low}}^{(m)}$ and $\mu_{\text{low}}^{(e)}$) requiring parameters *a* and *b*, where $\mu(a) = 1$, $\mu(b) = 0$, and $\mu[(a+b)/2] = 0.5$. For $\mu_{\text{high}}^{(m)}$ and $\mu_{\text{high}}^{(e)}$, $a = \rho_{\text{high}}$ and $b = -(0.25\rho_{\text{high}})$. Similarly, $a = \rho_{\text{low}}$ and $b = -(0.25\rho_{\text{low}})$ for μ_{low}^{m} and μ_{low}^{e}. If $\rho_{\text{high}} = -\rho_{\text{low}}$, then $\mu_{\text{high}}^{(m)}$ is a mirror image of $\mu_{\text{low}}^{(m)}$ (*y*-axis symmetry) and $\mu_{\text{high}}^{(m)} = \mu_{\text{low}}^{(m)}$ for $\rho^{(m)} = 0$. The overlap between $\mu_{\text{high}}^{(m)}$ and $\mu_{\text{low}}^{(m)}$ provides a soft transition between hypermethylation (up-regulation) and hypomethylation (down-regulation). It also represents regions of higher uncertainty, where samples receive low membership in both hypermethylation (up-regulation) and hypomethylation (down-regulation). The functions $\mu_{\text{high}}^{(m)}$ and $\mu_{\text{low}}^{(m)}$ ($\mu_{\text{high}}^{(e)}$ and $\mu_{\text{low}}^{(e)}$) are extended across $\rho^{(e)}$ ($\rho^{(m)}$).

The four methylation–expression relationships (R1 to R4) in Figure 7.4b are modeled using a fuzzy membership function μ_q in quadrants $q = \{1, 2, 3, 4\}$ defined as

$$\mu_q = \min(\mu_{\cdot}^{(m)}, \mu_{\cdot}^{(e)}), \tag{7.6}$$

where $\mu_{\cdot}^{(m)} \in \{\mu_{\text{low}}^{(m)}, \mu_{\text{high}}^{(m)}\}$ and $\mu_{\cdot}^{e} \in \{\mu_{\text{low}}^{(e)}, \mu_{\text{high}}^{(e)}\}$ appropriate for the quadrant, for example, R1 is modeled using $\mu_4 = \min(\mu_{\text{high}}^{m}, \mu_{\text{low}}^{e})$ (Fig. 7.4d). For clone *i* in array

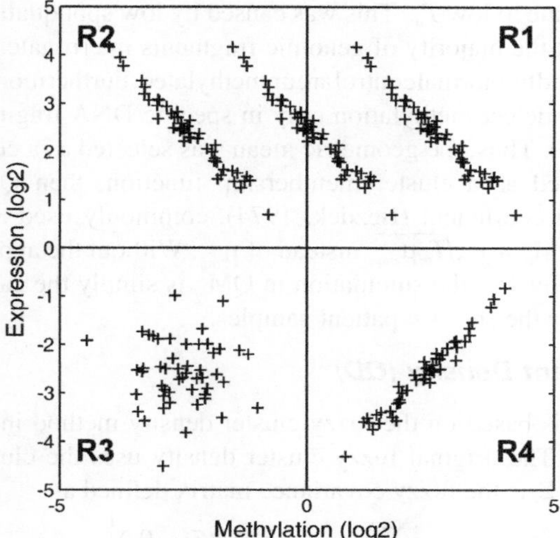

Figure 7.5 Correlation analysis on methylation–expression relationships. Data sets 1 to 5 are randomly generated from multivariate normal with $n = 50$. Data sets 1 to 3 have identical correlation coefficients (–0.98, –0.87, and –0.97 for Pearson, Kendall, and Spearman), although they differ considerably in locations and thus in their methylation–expression relationships. Data set 4 shows that the sign of correlation coefficient is not representative of methylation–expression relationships. With hyperspherical clusters (e.g., data set 5 with correlation coefficients of –0.017, 0.011, 0.015), correlation analysis shows little or no correlation between features, and it does not capture the methylation–expression relationship.

n with methylation and expression log ratios $\rho_{in}^{(m)}$ and $\rho_{in}^{(e)}$, $\mu_{inq} = \mu_q(\rho_{in}^{(m)}, \rho_{in}^{(e)})$ is the membership of clone i of sample n in the relationship represented by quadrant q. The fuzzy models μ_q along with the quality measure T_{in} serve as foundations for building the quadrant confident measures QM_{iq}, QD_{iq}, and QC_{iq} introduced next.

7.2.3.3 Quadrant Membership (QM)

For clone i across all N patients we define $X_i = \{\vec{x}_{in} | n = 1, \ldots, N\}$ as the universe of discourse, where $\vec{x}_{in} = [\rho_{in}^{(m)} \quad \rho_{in}^{(e)}]^{\mathrm{T}}$ is the methylation (m) and expression (e) data of clone i in patient n. Using Eq. (7.6), the memberships of $\vec{x}_{in}$ in the four quadrants are calculated as $\mu_{inq} = \mu_q(\vec{x}_{in})$ for $q = \{1, 2, 3, 4\}$. The measure $\mathrm{QM}_{iq} \in [0, 1]$ represents the total evidence that patient samples of clone i are present in quadrant q, defined as

$$\mathrm{QM}_{iq} = \frac{1}{N} \sum_{n=1}^{N} \sqrt{T_{in} \mu_{inq}}, \tag{7.7}$$

where T_{in} is incorporated as a weight to the cluster membership μ_{inq}. To aggregate μ_{inq} and T_{in} we considered the following fusion operators (in the order of most optimistic to most pessimistic for data in [0, 1]): maximum, arithmetic mean, geometric mean, minimum, and product. The product was not used since it produced very low aggregated value due to low T_{in}. This was caused by low spot quality in DMH arrays due to the fact that the majority of genomic fragments interrogated in DMH for any particular disease and the normal control are unmethylated. Furthermore, the methylation-sensitive enzymes detect methylation only in specific DNA fragments (CCGG and CGCG in our case). Thus, the geometric mean was selected as a compromise.

If μ_q is viewed as a cluster membership function, then QM_{iq} is similar to Bezdek's partition coefficient (Bezdek, 1974), commonly used as cluster validity measure, except QM_{iq} use $\sqrt{T_{in}\mu_{inq}}$ instead of μ^2_{inq}. Without the application of T_{in} and the normalization by $1/n$, the summation in QM_{iq} is simply the cardinality of fuzzy set μ_{iq} defined over the set of n patient samples.

7.2.3.4 Quadrant Density (QD)

The measure QD is based on the fuzzy cluster density method introduced by Gath and Geva (1989). The original fuzzy cluster density uses the cluster hypervolume $V_j = |\Sigma_j|^{1/2}$ where Σ_j is the fuzzy covariance matrix defined as

$$\Sigma_j = \frac{\sum_{n=1}^{N} u_{nj}^{\alpha}(\vec{x}_n - \vec{\theta}_j)(\vec{x}_n - \vec{\theta}_j)^{\mathrm{T}}}{\sum_{n=1}^{N} u_{uj}^{\alpha}} \tag{7.8}$$

where μ_{nj} are memberships and $\vec{\theta}_j$ are centroids, and the cluster density is defined as

$$\delta_j = \frac{1}{V_j}\sum_{n=1}^{N} u_{nj}. \tag{7.9}$$

Let the quadrant q be a cluster with membership function μ_q and its core area (μ_q = 1) as the cluster center. QD measures the fuzzy density of each quadrant. Data samples outside the core of μ_q increases the hypervolume and thus reduces QD. A high QD for a quadrant increases the confidence that the methylation–expression relationship represented by that quadrant can be used to describe the gene methylation and expression pattern. Instead of taking the cross product of $(\vec{x}_n - \vec{\theta}_j)$ as in Eq. 78, we calculate the fuzzy covariance matrix of quadrant q (Σ_q) by taking the cross product of $\vec{\Delta}_{inq}$ defined as a vector that originates from a point in the boundary of μ_q core area closest to $\vec{x}_{in}$ and extends to $\vec{x}_{in}$. The span of $\vec{\Delta}_{inq}$ is limited to the support area of μ_q only. If $\vec{x}_{in}$ is inside the core area, then $\vec{\Delta}_{inq} = [0 \quad 0]^{\mathrm{T}}$. Let $\vec{x}_{in} = [\rho_{in}^{(m)} \quad \rho_{in}^{(e)}]^{\mathrm{T}}$, and $\{a^{\mathrm{mq}}, b^{\mathrm{mq}}\}$ and $\{a^{\mathrm{eq}}, b^{\mathrm{eq}}\}$ be the parameters for $\mu_{\cdot}^{(m)}$ and $\mu_{\cdot}^{(e)}$ in quadrant q, then $\vec{\Delta}_{inq}$ is defined as (see Fig. 7.6a):

$$\vec{\Delta}_{inq} = \begin{cases} [0 \quad 0]^{\mathrm{T}} & \begin{array}{ll} \mu_{\cdot}^{(m)}(\rho_{in}^{(m)}) & = 1, \\ \mu_{\cdot}^{(e)}(\rho_{in}^{(e)}) & = 1, \end{array} \\ \left[0 \quad \left|\rho_{in}^{(e)} - a^{eq}\right| \wedge \left|b^{eq}\right|\right]^{\mathrm{T}}, & \begin{array}{ll} \mu_{\cdot}^{(m)}(\rho_{in}^{(m)}) & = 1, \\ 0 \le \mu_{\cdot}^{(e)}(\rho_{in}^{(e)}) & < 1, \end{array} \\ \left[\left|\rho_{in}^{(m)} - a^{mq}\right| \wedge \left|b^{mq}\right| \quad 0\right]^{\mathrm{T}}, & \begin{array}{ll} 0 \le \mu_{\cdot}^{(m)}(\rho_{in}^{(m)}) & < 1, \\ \mu_{\cdot}^{(e)}(\rho_{in}^{(e)}) & = 1, \end{array} \\ \left|\rho_{in}^{(m)} - a^{mq}\right| \wedge \left|b^{mq}\right| \quad \left|\rho_{in}^{(e)} - a^{eq}\right| \wedge \left|b^{eq}\right|^{\mathrm{T}}, & \begin{array}{ll} 0 \le \mu_{\cdot}^{(m)}(\rho_{in}^{(m)}) & < 1, \\ 0 \le \mu_{\cdot}^{(e)}(\rho_{in}^{(e)}) & < 1. \end{array} \end{cases} \tag{7.10}$$

Here, $\wedge$ and $\vee$ denote minimum and maximum operators.

We now define the modified fuzzy covariance matrix of clone i in quadrant q (Σ_{iq}) as:

$$\Sigma_{iq} = I + \sum_{n=1}^{N}\left(1-\sqrt{T_{in}\mu_{inq}}\,(\vec{\Delta}_{inq}\,\vec{\Delta}_{inq}^{T}\right). \tag{7.11}$$

Nonzero components in Σ_{iq} originate solely from variance produced by $\vec{x}_n$ located outside the core area of the μ_q. Similar to Gath and Geva's definition, we define the quadrant hypervolume $V_{iq} = |\Sigma_{iq}|^{1/2}$. We define the quadrant density QD_{iq} as

$$\mathrm{QD}_{iq} = \frac{\mathrm{QM}_{iq}}{V_{iq}} = \frac{1}{NV_{iq}}\sum_{n=1}^{N}\sqrt{T_{in}\mu_{inq}}. \tag{7.12}$$

In Eq. (7.11), the product $(\vec{\Delta}_{inq}\,\vec{\Delta}_{inq}^{T})$ is scaled by the factor $(1-\sqrt{T_{in}\mu_{inq}})$. This scale factor is equal to 0 if $\vec{x}_i$ is located within the core of the quadrant ($\mu_{inq} = 1$) with maximum sample quality ($T_{in} = 1$). If this condition is satisfied, diag(Σ_{iq}) is reduced resulting in a reduction of V_{iq} and, consequently, increasing QD_{iq}. To avoid singular Σ_{iq}, a 2×2 identity matrix I is used in Eq. (7.11) giving us $V_{iq} \geq 1$. This is possible because Σ_{iq} is not scaled by the sum of cluster membership as found in Gath and Geva's definition in Eq. (7.8). Since $V_{iq} \geq 1$, following Eq. (7.12), QM_{iq} becomes the ceiling of the quadrant density, i.e. $\mathrm{QD}_{iq} \leq \mathrm{QM}_{iq}$. The minimum hypervolume ($V_{iq} = 1$) is achieved when $\sum_{n=1}^{N}\sqrt{T_{in}\mu_{inq}} = N$, i.e. when all $\vec{x}_{in}$ are found within the core area of μ_q ($\mu_{inq} = 1$) with maximum spot quality ($T_{in} = 1$). This condition also maximizes QM_{iq} to 1, resulting in the maximum quadrant density of $\mathrm{QD}_{iq} = 1$. On the other hand, the lowest density $\mathrm{QD}_{iq} = 0$ is obtained when $\sum_{n=1}^{N}\sqrt{T_{in}\mu_{inq}} = 0$ ($\mathrm{QM}_{iq} = 0$); therefore QD_{iq} is bounded in [0, 1]. QD_{iq} is sensitive to μ_{inq} or T_{in} (due to negative correlation between QM_{iq} and V_{iq}), often QD_{iq} becomes too small in comparison to QM_{iq} and QC_{iq} (see next section). This becomes problematic if we are to fuse these measures together. To control this, we scale the hypervolume using an exponential power ($1/s$), $s \geq 1$. Let QD'_{iq} be the scaled quadrant density defined as:

$$\mathrm{QD}'_{iq} = \frac{\mathrm{QM}_{iq}}{(V_{iq})^{1/s}}. \tag{7.13}$$

The effects of scaling the hypervolume on QD'_{iq} for several values of s are illustrated in Figure 7.6b (assuming QM_{iq} is constant). Increasing s reduces the sensitivity of QD'_{iq} with respect to an increase in hypervolume V_{iq}.

7.2.3.5 Quadrant Measure Based on Clustering Analysis (QC)

The QM_{iq} is calculated based on the quadrant membership value and quality of data points within quadrant q. In QD_{iq}, the quadrant hypervolume capturing the distance of data points from the core of μ_q is taken as an additional factor. If the methylation and expression relationship for a gene across samples of a given disease is consistent, then we can expect data points to form one or more clusters in $\rho^{(m)} \times \rho^{(e)}$. Multiple clusters may suggest multiple subtypes within the disease. We now introduce a quadrant measure based on the cluster information (QC). QC is motivated by the notion

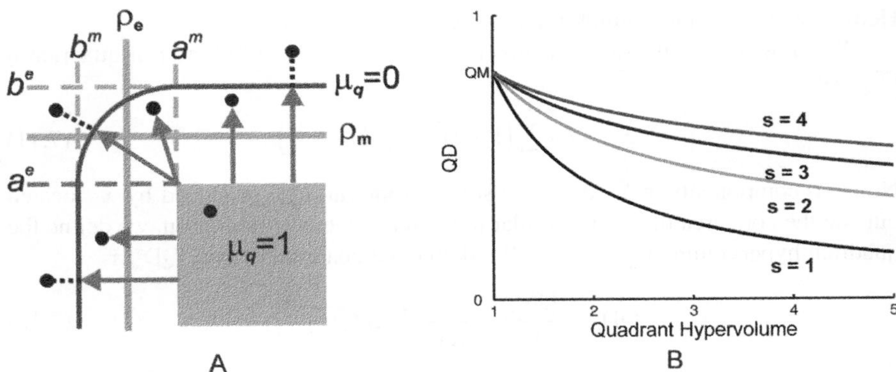

Figure 7.6 (a) Position vectors of data points (arrows) with respect to the quadrant center. **(b)** The effects of scale factor s applied to the quadrant hypervolume on the quadrant density QD_{iq}. Increasing s scales QD_{iq} up with QM_{iq} as the ceiling.

that having both individual data points and the cluster prototypes supporting the same quadrant provides more credibility to assign a higher confidence to that quadrant. If no a priori knowledge on the lymphoma subtypes is available, it is hard to determine the number of clusters to find. Therefore, clustering algorithms that do not require fixed number of clusters, such as PCM and the robust competitive agglomerative (RCA) algorithm (Frigui and Krishnapuram, 1999), are appropriate for this application since both are capable of finding an optimal number of clusters. Here, the PCM algorithm was selected due to its similarity to the fuzzy c-means (FCM) algorithm used in previous microarray studies (Asyali and Alci, 2005; Wang et al., 2003).

Briefly, the PCM clustering algorithm was introduced by Krishnapuram and Keller (1993, 1996) as a generalization of the FCM algorithm introduced earlier by Bezdek (1981). Let C be the number of clusters to find, $X = \{\vec{x}_n \mid n = 1, \ldots, N\}$ be the universe of discourse, and $\vec{\theta}_j$ be the prototype of the jth cluster. Then u_{nj} represents the membership value of $\vec{x}_n$ in cluster $j, u_{nj} \in (0, 1]$. While FCM requires $\sum_{j=1}^{C} u_{nj} = 1$, PCM has no such restriction giving PCM a "mode-seeking" property where each cluster seeks dense regions in X independent of other clusters. This allows C to be overspecified since different cluster prototypes $\{\vec{\theta}_j\}$ may converge on the same location if the natural number of cluster in X is actually smaller than C. On the contrary, FCM partitions X into C clusters regardless of the actual number of natural clusters in X. The definition of u_{nj} in PCM also effectively isolates outliers by assigning them low membership in all clusters. FCM cannot do this due to the constraint on u_{nj}. As in any clustering algorithm, success in PCM is influenced by the initialization of $\{\vec{\theta}_j\}$. If good initial $\{\vec{\theta}_j\}$ is not known, it is beneficial to run PCM for a number of times with different random initializations for $\{\vec{\theta}_j\}$ and evaluate the final results from all runs to determine the consistency of the final clusters.

We apply PCM to X_i of clone i. The number of clusters C is not known. But we should anticipate clusters in multiple quadrants and the possibility of having multiple clusters in the same quadrant. We randomly initialize the same number of cluster prototypes (P) in each quadrant. One prototype is initialized at the origin, yielding the total of $4P + 1$ prototypes. PCM drives prototypes $\{\vec{\theta}_j\}$ toward the dense regions. Upon convergence, the final $\{\vec{\theta}_j\}$ and an $N \times P$ membership matrix $U = [u_{nj}]$ are recorded. U contains membership values of each clone sample $\vec{x}_{in}$ in each cluster j. The membership of the final $\vec{\theta}_j$ in quadrant q is denoted as $\Psi_{jq} = \mu_q(\vec{\theta}_j)$. To reduce the effects of $\{\vec{\theta}_j\}$ initialization on the final clusters, Z-independent PCM runs are performed with different random initialization in each run. For each quadrant q, the quadrant measure QC_{iq} is defined as:

$$\mathrm{QC}_{iq} = \frac{1}{GN}\sum_{z=1}^{Z}\sum_{j=1}^{C} g_{zj}\Psi_{jqz}\sum_{n=1}^{N}(\mu_{inq}T_{in}u_{njz})^{1/3}. \tag{7.14}$$

At the zth run, the membership of $\vec{x}_{in}$ in quadrant q (μ_{inq}), the clone quality T_{in}, and the membership of $\vec{x}_{in}$ in the final cluster j (μ_{njz}) are aggregated using geometric means. The sum $\sum_{n=1}^{N}(\mu_{inq}T_{in}\mu_{njz})^{1/3}$ reflects the support for quadrant q provided by the data memberships in cluster j. This support is weighted by the membership of cluster prototype $\vec{\theta}_j$ in the quadrant q (Ψ_{jqz}), (that is, if the cluster prototype itself has a strong membership in quadrant q, then the support from its members for the quadrant is more likely to be real). Here, g_{zj} is a binary variable to filter out clusters that converge to outliers. A cluster j in the zth run is acceptable ($g_{zj} = 1$) if it contains significant membership values; in our case at least 10% of the data points in X_i must receive $u_{njz} > 0.2$ (such low thresholds are used due to the small number of samples and their variations due to individual patient biology, residual noise, and sample impurities). Alternatively, some cluster validity measures can be used to determine g_{zj}. The normalization factor G in Eq. (7.14) is the total number of acceptable clusters in all Z runs.

Higher QC_{iq} indicates higher likelihood that more compact clusters of $\vec{x}_{in}$ can be found in quadrant q. Assuming all $\vec{x}_{in}$ (with maximum quality T_{in}) are found in the core area of μ_q as a single cluster, then QC_{iq} approaches maximum value of 1 as the cluster tightens. In Figure 7.7, the four quadrants, R1 to R4, each contains a cluster of data points (marked with *) with increasing spread. The final cluster prototypes are marked with ♦. Each cluster is randomly created within a bounding rectangle of side 0.5, 1, 2, and 4 for R1 to R4, respectively. The dotted lines define the boundaries of the μ_q core area. All data points are located within the μ_q core area, thus QC_{iq} and QD_{iq} are equal to 1 for all quadrants. If we consider each quadrant independently (when processing one quadrant we assume the other three quadrants contain no data points) QC_{iq} for R1 to R4 are 0.999, 0.988, 0.9, and 0.804. Note that QC_{iq} decreases as the cluster expands. QD_{iq} and QM_{iq} do not capture this information.

In Figure 7.8, we illustrate the result of one PCM run on genes *CEP78* in FL and *RNPC2* in CLL. Each patient sample is marked as a circle whose intensity represents the locus quality T_{in} (darker means higher spot quality). We use $C = 21$; 5 prototypes for each quadrant ($P = 5$) and one at the origin. The cluster prototypes are initialized randomly and marked as *. The solid line starting from each initial cluster prototype outlines the convergence path of the prototype. The final position of cluster prototypes (convergence points) are marked as ♦. Encircled ♦ indicates acceptable final prototypes ($g_{zj} = 1$), that is, the cluster is not "trapped" by outliers.

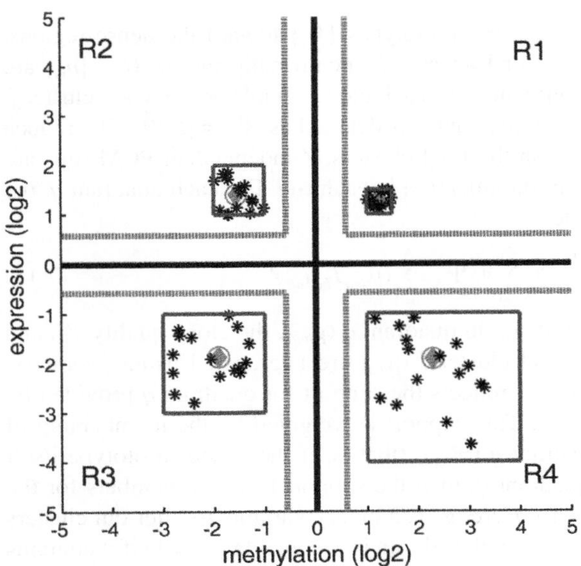

Figure 7.7 QC takes into account the cluster compactness. All data points are located in the quadrant core (outlined by dotted lines) giving QM and QD for R1 to R4 equal to 1 if each quadrant is calculated independently (when processing one quadrant, assume other quadrants are empty). QC for R1 to R4 are 0.999, 0.988, 0.9, and 0.804, respectively. If R1 to R4 are processed simultaneously, QM and QD are 0.25 and 0.125 for all quadrants, and QC are 0.083, 0.070, 0.043, and 0.038, for R1 to R4, respectively.

In both cases the 21 cluster prototypes converge to 1 acceptable final cluster and 1 rejected final cluster. In Figure 7.8a, QC_{i4} receives the only non-zero measure since the only acceptable final cluster is found inside the support area of μ_4. In Figure 7.8b, QC_{i2} receives the only nonzero measure. In both examples, the majority of data points are found within the proposed quadrants. However, not all of these data points are located in the quadrant's core areas [$\mu_{inq} \in [0, 1)$]. Thus, despite good hybridization quality, the values of QC_{i4} for *CEP78* (FL) and QC_{i2} for *RNPC2* (CLL) are relatively low. Based on maximum QC_{iq} alone, the most likely relationship to describe methylation and expression data in *CEP78* (FL) is {hypermethylation, down-regulation} while for *RNPC2* (CLL) it is {hypomethylated, up-regulated}.

The measures for quadrant 1 to 4 for Figure 7.8a are QC = {0, 0, 0, 0.424}, QD = {0.087, 0, 0, 0.46}, QM = {0.114, 0, 0, 0.536}, Q = {0.067, 0, 0, 0.473}, and for Figure 7.8b are QC = {0, 0.394, 0, 0}, QD = {0.171, 0.349, 0.013, 0.015}, QM = {0.228, 0.448, 0.019, 0.022}, Q = {0.133, 0.397, 0.011, 0.012}. All measures suggest that *CEP78* in FL is {hypermethylated, down-regulated} and *RNPC2* in CLL is {hypomethylated, up-regulated}. {Pearson, Kendall, Spearman} correlation coefficients for *CEP78* (FL) and *RNPC2* (CLL) are {0.037, 0.029, 0.046} and {0.666, 0.276, 0.39}. All coefficients suggest positive correlations, however, biologists would consider these as inverse relations.

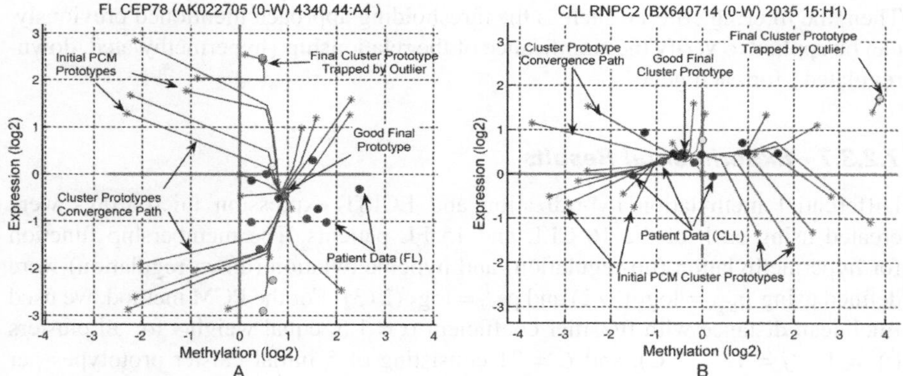

Figure 7.8 Examples of PCM analysis on methylation and expression data in FL and CLL.

7.2.3.6 Quadrant Aggregate Measure (Q)

QD, QM, and QC capture different aspects of the methylation–expression relationship. QC_{iq} is sensitive to inconsistencies in the relationship (data clusters are found in multiple quadrants). For instance, if the four clusters in Figure 7.7 are considered simultaneously $QC_{iq} = \{0.083, 0.07, 0.043, 0.038\}$ for R1 to R4, and $QD_{iq} = 0.125$, $QM_{iq} = 0.25$ for all relationships. Therefore, it is beneficial to fuse these measures into $Q_{iq} \in [0, 1]$ as the confidence that clone i can be described using a methylation–expression relationship represented by quadrant q. A number of fusion methods can be used, such as weighted averaging, the Choquet integral with fuzzy measures, and a special case of the Choquet integral called the linear combination of order statistics (LOS), which includes arithmetic mean. Here, QC_{iq}, QD_{iq}, and QM_{iq} are assumed to be of equal importance, thus Q_{iq} is calculated as the mean of QC_{iq}, QD_{iq}, and QM_{iq}. In Figure 7.7, $Q_{iq} = \{0.999, 0.996, 0.967, 0.935\}$ (independent quadrant processing) and $\{0.153, 0.149, 0.139, 0.137\}$ (simultaneous quadrant processing) for R1 to R4, respectively. The values of Q_{iq} for genes *CEP78* in FL and *RNPC2* in CLL are given in Figure 7.8. As the size of the data set grows and more information about the genes becomes available, more complex fusion algorithm can be used in place of the arithmetic mean.

We can use Q_{iq} to generate hypotheses on the methylation–expression relationship for a clone and use it to guide the next phase of the biology experiments. Additional filtering can be added, such as applying the minimum threshold to Q_{iq}, or $|Q_{iq1}- Q_{iq2}|$ where Q_{iq1} and Q_{iq2} are the two highest confidence values (a relationship is selected if it has *significantly* higher confidence than other relationships). If the criterion $|Q_{iq1}-Q_{iq2}| > 0$ is used (selecting the region with maximum confidence), then *CEP78* in FL is {hypermethylated, down-regulated} and *RNPC2* in CLL is {hypomethylated, up-regulated} (Fig. 7.8). Biologists commonly use microarrays as a discovery tool where candidate genes that best fit certain criteria were selected out of thousands of genes in the microarrays. For example, to find genes that are {hypermethylated, down-regulated} we can find genes having the highest Q_{i4} measures.

Then, the filtering criteria, such as the thresholding approach mentioned previously, can be applied to verify the significance of the relationship {hypermethylated, down-regulated} for each gene.

7.2.3.7 Experimental Results

Differential methylation hybridization and ECIST expression microarrays were created using each of the 16 CLL and 15 FL patients. The membership function for hypermethylation (up-regulation) and hypomethylation (down-regulation) were defined using $\rho_{\text{high}} = \log_2(3/2)$ and $\rho_{\text{low}} = \log_2(2/3)$. For the PCM method, we used Euclidean distance with fuzzifier coefficient $\alpha = 1.5$, equal weights for all clusters ($\eta_j = 1$; $\forall j = 1, \ldots, C$), and $C = 21$ consisting of 5 initial cluster prototypes per quadrant and one at the origin. PCM stopping criteria were the maximum number of iterations ($t_{\max} = 100$) and ($\max_{j=1\ldots c}\left\{\left\|\vec{\theta}_j(t) - \vec{\theta}_j(t-t_s)\right\|\right\} \leq \varepsilon$) for the last t_s iterations (t_s = 10, $\varepsilon = 0$). PCM was repeated for $Z = 100$ runs. Using these algorithms, we identified groups of genes that were {hypermethylated, down-regulated} or {hypomethylated, up-regulated} in CLL and FL. The top-10 loci whose Q_{iq} were the highest for {hypomethylated, up-regulated} ($q = 2$) are given in Table 7.3 and for {hypermethylated, down-regulated} ($q = 4$) are in given Table 7.4.

Only four genes proposed above are also listed in Tables 7.1 and 7.2: *MCF2L2* and *EFNA5* (CLL, hypermethylated, down-regulated), *LRP2* (FL, hypermethylated, down-regulated), and *HNRPA2B1* (FL, hypomethylated, up-regulated). Among these genes, only *MCF2L2* is designated as the top gene for the {hypermethylated, down-regulated} relationship in CLL by both the FRGS and the Q_{iq} measure. Observe that most of the top genes shown in Figure 7.3 contain low-quality samples, especially those in Figures 7.3b, 7.3c, and 7.3d. This is due to the fact that FRGS calculates relationship confidence based solely on the normalized ratios of methylation and expression, without taking into account the channel intensities (clone hybridization quality). On the contrary, the Q_{iq} measure takes both data into account. Assuming a gene follows a certain methylation–expression relationship, a higher Q_{iq} measure is given if the gene contains good-quality samples, as shown in Figure 7.8.

7.3 BIOLOGICAL SIGNIFICANCE

In this section, we look at a number of genes that have been identified as interesting in other studies. We represented these genes using the clones whose sequence alignment is closest to the gene's transcription start site (TSS). We compare the methylation–expression relationships found on these genes using the Q_{iq} measure algorithm with the biological findings reported in other studies. We want to show that the results of the Q_{iq} measure algorithm agree with the confirmation results obtained through biological experiments.

According to the Q_{iq} measure, *RCHY1* is {hypomethylated, up-regulated} in FL (Table 7.4). As reported in Leng et al. (2003), overexpression of *RCHY1* is correlated with the down-regulation of *TP53*, a known tumor suppressor gene whose down-

Table 7.3 Hypermethylated and Down-Regulated Genes in CLL and FL Found Using Q-measure approach

CLL		FL	
MCF2L2	Rho family guanine-nucleotide exchange factor	CEP78	Centrosomal protein 78 kDa
C15orf15	Chromosome 15 open reading frame 15	TTC4	Tetratricopeptide repeat domain 4
EFNA5	EPH-related receptor tyrosine kinase ligand 7	FAM90A1	Family with sequence similarity 90, member A1
ARF3	ADP-ribosylation factor 3	LRP2	Low-density lipoprotein-related protein 2
PSAT1	Phosphorine aminotransferase	LRRC25	Leucine-rich repeat containing 25
DKFZP586D0919	Hepatocellularcarcinoma-associated antigen HCA557a	SERPING1	Serpin peptidase inhibitor, clade G (C1 inhibitor), member 1
TTC4	Tetratricopeptide repeat domain 4	MNAT2	Eukaryotic translation elongation factor 1 alpha 1
KCNN2	Potassium intermediate/small conductance calcium-activated channel, subfamily N, member 2	TAFA2	Family with sequence similarity 19 (chemokine (C-C motif)-like), member A2
LHX2	LIM homeobox 2	SCAMP3	Secretory carrier membrane protein 3
KCNK2	Potassium channel, subfamily K, member 2	ACSS2	Acyl-CoA synthetase short-chain family member 2

regulation is often associated with cell malignancies including lymphomas. It was further shown that the expression of *RCHY1* was higher in malignant lung tissue as compared to normal, and that *TP53* was more ubiquitinated in the malignant tissue as compared to normal tissue provides strong evidence that an increase in the expression of *RCHY1* plays a role in lung tumorigenesis by reducing *TP53* activity (Duan et al., 2004). These findings along with the {hypomethylated, up-regulated} relationship of *RCHY1* found in FL suggest that further investigation is warranted to elucidate the relationship between *RCHY1* and *TP53* in FL.

Four histone genes are found to be {hypomethylated, up-regulated} in FL (Table 7.4). Furthermore, 11 of the top 20 genes in FL (not shown) in this relationship are histone genes. Histones are transcription regulatory proteins responsible for gene regulation, chromosome condensation, recombination, and replication. Santos-Rosa

Table 7.4 Hypomethylated and Up-Regulated Genes in CLL and FL Found Using *Q*-measure Approach

CLL		FL	
RNPC2	RNA-binding region	HNRPA2B1	Heterogeneous nuclear ribonucleoprotein A2/B1
RPL37A	Ribosomal protein L37a	PTMAP7	Prothymosin, alpha pseudogene 7
EEF1A1	Eukaryotic translation elongation factor 1 alpha 1	HIST1H3B	Histone cluster 1, H3b
EIF1AY	Eukaryotic translation initiation factor 1A, Y-linked	PSMB4	Proteasome beta 4 subunit
MLH1	MutL homolog 1, colon cancer, nonpolyposis type 2	HIST1H4E	H4 histone family, member J
RPL30	Ribosomal protein L30	HIST4H4	Histone cluster 4, H4
ZNF614	Zinc finger protein 614	HIST1H1D	H1 histone family, member 3
ELAVL1	Embryonic lethal, abnormal vision, Drosophila)-like 1	RCHY1	Androgen-receptor N-terminal-interacting protein
RBM8A	RNA binding motif protein 8A	RPL30	Ribosomal protein L30
TAF12	TAF12 RNA polymerase II, TATA box binding protein (TBP)-associated factor	AMD1	adenosylmethionine decarboxylase 1

and Caldas (2005) reported that transcription regulatory proteins are often identified in oncogenic chromosomal rearrangements and are overexpressed in a variety of malignancies. Methylation is one of the transcription repressors in histones. Thus hypomethylation may cause histone overexpression. The relationship found using our algorithm supported this notion in FL. Independent biological confirmation is needed to verify this relationship.

Rahmatpanah et al. (2006) investigated gene methylation patterns to classify CLL, FL, and mantle cell lymphoma (MCL). They claimed that *LHX2* and *LRP1B* were among the differentially methylated genes in CLL, FL, and MCL based on the cluster analysis of the DMH microarray data. They performed MSP to confirm the methylation status of *LHX2* and *LRP1B* on the FL and CLL data sets with 15 patients each. About 70% of the CLL and FL patient samples used in Rahmatpanah et al. (2006) are the same as those used in our CLL and FL data sets. Their MSP results showed that *LHX2* was methylated in 7/15 (47%) of CLL and 11/15 (73%) of FL, and *LRP1B* was methylated in 2/15 (13%) of CLL and 13/15 (87%) of FL patient samples. Bisulfite sequencing was performed on subsets of CLL and FL patients to satisfactorily confirm the MSP results. The reconfirmation results were satisfactory. RT-PCR was performed to assess the expression of *LHX2* and *LRP1B* on 8 CLL and 10 FL patients. Their RT-PCR results showed that the expression means for *LHX2* in both CLL and FL were lower by a factor of 3 ($\log_{10}$ scale) compared to the expression mean of the normal lymphocyte samples used as control. Similarly, the expression means for *LRP1B* in CLL and FL were lower by a factor of 1 and 1.5 compared to the normal control. All these results suggest that *LHX2* and *LRP1B* are hypermethylated and down-regulated in almost all FL samples. From the MSP results, *LHX2* was hypermethylated in almost half of the CLL samples and was down-regulated across all samples. *LRP1B* is hypermethylated only in a small number of CLL samples, but its expression is generally down-regulated. Between the two genes, FL shows more methylation than does CLL.

The Q_{iq} measures of *LHX2* and *LRP1B* are given in Table 7.5. For *LHX2*, Q_{i4} received the highest measures for both CLL and FL, suggesting that *LHX2* was likely to be {hypermethylated, down-regulated} as also suggested by the confirmation results from Rahmatpanah et al. (2006). Q_{i4} in FL was higher than that in CLL, suggesting that *LHX2* is hypermethylated and down-regulated in more FL samples than in CLL. This finding was in agreement with the MSP results in Rahmatpanah et al. (2006) where 47% of CLL and 73% of FL samples were hypermethylated. The Q_{i3} and Q_{i4} measures for *LRP1B* in CLL indicated that data points were found mostly in quadrant 3, although some were present in quadrant 4 as well. Both quadrants captured the notion of down-regulated gene expression, which was in agreement with the RT-PCR results in Rahmatpanah et al. (2006). Since $Q_{i3} > Q_{i4}$, it can be suggested that hypomethylated *LRP1B* were found in more CLL samples than the hypermethylated ones. The split in *LRP1B* methylation status among CLL samples conformed to the MSP results from Rahmatpanah et al. (2006) where only 13% of CLL samples were hypermethylated. This example suggested that gene down-regulation may be caused by factors other than promoter hypermethylation. For *LRP1B* in FL, Q_{i4} received the highest measure suggesting that {hypermethylated, down-regulated} as the proposed relationship. This choice of relationship was supported by the MSP and RT-PCR results from Rahmatpanah et al. (2006) where 87% of FL samples were methylated, and the mean expression level was down-regulated compared to that of the normal lymphocyte control.

Ephrin-A5, a member of the ephrin gene family encoded by *EFNA5*, was one of the novel epigenetic markers reported by Shi et al. (2007). The EPH and EPH-related receptors comprise the largest subfamily of receptor protein–tyrosine kinases and have been implicated in mediating developmental events, particularly in the

Table 7.5 Aggregate Quadrant Measures (Q_{iq}) for CLL and FL (the Largest Measure are in boldface)

Quadrant	CLL				FL			
(q)	1	2	3	4	1	2	3	4
LHX2	0.029	0.033	0.034	0.203	0.059	0	0.032	0.334
LRP1B	0.007	0.004	0.216	0.146	0.049	0.003	0.009	0.353
EFNA5	0.189	0.006	0.031	0.334	0.149	0.005	0.035	0.225
DDX51	0.287	0.012	0.053	0.124	0.439	0.039	0.002	0.074

nervous system. Himanen et al. (2004) found that ephrin-A5 binds to the *EphB2* receptor, a tumor suppressor gene (Huusko et al., 2004), leading to receptor clustering, autophosphorylation, and initiation of downstream signaling. In Shi et al. (2007), *EFNA5* was found to be differentially methylated using DMH on 6 cell lines originating from different subtypes of NHL. Using 75 primary NHL specimens comprised of 4 subtypes of different maturation stages, *EFNA5* was hypermethylated in 53% of the specimens. The COBRA method was used to confirm the methylation. Confirmation on gene expression for *EFNA5* was not performed. The Q_{iq} measures for *EFNA5* in CLL and FL are shown in Table 7.5. In both FL and CLL, Q_{i4} received the highest value suggesting {hypermethylation, down-regulation} as the most appropriate relationship. However, we found *EFNA5* was up-regulated in some CLL and FL patients despite hypermethylation.

DDX51 was one of the genes studied in ALL (acute lymphoblastic leukemia) patients and cell lines by Taylor et al. (2006). They found *DDX51* was methylated in 70% of the B-ALL and none in the T-ALL patients, hence making it a potential biomarker for separating the two ALL subtypes. Taylor et al. (2006) treated the cell lines using a demethylating agent and conducted an independent confirmation for *DDX51* gene expression on both treated and untreated cell lines using RT-PCR. Despite the presence of methylation, the untreated *DDX51* still showed moderate level of gene expression. The treated cell lines showed an increase in expression level by at least twofold compared to untreated ones. We found *DDX51* was hypermethylated in CLL and FL patient samples with moderate level of gene expression indicated by Q_{i1} being the highest measures in both CLL and FL (Table 7.5). This condition is similar to that reported in Taylor et al. (2006). Overall, we have excellent locus quality with 12/15 (80%) of the samples found to be hypermethylated. The methylation percentage is roughly equal to that of ALL found in Taylor et al. (2006). We also notice that PCM prototypes converge to a single final cluster whose location suggesting significant hypermethylation with low to moderate level of up-regulation in the gene expression. These results imply that hypermethylation does not always lead to down-regulated gene expression. Our results agree with the findings reported in Taylor et al. (2006) for ALL. To confirm this apparent similarity, *DDX51* in FL and CLL need to be independently verified for methylation using MSP or COBRA, and for expression using RT-PCR. The COBRA method is preferred for confirming the gene methylation status since it can be designed to target the same cut sites as those used in DMH.

7.4 CONCLUSIONS

Aberrant gene methylation is suspected to have links with tumorigenesis through the silencing of tumor suppressor genes or overexpression of oncogenes. Better understanding of the relationships between methylation and expression of important genes in NHL may lead to better cancer diagnosis and treatments. In this chapter we propose two fuzzy algorithms for simultaneous analysis of the methylation–expression relationship from a large volume of methylation and expression microarray data. The fuzzy models allow us to better manage common uncertainties in microarray experiments, such as the probe hybridization quality and level of expression/methylation.

The first algorithm, FRGS, employs a fuzzy rule approach with linguistic values defined for the normalized ratios of methylation and expression. The advantage of this approach is simplicity. Using this approach we identified several genes previously mentioned in the literature as connected to NHL, such as *DDX51*, *LHX2*, and *EFNA5*. However, this method produces many false positives because it does not account for the channel intensities (hybridization quality) in the loci.

The second algorithm utilizes a compound quality measure, Q_{iq}, for the clone hybridization levels in methylation and expression arrays to reduce the number of false positives. Unlike the approach in FRGS where fuzzy sets are used to capture the level of methylation and expression separately, here we use fuzzy sets to model the methylation–expression relationships directly. As a result about 85% of the genes found significant by FRGS were removed by the Q-measure algorithm as false alarms.

To validate our models, we analyze some of the genes that were found by both algorithms, such as *LHX2*, *EFNA5* and *DDX51*. These genes have been also found to be related to NHL in other studies (Rahmatpanah et al., 2006; Shi et al., 2007; Taylor et al., 2006). In all three genes, our algorithms select methylation–expression relationships that matched the confirmation results obtained using MSP, COBRA, bisulfite sequencing, and RT-PCR reported in Rahmatpanah et al. (2006), Shi et al. (2007), and Taylor et al. (2006).

The new computation models proposed in this chapter enables us to analyze the methylation–expression relationship from a large volume of DMH and ECIST cDNA microarray data simultaneously. Our approaches enables investigators to build initial hypothesis of the relationship between promoter methylation and gene expression in various genes.

Acknowledgments

This work is supported by National Cancer Institute grants CA100055 and CA09780 and National Library of Medicine grant LM07089 (C. W. Caldwell).

REFERENCES

Adorjan, P., J. Distler, E. Lipscher, F. Model, J. Muller, C. Pelet, A. Braun, A. R. Florl, D. Gutig, G. Grabs, A. Howe, M. Kursar, R. Lesche, E. Leu, A. Lewin, S. Maier, V. Muller, T. Otto, C. Scholz, W. A. Schulz, H. H. Seifert, I. Schwope, H. Ziebarth, K. Berlin, C. Piepenbrock, and A. Olek (2002). "Tumour class prediction and discovery by microarray-based DNA methylation analysis," *Nucleic Acids Res.*, Vol. 30, p. e21.

Asyali, M. H. and M. Alci (2005). "Reliability analysis of microarray data using fuzzy c-means and normal mixture modeling based classification methods," *Bioinformatics*, Vol. 21, pp. 644–649.

Bezdek, J. C. (1974). Cluster validity with fuzzy sets, *J. Cybernet.*, Vol. 3, pp. 58–72.

Bezdek, J. C. (1981). *Pattern Recognition with Fuzzy Objective Function Algorithms*. Plenum, New York.

Cameron, E. E., S. B. Baylin, and J. G. Herman (1999). "p15(INK4B) CpG island methylation in primary acute leukemia is heterogeneous and suggests density as a critical factor for transcriptional silencing," *Blood*, Vol. 94, pp. 2445–2451.

Cleveland, W. S. (1979). "Robust locally-weighted regression and smoothing scatterplots," *J. Am. Statist. Assoc.*, Vol. 74, pp. 829–836.

Cleveland, W. S., S. J. Devlin, and E. Grosse (1988). "Regression by local fitting," *J. Econometr.*, Vol. 37, pp. 87–114.

Cleveland, W. S. and E. Grosse (1991). "Computational methods for local regression," *Statist. Comput.*, Vol. 1, pp. 47–62.

Costello, J. F., M. C. Fruhwald, D. J. Smiraglia, L. J. Rush, G. P. Robertson, X. Gao, F. A. Wright, J. D. Feramisco, P. Peltomaki, J. C. Lang, D. E. Schuller, L. Yu, C. D. Bloomfield, M. A. Caligiuri, A. Yates, R. Nishikawa, H. H. Su, N. J. Petrelli, X. Zhang, M. S. O'Dorisio, W. A. Held, W. K. Cavenee, and C. Plass (2000). "Aberrant CpG-island methylation has N-random and tumour-type-specific patterns," *Nat. Genet.*, Vol. 24, pp. 132–138.

Duan, W., L. Gao, L. J. Druhan, W. G. Zhu, C. Morrison, G. A. Otterson, and M. A. Villalona-Calero (2004). "Expression of Pirh2, a newly identified ubiquitin protein ligase, in lung cancer," *J. Natl. Cancer Inst.*, Vol. 96, pp. 1718–1721.

Egger, G., G. Liang, A. Aparicio, and P. A. Jones (2004). "Epigenetics in human disease and prospects for epigenetic therapy," *Nature*, Vol. 429, pp. 457–463.

Eisenberg, E. and E. Y. Levanon (2003). Human housekeeping genes are compact, *Trends Genet.*, Vol. 19, pp. 362–365.

Frigui, H. and R. Krishnapuram (1999). "A robust competitive clustering algorithm with applications in computer vision," *IEEE Trans. PAMI*, Vol. 21, no. 5, pp. 450–465.

Gath, I. and A. B. Geva (1989) "Unsupervised optimal fuzzy clustering," *IEEE Trans. Pattern Anal. Mach. Intell.*, Vol. 11, pp. 773–847.

Glas, A. M., M. J. Kersten, L. J. delahaye, A. T. Witteveen, R. E. Kibbelaar, A. Velds, L. F. Wessels, P. Joosten, R. M. Kerkhoven, R. Bernards, J. H. van Krieken, P. M. Kluin, L. J. Van't Veer, and D. de Jong (2005). "Gene expression profiling in follicular lymphoma to assess clinical aggressiveness and to guide the choice of treatment," *Blood*, Vol. 105, pp. 301–307.

Herman, J. G., J. R. Graff, S. Myohanen, B. D. Nelkin, and S. B. Baylin (1996). "Methylation-specific PCR: A novel PCR assay for methylation status of CpG islands," *Proc. Natl. Acad. Sci., USA*, Vol. 93, pp. 9821–9826.

Himanen, J. P., M. J. Chumley, M. Lackmann, C. Li, W. A. Barton, P. D. Jeffrey, C. Vearing, D. Geleick, D. A. Feldheim, A. W. Boyd, M. Henkemeyer, and D. B. Nikolov (2004). "Repelling class discrimination: Ephrin-A5 binds to and activates EphB2 receptor signaling," *Nat. Neurosci.*, Vol. 7, pp. 501–509.

Huang, T. H., M. R. Perry, and D. E. Laux (1999). "Methylation profiling of CpG islands in human breast cancer cells," *Hum. Mol. Genet.*, Vol. 8, pp. 459–470.

Husson, H., E. G. Carideo, D. Neuberg, J. Schultze, O. Munoz, P. W. Marks, J. W. Donovan, A. C. Chillemi, P. O'Connell, and A. S. Freedman (2002). "Gene expression profiling of follicular lymphoma and normal germinal center B cells using cDNA Arrays," *Blood*, Vol. 99, pp. 282–289.

Huusko, P., D. Ponciano-Jackson, M. Wolf, J. A. Kiefer, D. O. Azorsa, S. Tuzmen, D. Weaver, C. Robbins, T. Moses, M. Allinen, S. Hautaniemi, Y. Chen, A. Elkahloun, M. Basik, G. S. Bova, L. Bubendorf, A. Lugli, G. Sauter, J. Schleutker, H. Ozcelik, S. Elowe, T. Pawson, J. M. Trent, J. D. Carpten, O. P. Kallioniemi, and S. Mousses (2004). "Nonsense-mediated decay microarray analysis identifies mutations of EPHB2 in human prostate cancer," *Nat. Genet.*, Vol. 36, pp. 979–983.

Katzenellenbogen, R. A., S. B. Baylin, and J. G. Herman (1999). "Hypermethylation of the DAP-kinase CpG Island is a common Alteration in B-cell malignancies," *Blood*, Vol. 93, pp. 4347–4353.

Krishnapuram, R., and J. M. Keller (1993). A possibilistic approach to clustering, *IEEE Trans. Fuzzy Syst.*, Vol. 1, pp. 98–110.

Krishnapuram, R., and J. M. Keller (1996). "The possibilistic C-means algorithm: Insights and recommendations," *IEEE Trans. Fuzzy Syst.*, Vol. 4, pp. 385–393.

Lee, M-L. T. (2004). *Analysis of Microarray Gene Expression Data*. Kluwer Academic, Boston.

Leng, R. P., Y. Lin, W. Ma, H. Wu, B. Lemmers, S. Chung, J. M. Parant, G. Lozano R. Hakem, and S. Benchimol (2003). "Pirh2, a p53-induced ubiquitin-protein ligase, promotes p53 degradation," *Cell*, Vol. 112, pp. 779–791.

Li, Y., H. Nagai, T. Ohno, M. Yuge, S. Hatano, E. Ito, N. Mori, H. Saito, and T. Kinoshita (2002). "Aberrant DNA methylation of p57(KIP2) gene in the promoter region in lymphoid malignancies of B-cell phenotype," *Blood*, Vol. 100, pp. 2572–2577.

Nagasawa, T., Q. Zhang, P. N. Raghunath, H. Y. Wong, M. El-Salem, A. Szallasi, M. Marzec, P. Gimotty, A. H. Rook, E. C. Vonderheid, N. Odum, and M. A. Wasik (2006). "Multi-gene epigenetic silencing of tumor suppressor genes in T-cell lymphoma cells; delayed expression of the p16 protein upon reversal of the silencing," *Leukemia Res.*, Vol. 30, pp. 303–312.

Rahmatpanah, F., S. J. Carstens, J. Guo, O. Sjahputera, K. H. Taylor, D. J. Duff, H. Shi, J. W. Davis, S. I. Hooshmand, R. Chitma-Matsiga, C. W. Caldwell (2006). "Differential DNA methylation patterns of small B-cell lymphoma subclasses with different clinical behavior," *Leukemia*, Vol. 20, pp. 1855–1862.

Santos-Rosa, H. and C. Caldas (2005). "Chromatin modifier enzymes, the histone code and cancer," *Eur. J. Cancer*, Vol. 41, pp. 2381–402.

Shi, H., J. Guo, D. J. Duff, F. Rahmatpanah, R. Chitma-Matsiga, M. Al-Kuhlani, K. H. Taylor, O. Sjahputera, M. andreski, J. E. Wooldridge, and C. W. Caldwell (2007). Discovery of novel epigenetic markers in non-hodgkin's lymphoma, *Carcinogenesis*, Vol. 28, No. 1, pp. 60–70.

Shi, H., P. S. Yan, C. M. Chen, F. Rahmatpanah, C. Lofton-Day, C. W. Caldwell, and T. H. Huang (2002). "Expressed CpG island sequence tag microarray for dual screening of DNA hypermethylation and gene silencing in cancer cells," *Cancer Res.*, Vol. 62, pp. 3214–3220.

Simon, R. M., E. L. Korn, L. M. McShane, M. D. Radmacher, G. W. Wright, and Y. Zhao (2004). *Design and Analysis of DNA Microarray Investigations*. Springer, New York.

Sjahputera, O., J. M. Keller, J. W. Davis, K. H. Taylor, F. Rahmatpanah, H. Shi, D. T. Anderson, S. N. Blisard, R. H. Luke III, M. Popescu, G. C. Arthur, and C. W. Caldwell (2007). Relational analysis of CpG islands methylation and gene expression in human lymphomas using possibilistic C-means clustering and modified cluster fuzzy density, *IEEE. Trans. Comp. Biol. Bioinform.*, Special Issue on Computational Intelligence,Vol. 4, No. 2, pp. 176–189.

Taylor, K. H., D. J. Duff, D. Durtschi, K. E. Pena-Hernandez, H. Shi, O. Sjahputera, and C. W. Caldwell (2006). "Differentially methylated genes in acute lymphoblastic leukemia using differential methylation hybridization," unpublished manuscript, University of Missouri–Columbia.

Wang, J. and T. H. Bo, I. Jonassen, O. Myklebost, and E. Hovig (2003). "Tumor classification and marker gene prediction by feature selection and fuzzy c-means clustering using microarray data," *BMC Bioinformatics*, Vol. 4, p. 60.

Woolf, P. J. and Y. Wang (2000). "A fuzzy logic approach to analyzing gene expression data," *Physiol. Genomics*, Vol. 3, pp. 9–15.

Xiong, Z. and P. W. Laird (1997). "COBRA: A sensitive and quantitative DNA methylation assay," *Nucleic Acids Res.*, Vol. 25, pp. 2532–2534.

Yager, R. R. (1993). "Families of OWA operators," *Fuzzy Sets Syst.* Vol. 57, pp. 125–148.

Yan, P. S., C. M. Chen, H. Shi, F. Rahmatpanah, S. H. Wei, C. W. Caldwell, and T. H. Huang (2001). "Dissecting complex epigenetic alterations in breast cancer using CpG island microarrays," *Cancer Res.*, Vol. 61, pp. 8375–8380.

Yang, Y. H., S. Dudoit, P. Luu, D. M. Lin, V. Peng, J. Ngai, and T. P. Speed (2002). "Normalization for cDNA microarray data: A robust composite method for addressing single and multiple slide systematic variation," *Nucleic Acid Res.*, Vol. 30, p. e15.

Zadeh, L. (1965). "Fuzzy sets," *J. Info. Control*, Vol. 8, pp. 338–353.

Part Three

Molecular Structure and Phylogenetics

Chapter 8

Protein–Ligand Docking with Evolutionary Algorithms

René Thomsen

8.1 INTRODUCTION

Proteins found in nature have evolved by natural selection to perform specific functions. The biological function of a protein is linked to its three-dimensional structure or conformation. Consequently, protein functionality can be altered by changing its structure. The idea in molecular docking is to design pharmaceuticals computationally by identifying potential drug candidates targeted against proteins. The candidates can be found using a docking algorithm that tries to identify the bound conformation of a small molecule (also referred to as a ligand) to a macromolecular target at its active site, which is the region of an enzyme[1] where the natural substrate[2] binds. Often, the active site is located in a cleft or pocket in the protein's tertiary structure. The structure and stereochemistry[3] at the active site complements the shape and physical/chemical properties of the substrate so as to catalyze a particular reaction. The purpose of drug discovery is thus to derive drugs or *ligands* that bind stronger to a given protein target than the natural substrate. By doing this, the biochemical reaction that the enzyme catalyzes can be altered or prevented.

Until recently, drugs were discovered by chance via a trial-and-error manner using high-throughput screening methods that experimentally test a large number of compounds for activity against the target in question. This process is very expensive and time consuming. If a three-dimensional structure of the target exists, simulated molecular docking can be a useful tool in the drug discovery process. This so-called

[1] In short, enzymes are large proteins that catalyze chemical reactions.

[2] The specific molecule an enzyme acts upon.

[3] The spatial arrangement of the atoms.

Computational Intelligence in Bioinformatics. Edited by G. B. Fogel, D. Corne, and Y. Pan

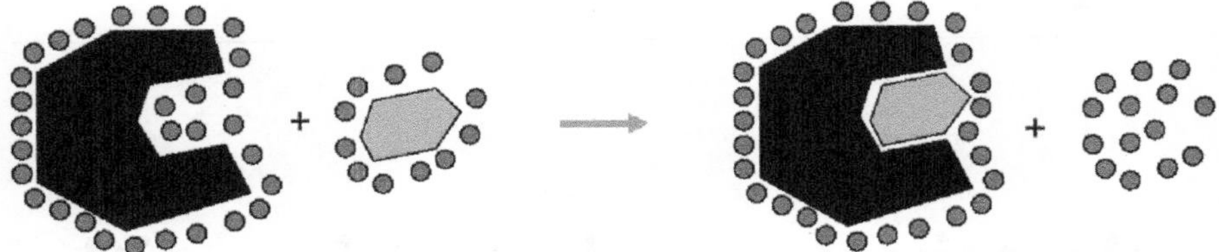

Figure 8.1 Docking of a small molecule into a protein's active site. Solvent molecules are indicated by filled circles.

in silico approach allows for many possible lead candidates to be tested before committing expensive resources for wet lab experiments (synthesis), toxicological testing, bioavailability, clinical trials, and so forth.

Molecular docking algorithms try to predict the bound conformations of two interacting molecules, such as protein–ligand and protein–protein complexes. The task of predicting these interactions is also commonly referred to as *the docking problem*. More specifically, given two molecules of known initial structure, the task is to determine their three-dimensional structure when bound together as a *complex*. These molecules can be proteins, DNA (deoxyribonucleic acid), RNA (ribonucleic acid), or small molecules. Figure 8.1 illustrates the docking of a small molecule to an active site of a protein.

The focus of this chapter will be on docking proteins and ligands, which is commonly referred to as *protein–ligand docking*. This chapter contains an overview of protein–ligand docking algorithms currently used by most practitioners as well as a short introduction to the biochemical background and the scoring functions used to evaluate docking solutions. The chapter concludes with an introduction to the authors own research contributions and an overview of future work topics. For a more detailed introduction to the biochemical background and the molecular docking field see Alberts et al. (2002), Creighton (1997), Cotterill (2002), Morris et al. (2000), Muegge and Rarey (2001), and Höltje et al. (2003).

8.2 BIOCHEMICAL BACKGROUND

8.2.1 Amino Acids, Peptides, and Proteins

Proteins are large molecules that have many important biological functions, and they can act as both structural components or active agents. Examples of structural components are proteins that constitute skin, hair, and muscles. Examples of active agents are enzymes that catalyze most intracellular chemical reactions and transport proteins, such as hemoglobin, that carry oxygen to our tissues.

The basic building blocks of proteins are amino acids. An amino acid consists of a central carbon atom (denoted by C_α) that binds to an amino group (NH_2), a carboxyl group (COOH), and a side chain (denoted by R). There are 20 common amino acids in living organisms. The amino acids differ in the chemical composition of the side chain R, which contains between 1 and 18 atoms (glycine and arginine,

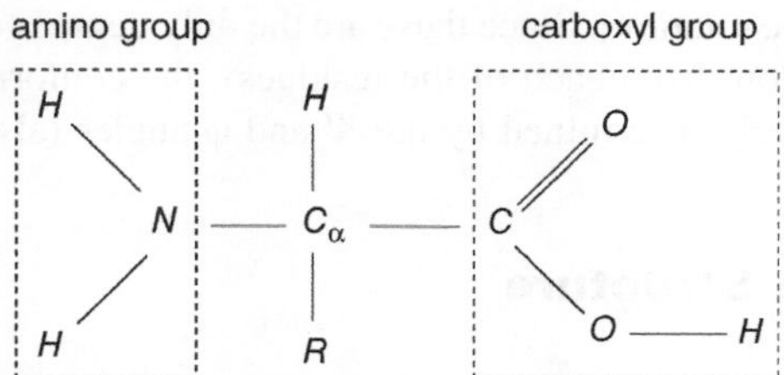

Figure 8.2 General structure of amino acids (displayed in its uncharged state).

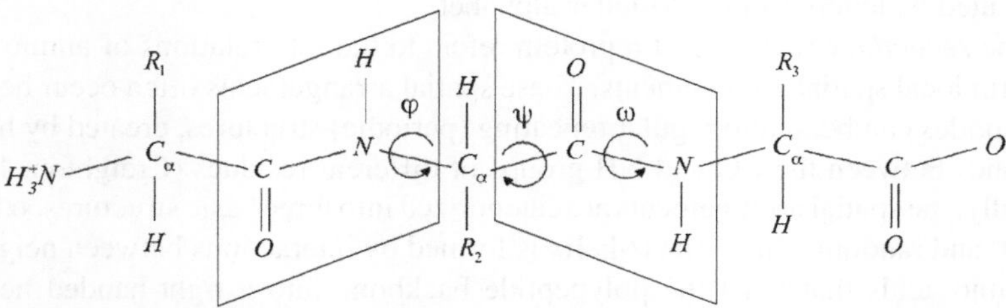

Figure 8.3 Schematic of backbone structure of a polypeptide. Atoms inside each quadrilateral are in the same plane, which may rotate according to the torsional angles φ and Ψ. The ω angle is fixed at 180° since the peptide group is planar.

respectively) (Branden and Tooze, 1998). Figure 8.2 shows the general form of amino acids, which is the same for all except proline.[4]

Amino acids are linked together by covalent bonds called *peptide bonds*. A peptide bond (C—N) is formed through a condensation reaction, where the carboxyl group of the first amino acid condensates with the amino group of the next and eliminates water in the process (Griffiths et al., 1999). Several amino acids linked together by peptide bonds form a molecule called a *polypeptide*. When the peptide bond is formed, the amino acid is changed (losing two hydrogen atoms and an oxygen atom), so the portion of the original amino acid integrated into the polypeptide is often called a *residue*. The formation of peptide bonds generates a backbone (also referred to as the main chain), consisting of the common repeating unit, N—C_α—C, from which the side chains (R) extend. A schematic illustration of the backbone structure of a polypeptide is shown in Figure 8.3.

The two ends of a protein chain are different (see Fig. 8.3). One end has a free protonated amino group and is called the *N terminus*, whereas the other end has a free deprotonated carboxyl group and is called the *C terminus*. Since the peptide units are effectively rigid groups that are linked into a chain by covalent bonds through the C_α atoms, the only degrees of freedom they have are rotations around these bonds (Branden and Tooze, 1998). Each unit can rotate around two such bonds: the C_α—C [referred to as the Ψ (psi) angle rotation] and the N—C_α bonds [referred

[4] Proline has a bond between the side chain and the amino group.

to as the φ (phi) angle rotation]. Since these are the only degrees of freedom[5] (besides the degrees of freedom from each of the residues), the conformation of the whole backbone is completely determined by the Ψ and φ angles (also referred to as *torsional angles*).

8.2.2 Protein Structure

Proteins have a complex structure that is traditionally thought of as having four levels (Branden and Tooze, 1998). The linear sequence of the amino acids in the polypeptide chain is called the *primary structure*. Typically, the amino acids are represented as letters from a 20-letter alphabet.

The *secondary structure* of a protein refers to the interrelations of amino acids that form local spatial arrangements. These spatial arrangements often occur because polypeptides can bend into regular repeating (periodic) structures, created by hydrogen bonds between the CO and NH groups of different residues (Creighton, 1997). Typically, the spatial arrangements are categorized into three basic structures, α helix, β sheet, and random coils.[6] The α-helix is formed by interactions between neighboring amino acids that twist the polypeptide backbone into a right-handed helix in which the CO group in each amino acid residue is hydrogen bonded to the NH group in the fourth amino acid toward the G-terminus. In contrast, the β sheet involves extended protein chains, packed side by side, that interact by hydrogen bonding (Creighton, 1997). The packing of the chains creates the appearance of a pleated sheet. Illustrations showing these arrangements are depicted in Figures 8.4 and 8.5. Another example of a secondary structure belonging to the coil category is the *turn*, which connects α-helix and β-sheet elements in a protein.

The overall three-dimensional shape of a polypeptide created by electrostatic, hydrogen, sulfur bridges, and *van der Waals* bonds is its *tertiary structure* (Creighton, 1997). Many proteins fold up in a roughly spherical or globular structure. In many

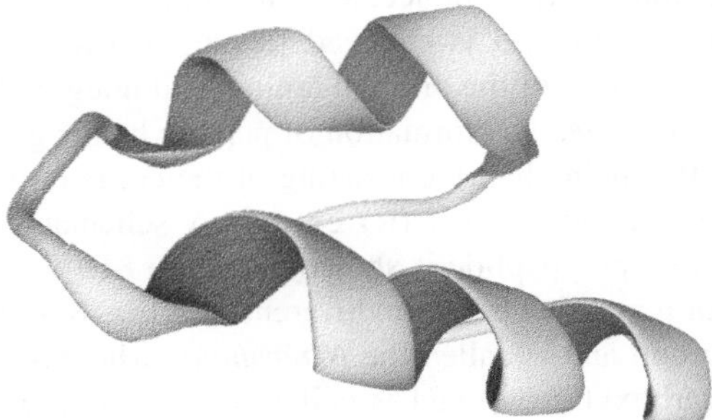

Figure 8.4 Alpha helices in crambin.

[5] The peptide group is planar because the additional electron pair of the *C=O* bond is delocalized over the peptide group such that rotation around the *C—N* bond is prevented by a high-energy barrier. Therefore the ω angle is fixed at 180°.

[6] Random coils are structures such as loop regions not covered by other structural classes. They are only random in the sense that they are not periodic.

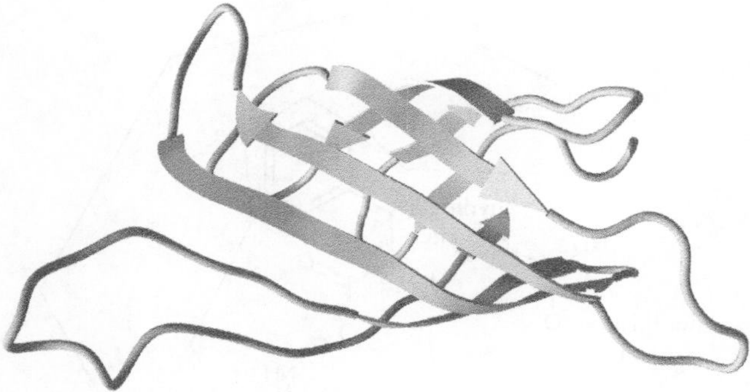

Figure 8.5 Beta sheets in streptavidin.

cases, amino acids that are far apart in the linear sequence are brought close together in the tertiary structure. The protein surface usually has one or more cavities (active sites), where specific molecules (substrates) can catalyze chemical reactions by binding to the active site.

Some proteins contain several polypeptide chains (e.g., hemoglobin). When a protein contains more than one chain, each chain is called a subunit, and the structure formed between subunits is called the *quaternary* structure.

8.2.3 Molecular Interactions

Two types of bonds occur between atoms and molecules: *covalent* and *noncovalent*. In short, covalent bonds provide the "glue" that holds the atoms in a molecule together by sharing one or more electron pairs[7] between atoms. For instance, when two cysteine amino acids are close together in three-dimensional space, a covalent disulfide bond S—S can be formed between different regions of the protein chain through oxidation.

An example of a noncovalent bond of particular importance in biological systems is the *hydrogen bond*. A hydrogen bond can be regarded as a relatively strong intermolecular interaction between a positively charged hydrogen atom H and an electronegative acceptor atom A and occurs when two electronegative atoms compete for the same hydrogen atom: *—D—H ··· A—*. Formally, the hydrogen atom is bonded covalently to the donor atom D, but it also interacts favorably with the acceptor atom A. In contrast to other noncovalent forces (e.g., van der Waals interactions), the hydrogen bonding interaction is directional (i.e., it depends on the propensity and orientation of the unpaired electron pairs of the acceptor atom and is close to linear around H).

Other examples of noncovalent bonds are *salt bridges* (ion–ion bonds) and *coordination bonds* (metal–ligand bonds). Figure 8.6 illustrates the most common

[7] Sharing of one pair, two pairs, or three pairs of electrons is referred to as single bond, double bond, and triple bond, respectively.

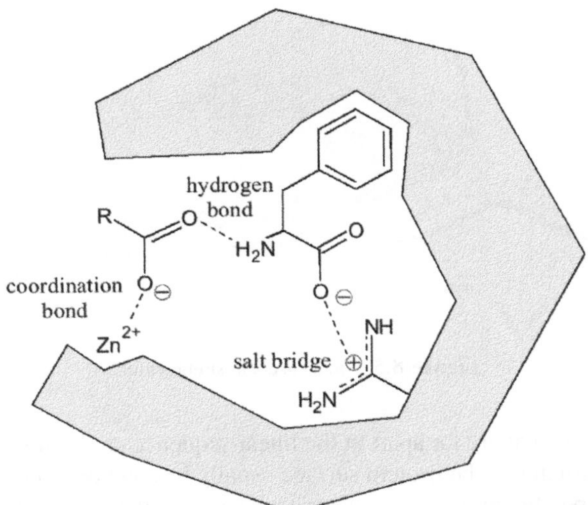

Figure 8.6 Examples of different binding interactions.

intermolecular interactions between a ligand situated in an active site. Generally, noncovalent bonds consist of both short- and long-range interactions that mediate molecular recognition between complementary molecular surfaces and are responsible for the folding and unfolding of proteins (e.g., determining their three-dimensional structure).

The nonbonded terms (i.e., interactions acting between atoms that are not linked by covalent bonds) that are used in docking scoring functions, such as force fields, are usually considered in two groups, one comprising electrostatic interactions and the other van der Waals interactions. A short introduction to these interactions is provided in the following two sections. For a more detailed introduction to molecular forces and interactions and a description of other forces acting between atoms and molecules see Creighton (1997), Cotterill (2002), Fersht (2000), and Leach (2001).

8.2.4 Electrostatic Interactions

All intermolecular forces are essentially electrostatic in nature and the most fundamental noncovalent interaction is the attraction between atoms caused by electrostatic charges. Coulomb's law states that the energy of the electrostatic interaction between two atoms i and j is the product of their two charges (q_i and q_j, respectively) divided by the medium's (e.g., liquid water) permittivity ε and the distance r_{ij} between them:

$$E_{\text{elec}} = \frac{q_i q_j}{4\pi\varepsilon r_{ij}}. \tag{8.1}$$

Consequently, negatively and positively charged atoms attract each other because the energy E_{elec} decreases as they approach each other making the interaction favorable. If the charged atoms are of the same sign, they repel each other because the interaction becomes energetically unfavorable. Electrostatic interactions are important due to their long range of attraction between ligand and protein atoms. The Coulomb equation (8.1) is widely used for the calculation of the electrostatic term (E_{elec}) with the addition of a distance-dependent dielectric constant, although it is a simplification of the physical reality (Leach, 2001).

8.2.5 Van der Waals Interactions

Van der Waals discovered that the electrostatic interactions do not account for all nonbonded interactions occurring in a system (Leach, 2001). Thus, the other interactions occurring are often called *van der Waals* interactions. The van der Waals interaction energy can be regarded as a combination of long-range attractive forces and short-range repulsive forces between nonbonded atoms. The attractive contribution is due to so-called dispersive forces discovered by London (1930) (also sometimes called London forces) in which the correlated motion of electrons around the nuclei results in induced dipole interactions. Van der Waals interactions are often represented by an energy potential as a function of the distance r that includes both the attractive force and the repulsion at close range. The most well-known of these is the Lennard-Jones potential of the form ($n > 6$):

$$V(r) = \frac{A}{r^n} - \frac{B}{r^6}, \tag{8.2}$$

where A and B are van der Waals repulsion and attraction parameters depending on the actual atoms in question and r is the distance between them. The first term gives the repulsions and the second term the attractions. The most common potential uses $n = 12$, which is known as the Lennard-Jones 12–6 (or 6–12) potential. Figure 8.7 illustrates the Lennard-Jones 12–6 potential using $A = 4$ and $B = 4$.

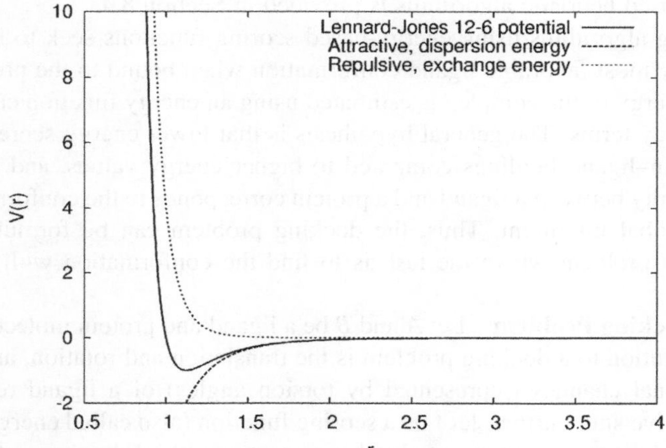

Figure 8.7 Example of Lennard-Jones 12–6 potential using $A = 4$ and $B = 4$.

8.3 THE DOCKING PROBLEM

Protein–ligand docking is a geometric search problem where the task is to find the best ligand conformation within the protein active site or binding pocket. Unfortunately, since the relative orientation of the two molecules (protein and ligand) and their conformations have to be taken into account, docking is not an easy problem. Usually, the protein is treated as fixed in a three-dimensional coordinate system, whereas the ligand is allowed to reposition and rotate resulting in 6 degrees of freedom (3 for translations and 3 for the orientation). If either the protein or the ligand (or both) are allowed to be flexible (e.g., change conformation during the docking), the problem becomes even more difficult. The flexibility is often measured as the number of bonds that are allowed to rotate during the docking run (simulation).

Typically, the docking problem can be classified according to the following categories listed by increasing complexity: (1) both protein and ligand remain rigid (also called rigid-body docking), (2) rigid protein and flexible ligand, (3) flexible protein and rigid ligand, and (4) both protein and ligand are flexible. Category 4 is further divided in two cases: (4.a) partly flexible, protein backbone is rigid, selected side chains are flexible and (4.b) fully flexible.

Docking problems have huge search spaces, and the number of possible conformations to consider increases dramatically when flexibility is taken into account. Unfortunately, the search space is not unimodal but rather highly multimodal due to the numerous local optima caused by the energy function used. Moreover, the complexity of the docking problem is influenced by the size, shape, and bonding topology of the actual ligand being docked. Despite great improvements in computational power, docking remains a very challenging problem, generally considered to be NP-hard (although no formal proof exists). Thus, a brute-force approach looking at all possible docking conformations is impossible for all but the simplest docking problems. To handle docking flexibility in an efficient manner, search heuristics that sample the search space of possible conformations are required. A survey of the most commonly used heuristic algorithms is provided in Section 8.4.

Docking algorithms using energy-based scoring functions seek to identify the energetically most favorable ligand conformation when bound to the protein molecule. The energy of the complex is estimated using an energy function composed of several energy terms. The general hypothesis is that lower energy scores represent better protein–ligand bindings compared to higher energy values, and the highest binding affinity between a ligand and a protein corresponds to the conformation near or at the global minimum. Thus, the docking problem can be formulated as an optimization problem where the task is to find the conformation with the lowest energy:

The Docking Problem Let A and B be a ligand and protein molecule, respectively. A solution to a docking problem is the translation and rotation, and possibly conformational changes (represented by torsion angles) of a ligand relative to a protein's active site. Further, let f be a scoring function (also called energy function) that ranks solutions with respect to binding energy, and let C be the conformational search space of all possible conformations (docking solutions) between A and B. The *docking problem* is a computational problem in which the task is to find an $x \in C$ satisfying

$$f(x) \leq f(y)\ \forall y \in C \text{(minimization problem)}.$$

The next two sections will describe in more detail different ways to (1) score protein–ligand complexes and (2) search for the protein–ligand combination that corresponds to the lowest energy conformation.

8.3.1 Scoring Protein–Ligand Complexes

The binding free energy of a protein–ligand complex is the energetic difference relative to the uncomplexed state and intuitively indicates how much the ligand likes to be situated at the active site of the protein as opposed to be placed somewhere in the solvent (usually aqueous solution) far away from the protein. Theoretically, the free energy of binding is given by the Gibbs free energy according to the laws of thermodynamics in which a negative value indicates that a reaction is being favored. Lower free energy values correspond to more favorable reactions.

Docking algorithms using energy-based scoring functions seek to identify the energetically most favorable ligand conformation when bound to the protein molecule. However, because of high computational costs, most energy functions used in docking programs do not estimate the free energy but merely try to rank the different candidate solutions using a simplified energy function. Thus, the term *scoring function* is often used instead of *energy function*. Generally, the fact that all docking programs use different scoring functions underscores the notion that an optimal scoring function still awaits discovery.

There is often a trade-off between the sophistication of a scoring function in terms of accuracy and its computational cost. Usually, the scoring functions are grouped into the following categories: (1) geometric or steric scoring functions, (2) force-field-based scoring functions, (3) empirical scoring functions, and (4) knowledge-based scoring functions. The geometric scoring functions are typically used for protein–protein docking or for performing fast screening of compound databases containing hundreds of thousands of rigid ligands. However, a major concern is that they may miss important ligands or binding modes because of poor chemical selectivity. The next three sections will focus on the other three types of scoring functions. For a detailed overview of scoring functions used for protein–ligand docking see Muegge and Rarey (2001), Halperin et al. (2002), and Wang and Wang (2003).

8.3.2 Force-Field-Based Scoring Functions

Several empirically based molecular mechanics models have been developed for protein systems, such as CHARMM (Brooks et al., 1983), AMBER (Weiner et al., 1981), GROMOS (Hermans et al., 1984), and OPLSAA (Jorgensen et al., 1996). These models, also known as force fields, are typically expressed as summations of several potential energy components. Usually, a total energy equation (8.3) includes terms for bond stretching (E_{bond}), angle bending (E_{angle}), torsion (E_{tor}), nonbonded (E_{nb}), and coupled (E_{cross}) interactions:

$$E_{energy} = E_{bond} + E_{angle} + E_{tor} + E_{nb} + E_{cross}. \tag{8.3}$$

Force-field scoring is based on the idea of using only the enthalpic contributions to estimate the free energy of the binding. Most often, nonbonded interaction energy terms of standard force fields are used, such as the Lennard-Jones potential describing van der Waals interactions and Coulomb energy describing the electrostatic components of the interactions. The nonbonded interaction energy typically takes the following form:

$$E_{binding} \approx E_{nb} = \sum_{i}^{\text{ligand}} \sum_{j}^{\text{protein}} \left(\frac{A_{ij}}{r_{ij}^{12}} - \frac{B_{ij}}{r_{ij}^{6}} + C\frac{q_i q_j}{\varepsilon \cdot r_{ij}} \right), \tag{8.4}$$

where A_{ij} and B_{ij} are van der Waals repulsion and attraction parameters between two atoms i and j at a distance r_{ij}, q_i and q_j are the point charges on atoms i and j, ε is the medium's permittivity (e.g., liquid water), and C (equal to 332) is a constant factor that converts the electrostatic energy into kilocalories per mole. Additionally, intraligand interactions are added to the score if flexibility is allowed.

A disadvantage of force-field-based scores is that only potential energies rather than free energies are taken into account. Therefore, solvation or entropy terms are sometimes added to provide more accurate energy estimates. Examples of algorithms that use force-field-based scoring functions include DOCK (Kuntz et al., 1982) and old versions of AutoDock (Goodsell and Olson, 1990).

8.3.3 Empirical Scoring Functions

Empirical scoring functions are similar to the force-field-based scoring functions but often include entropic terms, taking loss of conformational degrees of freedom and changes in solvation of the ligand upon binding into account. Moreover, an estimate of the free energies of binding is usually provided.

Empirical scoring functions use several terms that describe properties known to be important in molecular docking when constructing a master equation for the prediction of the binding affinity. These terms generally describe polar interactions, such as hydrogen bonds and ionic interactions, apolar and hydrophobic interactions, loss of ligand flexibility (entropy), and sometimes also desolvation effects. Linear regression analysis is used to derive optimal coefficients to weight each of the terms. The training set used by the linear regression method consists of protein–ligand complexes for which both experimentally determined three-dimensional structures and binding affinities are known. Because the empirical score is an approximation of free energy, lower scores represent greater stability, and the lowest score should ideally correspond to the native cocrystalized conformation.

A disadvantage of the empirical scoring functions is that they will only work well on protein–ligand complexes that are similar to the ones used for the linear regression analysis. Thus, choosing a representative set of protein–ligand complexes for the regression analysis is of great importance. Examples of empirical scoring functions include the AutoDock 3.0 scoring function (Morris et al., 1998), ChemScore (Eldridge et al., 1997), LUDI (Böhm, 1994), and X-CSCORE (Wang and Wang, 2002).

8.3.4 Knowledge-Based Scoring Functions

Knowledge-based scoring functions are based on a statistical analysis of known structures of protein–ligand complexes. The purpose is to establish a score by extracting forces and potentials from a database of known protein–ligand complexes assuming that an experimentally derived complex represents the optimum binding between the protein and the ligand. Usually, complexes determined by X-ray crystallography, such as the ones deposited in the Protein Data Bank (PDB) (Berman et al., 2000) are used.

The structural information gathered from the protein–ligand X-ray coordinates is converted into pair potentials accounting for interaction energies of protein–ligand atom pairs. It is assumed that the more often protein atoms of type i and j are found in a certain distance r, the more favorable this interaction is. Thus, pair potentials $\Delta W_{ij}(r)$ are made for each interaction type between protein atom of type i and a ligand atom of type j in a distance r depending on its frequency:

$$\Delta W_{ij}(r) \propto -\ln \frac{g_{ij}(r)}{g_{ref}}, \tag{8.5}$$

where $g_{ij}(r)$ is the atom pair distribution function for a protein–ligand atom pair ij. The distribution function is calculated from the number of occurrences of that pair ij at a certain distance r in the entire database of complexes used. The term g_{ref} corresponds to a reference distribution incorporating all nonspecific information common to all atom pairs in the protein environment. The final score of the protein–ligand binding is then calculated as the sum of all the interatomic interactions of the protein–ligand complex using the corresponding pair potentials $\Delta W_{ij}(r)$. Examples of knowledge-based scoring functions include the potential of mean force (PMF) function (Muegge and Martin, 1999) and DrugScore (Gohlke et al., 2000).

8.4 PROTEIN–LIGAND DOCKING ALGORITHMS

During the last two decades, numerous docking methods have been introduced. This section provides a brief survey of methods for protein–ligand docking. The main focus is on explaining the algorithms and highlighting their similarities and differences, not to provide a general comparison between them. Several review studies have been published providing more detailed description of each method; see Muegge and Rarey (2001), Morris et al. (2000), Höltje et al. (2003), and Taylor et al. (2002). Furthermore, comparisons have been made between some of the introduced methods; see, for instance, Vieth et al. (1998), Diller and Verlinde (1999), Westhead et al. (1997), and Kontoyianni et al. (2004).

Protein–ligand docking algorithms can be grouped into three main categories: (1) incremental construction algorithms, (2) stochastic algorithms, and (3) molecular dynamics. Other methods have been suggested based on shape matching, distance geometry, or even exhaustive search. However, only methods from the three categories mentioned will be given in the following sections since they represent the majority of docking approaches.

8.4.1 Incremental Construction Algorithms

For *incremental construction algorithms* (also referred to as fragment-based algorithms), the ligand is not docked completely at once but is instead divided into fragments and incrementally reconstructed inside the active site of the protein.

The first program using the incremental construction strategy was DOCK introduced in 1982 by Kuntz et al. DOCK works in the following manner. First, the active site is filled up with overlapping spheres representing potential ligand atoms. The purpose of using spheres is to identify shape characteristics of the active site. Second, the ligand is partitioned into rigid segments (along flexible bonds). Third, an anchor point (found manually or automatically) is chosen from the ligand fragments and positioned inside the active site by matching its atoms with the sphere centers. Fourth, all possible anchor placements are scored using the DOCK scoring function, and the most favorable anchor(s) is used for the subsequent "growth" of the ligand. Finally, the fully placed ligand with the highest score is returned as the docking result.

Another widely used program is FlexX (Rarey et al., 1995, 1996, 1999; Kramer et al., 1999), which is very similar to DOCK. FlexX docks a fragment into the active site followed by remaining fragments being attached. However, FlexX defines interaction sites for each possible interacting group of the ligand and the active site instead of identifying possible ligand atom locations using spheres. The interaction sites are assigned different types (hydrogen bond acceptor, hydrogen bond donor, etc.) and are modeled by an interaction geometry consisting of an interaction center and a spherical surface. The anchor (base) fragment is oriented by searching for positions where interactions between ligand and active site can occur. Afterwards, the remaining fragments are added to the anchor one at a time. At each step of growth, a list of preferred torsional angle values is probed and the most energetic favorable conformation is selected for further "growing" of the ligand. Finally, the fully grown ligand with the highest energy score is selected. FlexX has been extended in several ways to account for receptor flexibility [FlexE (Claussen et al., 2001)] and pharmacophore constrained docking [FlexX-Pharm (Hindle et al., 2002)].

DOCK and FlexX are the most widely used representatives of the fragment-based approach. Other programs using the same strategy, such as Hammerhead (Welch et al., 1996) and Slide ("screening for ligands by induced-fit docking, efficiently") (Schnecke and Kuhn, 2000), differ mainly by the scoring function used.

8.4.2 Stochastic Algorithms

Different variants of *stochastic algorithms* have been applied to the docking problem, such as simulated annealing (SA), tabu search (TS), and evolutionary algorithms (EAs). This section will briefly describe some of the docking methods that use these algorithms.

Several docking methods exist that make use of the Metropolis (Metropolis et al., 1953) or SA (Kirkpatrick et al., 1983) algorithms, such as Affinity (http://www.accelrys.com), QXP (McMartin and Bohacek, 1997), MCDOCK (Liu and Wang, 1999), ICM (Abagyan et al., 1994; Totrov and Abagyan, 1997), and Glide (Eldridge et al., 1997; Friesner et al., 2004). A few examples are described below.

The internal coordinates mechanics (ICM) method by Abagyan et al. (1994; Totrov and Abagyan, 1997) allows for both ligand and protein flexibility. The program uses a traditional Metropolis search algorithm in which the internal variables are subjected to random variation (moves). The internal variables to optimize are: (1) position and orientation of ligand (translation and rotation), (2) ligand torsion angles, and (3) torsion angles of protein active site side-chains. However, the sampling of the side-chain torsion angles is biased according to their frequency of occurrence based on a rotamer library comprising commonly observed side-chain angles. After a random move is made, a local energy minimization is performed and the conformation is accepted or rejected based on the Boltzmann criterion [$e^{|-f(x')-f(x)|/T}$, where x is the old solution, x' is the new solution, and T is a fixed temperature].

Glide (grid-based ligand docking with energetics) (Eldridge et al., 1997; Friesner et al., 2004) uses the Metropolis algorithm. A key feature of Glide is the use of the so-called glide funnel, which is a series of filters that is gradually applied to lower the number of ligand conformations being examined by weeding out inferior ligand conformations.

Another popular docking program is AutoDock (Goodsell and Olson, 1990; Morris et al., 1998; Hart et al., 2000). AutoDock used SA prior to version 3.0, but SA was later superseded by EAs since it was unable to handle ligands with more than eight rotatable torsion angles.

Docking methods using TS typically start with an initial random solution and create new solutions by random moves (similar to random moves in SA). During the optimization process (iterations), a list (the tabu list) is maintained containing the best and most recently visited solutions (conformations). Moves resulting in conformations close to the ones present in the tabu list are ignored, unless they improve the overall best found solution. Using the tabu list scheme makes the algorithm less prone to prematurely converge to local optima. An example of a TS-based docking method is Pro_Leads (Baxter et al., 1998).

Several hybrid algorithms have been introduced. Examples of these are SFDock (Hou et al., 1999), which combines a genetic algorithm (GA) with TS, and the Lamarckian GA (Morris et al., 1998; Hart et al., 2000) provided by AutoDock, which combines a GA with a local search method. Another approach is to apply molecular dynamics to the final solution obtained by the stochastic-based methods to fine-tune the found conformation. The next section gives a brief introduction to the molecular dynamics approach.

8.4.3 Molecular Dynamics

In contrast to the docking algorithms presented in the previous sections, which use a combinatorial approach to sample different conformational "snapshots," *molecular dynamics* (MD) simulates the entire docking process. Thus, the entire process of the ligand binding to the protein's active site is simulated in small time steps by calculating the forces acting on each atom using a force field. Starting from an initial configuration, a sampling of the conformational space of the protein and the ligand is performed using constant temperature (or energy). After each time step new velocities and accelerations are calculated using Newton's laws of motion and the positions of all atoms are updated accordingly. To make the simulation realistic, very small time steps need to be taken, making the entire simulation very time consuming. Although several modifications can be made to lower the total run time, MD is typically not used as a stand-alone method for molecular docking. Another disadvantage of MD is the dependency of good starting conformations in order to obtain good docking results because of the ruggedness (multiple optima) of the energy surface caused by the force field. Often, MD will not be able to cross high-energy conformational barriers making it prone to get stuck in local minima. Despite these deficiencies, MD has successfully been used in molecular docking studies (Given and Gilson, 1998; Mangoni et al., 1999) and is often combined with other algorithms, such as SA or EAs, to fine-tune the final solution obtained. Moreover, different low-energy ensembles of protein active sites can be generated using MD and subsequently be used by faster docking methods to try out alternative protein conformations. For a more detailed introduction to the MD field see Leach (2001).

8.5 EVOLUTIONARY ALGORITHMS

Stochastic search methods such as EAs can be used to sample large search spaces efficiently and are some of the most commonly used methods for flexible ligand docking (Morris et al., 2000). The application of EAs to protein–ligand docking has been tackled in many ways. The main differences can be summarized in the following categories: (1) problem representation (encoding), (2) protein and ligand flexibility, (3) variation operators, (4) hybrid methods, and (5) scoring functions used.

One of the first applications of EAs to molecular docking was introduced by Dixon in 1992 (Dixon, 1993). Dixon used a simple GA with binary encoding representing a matching of ligand atoms and spheres placed inside the active site, similar to DOCK (Kuntz et al., 1982). A binary string was used to represent the torsion angles for the rotatable bonds in the ligand. Unfortunately, the initial study by Dixon did not provide a detailed description of the GA and only limited experimental results were presented. Later on, Oshiro, Kuntz, and Dixon (Oshiro et al., 1995) introduced a similar GA using the DOCK scoring function. Here, an alternative representation was suggested that encoded the relative position of the ligand (translation and rotation) in a bit string, substituting the sphere-matching scheme. Both GA variants were tested on three protein–ligand complexes and compared regarding their ability to screen a database of compounds. The experimental results indicated that the explicit representation outperformed the sphere-matching approach and resulted in overall lower root-mean-square deviation (RMSD) values.

Genetic algorithm for minimization of energy (GAME) by Xiao and Williams (1994a,b) was another early attempt to molecular docking. GAME was designed to dock one or more rigid ligands to a rigid protein simultaneously. Candidate solutions were encoded as binary strings representing three rotational angles and three translational coordinates. The algorithm was tested on one protein–ligand complex trying different search strategies: (1) allowing the ligand to move with fixed rotation, (2) allowing the ligand to rotate with fixed translation, and (3) allowing both translations and rotations thus exploring all 6 degrees of freedom.

Judson et al. (1994, 1995) introduced yet another approach using a so-called pivot atom. The pivot atom served as the origin of translation and rotation of the ligand restraining it to positions near the known crystallographic position. Initially, only a small portion of the ligand (the pivot atom and adjacent atoms) was docked. As the search process proceeded, atoms were gradually added to the initial submolecule until all the ligand atoms were assembled. This so-called growing process is very similar to the fragment-based approach made by the incremental construction methods introduced in Section 8.4.1. The protein remained rigid whereas the ligand was flexible. The ligand position (translation and rotation) was represented by Gray encoding[8] and the variation operators used were one-point crossover and bit-flip mutation. Moreover, the GA used a niching scheme (similar to the classic "island model"; Martin et al., 1997) where all individuals were divided into four subpopulations with occasional migration of the best individual from each subpopulation to the others. Finally, elitism was applied to save the overall best found solution from deletion.[9] The GA was evaluated on 10 protein–ligand complexes using the CHARMM force field as scoring function and was able to obtain good docking conformations resembling known structures.

DIVALI (Clark and Ajay, 1995), short for docking with evolutionary algorithms was introduced by Clark and Jain.[10] Similar to the previously introduced GAs, DIVALI used a Gray-coded binary representation, bit-flip mutation, and an elitism scheme. Two-point crossover was used instead of one-point crossover because it was believed to perform better. The protein remained rigid during the docking process whereas the ligand was flexible. The key feature of DIVALI was the *masking*

[8] Gray decoding ensures that adjacent binary numbers correspond to neighboring integer numbers, thereby lowering the disruptive effects caused by bit-flip mutation.

[9] Elitism is used to keep the k best performing individuals in the population by protecting them from, for example, crossover, mutation, and selection. Without the elitism scheme the best individuals could accidentally be removed from the population forcing the EA to rediscover the solutions once more. Typically, only the best individual is protected from deletion ($k = 1$).

[10] The published paper omitted A. N. Jain's middle- and last name, thus only Ajay occurs in the paper.

operator, which divides the search space into eight parts. In principle, the masking operator is nothing but a simple subpopulation strategy restraining individuals to certain areas in the search space. The masking operator worked as follows: Eight distinct subpopulations were created by fixing the most significant bit in each of the three translational parts of the binary encoding. Thus, the translational search space was divided into eight parts and individuals were assigned to each of these parts according to their binary encoding. Additionally, new individuals were restrained to the area where their parents were residing. For instance, the first bit of the corresponding translation string was changed (masked) if an offspring was accidentally relocated to another area by mutation. DIVALI was evaluated on four protein–ligand complexes using an AMBER-like force field as scoring function and was able to obtain good docking conformations similar to the crystal structures.

Gehlhaar et al. (1995a,b, 1998) developed the EPDOCK algorithm based on evolutionary programming (Fogel, 1962; Fogel et al., 1966). In contrast to many GA-based docking methods, EPDOCK used a real-valued encoding. A self-adaptive scheme was used to evolve the variance used in the Gaussian mutation operator.[11] Moreover, the overall best individual was subjected to conjugate gradient energy minimization. A simple scoring function based on hydrogen bonding and a piecewise-linear potential instead of the usual Lennard-Jones potential was used. EPDOCK was evaluated on several protein–ligand complexes and used to screen a database for promising compounds. In both cases, EPDOCK generally performed well and was shown to be both efficient and accurate.

The docking program GOLD, short for genetic optimization for ligand docking, developed by Jones et al. (1995, 1997, 1999) used a mixed encoding combining binary string and real-valued numbers. Two binary strings were used to encode torsion angles of the ligand and selected side-chain angles of the protein providing partial protein flexibility. In addition, two integer strings were used to represent mappings of hydrogen bonding sites between the ligand and the protein. When evaluating a candidate conformation, least-square fitting was used to place the ligand into the active site, such that as many as possible of the hydrogen bonds defined by the mapping were satisfied. The fitness was defined as the number and strength of the hydrogen bonds plus the van der Waals energy. The variation operators used in GOLD were standard one-point crossover and bit-flip mutation. Furthermore, a niching strategy was used to maintain population diversity. When new candidate solutions were added to the population, they were compared to all individuals in the population. All individuals were divided into distinct niches using the RMSD between all donors and acceptors of two individuals as a similarity measure. If the RMSD between two individuals was less than 1 Å, they were determined to belong to the same niche. The new individual replaced the least fit individual in its niche if more than one was present. Finally, they applied a subpopulation scheme that distributed the individuals into distinct subpopulations with occasional migration. Although no improvements were observed regarding effectiveness, the subpopula-

[11] The Gaussian mutation operator adds a randomly generated vector $m = (m_1, m_2, \ldots, m_n)$ to the genome of the individual. The random numbers are taken from a Gaussian $N(0,\sigma^2)$ distribution.

tion scheme generally improved the efficiency of the algorithm. The method was extensively tested on numerous protein–ligand complexes and obtained good results (Jones et al., 1997).

The family competition EA (FCEA) by Yang and Kao (2000) used several variation operators to balance between global exploration and local exploitation. In short, FCEA used three mutation operators: (1) decreasing-based Gaussian mutation, (2) self-adaptive Gaussian mutation, and (3) self-adaptive Cauchy mutation, and two crossover operators: a modified version of discrete (uniform) recombination and intermediate recombination [see Yang and Kao (2000) for more details]. Each of the mutation operators was applied in turn to the entire population, creating a new population that again was subjected to another mutation operator, and so forth. However, for each mutation operator, each parent in the current population created a number L of offspring by mutation and recombination and only the best solution survived, hence the name family competition. FCEA has been validated on several docking benchmarks with good results. Recently, FCEA has been extended with the variation operator used in the differential evolution (DE) algorithm (Yang et al., 2001) showing good performance, although no comparison was made to the FCEA without the DE operator or DE alone. FCEA has later been modified and renamed to generic evolutionary method for molecular docking (GEMDOCK) and evaluated on a set of 100 benchmark complexes (Yang and Chen, 2004).

AutoDock (Goodsell and Olson, 1990; Rosin et al., 1997) is another popular docking program that originally used SA to search for promising docking conformations. The SA was later superseded by a GA, which was again replaced by a hybrid algorithm combining a GA and a local search method, termed Lamarckian GA (LGA) (Morris et al., 1998; Hart et al., 2000). LGA uses a real-valued representation, two-point crossover, and Cauchy mutation with fixed α and β values (defining the mean and spread of the Cauchy distribution). Each candidate solution is represented by three Cartesian coordinates for the ligand translation, four variables defining a quaternion determining the orientation of the ligand, and one variable for each flexible torsion angle. Further, in each generation a local search method is applied to a subset of randomly chosen individuals with a fixed probability. When local search is applied, the genotype of the individual is replaced with the new best solution found by the local search process, thus mimicking Lamarckian evolution (hence the name of the algorithm). Compared to other EAs, the parameter settings used in AutoDock makes the application of the mutation (using fixed variance) and crossover operators very disruptive similar to random search relying solely on the local search method to fine-tune the solutions.

8.6 EFFECT OF VARIATION OPERATORS

During the last decade, numerous EAs for protein–ligand docking have been introduced using different representations, variation operators, scoring functions, and the like. Apart from a study by Gehlhaar and Fogel (1996) that investigated different tournament schemes, self-adaptive strategies, and the effect of varying the population sizes using the EPDOCK method, few investigations have compared the effectiveness of variation operators for docking.

In AutoDock, the variation operators used by LGA were focused on providing high exploration using one-point crossover and a Cauchy mutation operator followed by a specialized local search method to do the fine-tuning. The local search method uses a lot of "expensive" fitness evaluations to probe the neighborhood for potential search directions. Alternatively, it might be possible to replace the local search method using more advanced EA operator strategies without sacrificing the docking accuracy.

This particular issue was further investigated in a study using the energy function provided by AutoDock (Thomsen, 2003a). Different EA settings, such as variation operators (mutation and recombination), population size, and usage of local-search methods, were investigated. Each EA setting was evaluated using a suite of six docking problems previously used to benchmark the performance of search algorithms provided with the AutoDock program package. The experimental results demonstrated that a good choice of variation operators can lead to substantial improvements in performance without the need of specialized local search techniques. Based on these findings, a simple EA termed DockEA using annealed Gaussian mutation and arithmetic crossover was introduced [see Thomsen (2003a) for more details].

Overall, DockEA outperformed LGA in terms of mean energy score on the two most flexible test cases and obtained the overall best docking solutions compared to the LGA provided with AutoDock. Furthermore, the DockEA proved to be more robust than the LGA (in terms of reproducing the results in several runs) on the more difficult problems with a high number of flexible torsion angles.

Generally, the energy function provided by AutoDock was able to discriminate between "good" and "bad" solutions, making the overall docking simulation successful. However, the correspondence between energy and similarity between found ligand conformations and crystallized ligand structures breaks down in the low-energy regions (see Fig. 8.8). Thus, the energy function was not able to distinguish between near-optimal and nonoptimal solutions in low-energy regions of the energy landscape (search space). These results indicate that using local search to refine the final solution is pointless as long as the solution is located in the low-energy region.

8.7 DIFFERENTIAL EVOLUTION

Differential evolution (DE) was introduced by Storn and Price (1995). Compared to more widely known EA-based techniques (e.g., genetic algorithms, evolutionary programming, and evolution strategies), DE uses a different approach to select and modify candidate solutions (individuals). The main innovative idea in DE is to create offspring from a weighted difference of parent solutions.

The DE works as follows: First, all individuals are initialized and evaluated according to the fitness function (here we consider this as a docking scoring func-

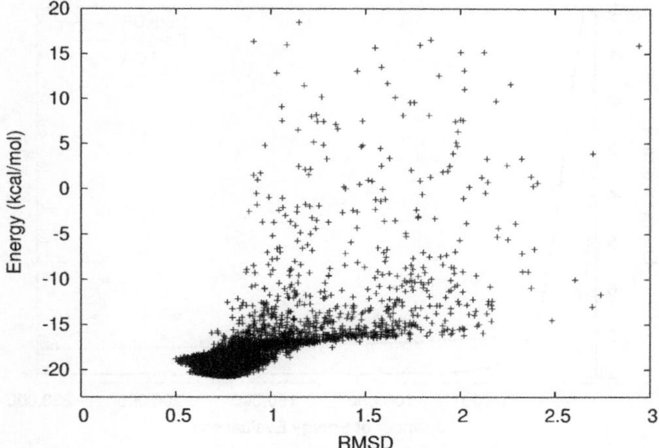

Figure 8.8 Energy and corresponding RMSD of observed ligand conformations of the 1hvr complex during a run of the DockEA.

tion). Afterward, the following process will be executed as long as the termination condition is not fulfilled: For each individual in the population, an offspring is created by adding a weighted difference of the parent solutions, which are randomly selected from the population. Afterward, the offspring replaces the parent, if and only if it has higher fitness. Otherwise, the parent survives and is passed on to the next generation. The termination condition used was to stop the search process when the current number of fitness (energy) evaluations performed exceeded the maximum number of evaluations allowed.

The DE algorithm has recently demonstrated superior performance, outperforming related heuristics, such as GAs and *particle swarm optimization* on both numerical benchmark functions (Vesterstrøm and Thomsen, 2004) and in several real-world applications (Ursem and Vadstrup, 2003; Paterlini and Krink, 2005). Typically, DE has a fast convergence, high accuracy, and good robustness in terms of reproducing the results. These properties are highly desirable, particularly for protein–ligand docking because the energy functions used to evaluate candidate solutions (ligand conformations) are very time consuming.

Based on the great success of DE, a novel application of DE to flexible ligand docking (termed DockDE) was introduced (Thomsen, 2003b). Similar to the DockEA algorithm, DockDE used the energy function provided by AutoDock. The DockDE algorithm was compared with the LGA, and the DockEA previously found to outperform LGA. The comparison was performed on a suite of six commonly used docking benchmark problems.

Overall, the DockDE obtained the best energy for all six docking problems. Furthermore, the DockDE had the best mean value for four of the docking problems compared to the DockEA and the LGA algorithms (Fig. 8.9 shows the result for the 1stp complex). Moreover, the standard deviations were much lower for DockDE

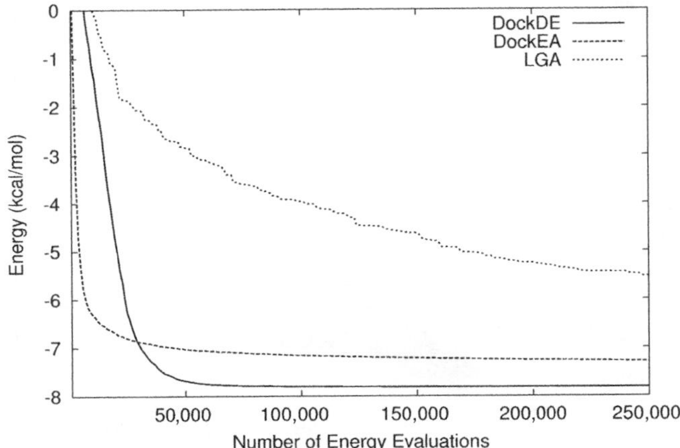

Figure 8.9 Average energy of best observed ligand conformations for the 1stp complex using the DockDE, DockEA, and LGA algorithms (mean of 30 docking runs).

compared to DockEA and LGA indicating the robustness of DockDE (in terms of reproducing the docking results).

In a recent study, a new docking algorithm called MolDock was introduced (Thomsen and Christensen, 2006). MolDock is based on a hybrid search algorithm that combines the DE optimization technique with a cavity prediction algorithm. The use of predicted cavities during the search process allows for a fast and accurate identification of potential binding modes (docking solutions) since predicted conformations (poses) are constrained during the search process. More specifically, if a candidate solution is positioned outside the cavity, it is translated so that a randomly chosen ligand atom will be located within the region spanned by the cavity.

MolDock used a real-valued representation. Thus, a candidate solution was encoded by an array of real-valued numbers representing ligand position, orientation, and conformation as Cartesian coordinates for the ligand translation, four variables specifying the ligand orientation (encoded as a rotation vector and a rotation angle), and one angle for each flexible torsion angle in the ligand (if any).

The docking scoring function of MolDock that was used is based on a *piecewise linear potential* (PLP) introduced by Gehlhaar et al. (1995a,b) and further extended in GEMDOCK by Yang and Chen (2004). In MolDock, the docking scoring function is extended with a new term taking hydrogen bond directionality into account [see Thomsen and Christensen (2006) for more details]. The MolDock algorithm has been shown to yield higher docking accuracy than other state-of-the-art docking products (see Section 8.9 for more details).

One of the reasons DE works so well is that the variation operator exploits the population diversity in the following manner: Initially, when the candidate solutions in the population are randomly generated, the diversity is large. Thus, when offspring are created, the differences between parental solutions are high, resulting in large step sizes being used. As the algorithm converges to better solutions, the population diversity is lowered, and the step sizes used to create offspring are lowered correspondingly. Therefore, by using the differences between other individuals in the population, DE automatically adapts the step sizes used to create offspring as the search process converges toward good solutions.

8.8 EVALUATING DOCKING METHODS

Docking methods are usually evaluated by trying to reproduce the crystal structure of one or more protein–ligand complexes. Thus, to reproduce a known complex using a docking algorithm, the ligand and protein are separated. Afterward, the ligand is randomly translated, reorientated, and conformationally changed. The task is then to redock the ligand back into the active site of the protein. For each docking run, the RMSD of the atomic positions in the found candidate structure and the corresponding atoms in the crystallographic structure is calculated as follows:

$$\mathrm{RMSD} = \sqrt{\frac{\sum_{i=1}^{n}(dx_i^2 + dy_i^2 + dz_i^2)}{n}}, \tag{8.6}$$

where n is the number of heavy atoms in the comparison and dx_i, dy_i, and dz_i are the deviations between the experimental structure and the corresponding coordinates from the predicted structure on Cartesian coordinate i.

Typically, only the ligand is considered in the comparison since the protein normally remains rigid during the entire docking process. Generally, lower RMSD values indicate closer resemblance between observed and predicted structures, and RMSD values below 2 Å usually indicate high agreement with the known complex, whereas values higher than 2 Å indicate poorer correspondence.

An important aspect to consider when using the RMSD measure to benchmark docking algorithms is to take potential structural symmetries into account (e.g., by enumerating all automorphisms of the molecule). Not taking symmetries into account can result in anomalously high RMSD values being reported, which do not reflect the actual performance of the docking algorithm.

As pointed out by Cole et al. (2005) using the RMSD measure as an indication of overall success can be problematic (e.g., small ligands that are randomly placed can result in low RMSD values or docked solutions with RMSD values above 2 Å can still represent good binding modes). The study by Cole et al. (2005) also contains a general discussion on common problems and difficulties when comparing docking programs that can be used as guidance when conducting docking experiments and comparisons.

8.9 COMPARISON BETWEEN DOCKING METHODS

In general, it is a difficult task to compare the performance of different docking programs because they do not use the same energy function or search algorithm. For this reason it is hard to determine if their success or failure was due to the docking algorithm or the energy function (or both). Moreover, when reporting the performance of a docking program, important details are sometimes omitted: (1) the total number of energy evaluations used (or total run time), (2) how the ligands and proteins were prepared (e.g., preoptimized) and randomized (to avoid any bias), (3) if the results presented are the mean of several runs or the best found solution, (4) the number of rotatable angles and which ones were allowed to be flexible (determining the size of the search space), and (5) handling of water and cofactor molecules.

Despite these shortcomings, a comparison is provided below to give an indication of the performance of a DE-based docking method (MolDock) compared with four commercial docking programs that are considered to be state of the art for protein–ligand docking and representing different types of docking algorithms (incremental search, simulated annealing, and genetic algorithms).

To evaluate the docking accuracy of MolDock, a selection of 77 protein–ligand complexes from the publicly available GOLD data set (Nissink et al., 2002) was selected. This data set represents a diverse set of druglike ligands containing between 6 and 48 heavy atoms and between 0 and 15 rotatable bonds and have previously been used to evaluate the docking programs Surflex and Glide (Jain, 2003; Friesner et al., 2004). Further, to avoid any bias in the optimization process, the ligands were energy minimized and randomized beforehand.

For each benchmark complex, 10 independent runs with the MolDock algorithm were conducted, and the highest ranked solution was compared with the known experimental structure using the standard RMSD measure (between similar atoms in the solution found and the experimental structure). A solution was considered successful if the RMSD between the pose and the experimentally known ligand was less than 2.0 Å. The same RMSD threshold was used to estimate the accuracy of Glide, Surflex, FlexX, and GOLD.

The experimental results were compared with published results from Glide, GOLD, FlexX, and Surflex. Table 8.1 provides a summary of the docking accuracy for each of the docking programs. The RMSD results for Glide, Surflex, FlexX, and GOLD are adopted from previously published comparisons (Jain, 2003; Friesner et al., 2004).

The docking accuracy of MolDock on the 77 complexes is 87%, outperforming the other docking programs on the benchmark data set. The detailed results are further described in the article by Thomsen and Christensen (2006). In conclusion, the results strongly suggest that the DE algorithm has a great potential for protein–ligand docking.

8.10 SUMMARY

Molecular docking is a topic of great importance because it aids to the understanding of how molecules interact and can be used to design new drugs. Docking tools are valuable to pharmaceutical companies because they can be used to suggest promising drug candidates at a low cost compared to making real-world experiments. As a result of the need for methods that can predict the binding of two molecules when complexed, numerous algorithms have been suggested during the last three decades.

Table 8.1 Accuracy of Selected Docking Programs

Docking Program	Accuracy (%)
MolDock	87
Glide	82
Surflex	75
GOLD[a]	78
FlexX[b]	58

[a]The docking accuracy is based on 55 out of the 77 complexes.
[b]The docking accuracy is based on 76 out of the 77 complexes.

In this chapter, a review of the most representative algorithms for protein–ligand docking has been presented. In particular, docking methods utilizing differential evolution have demonstrated to have superior performance to other well-studied approaches regarding accuracy and speed, issues that are important when docking hundreds of thousands of ligands in virtual screening experiments.

8.11 FUTURE RESEARCH TOPICS

Several important research topics within the molecular docking field call for further investigation. First, incorporating flexible side chains at the active site is a topic of current research and extending existing docking methods in that respect is important when trying to dock molecules to proteins where part of the receptor needs to be flexible in order to allow access (Najmanovich et al., 2000; Österberg et al., 2002; Cavasotto and Abagyan, 2004). Second, solvent modeling is important when trying to compare dissimilar molecules based on their binding energies since the inclusion of solvation terms has shown to improve the overall ranking of known ligands (Shoichet et al., 1999). Taking structural water molecules into account in the docking simulation can further improve the docking accuracy (Thomsen and Christensen, 2006). Therefore, including models for implicit and explicit water would be worth looking into. Finally, a better estimation of energy and binding affinity is one of the major determinants for increasing performance in virtual screening of compounds using computational docking methods, and further research in developing and testing energy and binding affinity scoring functions is of great importance.

REFERENCES

ABAGYAN, R., M. N. TOTROV, and D. A. KUZNETSOV (1994). "ICM: A new method for structure modeling and design: Applications to docking and structure prediction from the distorted native conformation," *J. Comput. Chem.*, Vol. 15, pp. 488–506.

ALBERTS, B., A. JOHNSON, J. LEWIS, M. RAFF, K. ROBERTS, and P. WALTER (2002). *Molecular Biology of the Cell*, 4th ed. Garland Science, New York.

BAXTER, C. A., C. W. MURRAY, D. E. CLARK, D. R. WESTHEAD, and M. D. ELDRIDGE (1998). "Flexible docking using tabu search and an empirical estimate of binding affinity," *Proteins*, Vol. 33, pp. 367–382.

BERMAN, H. M., J. WESTBROOK, Z. FENG, G. GILLILAND, T. N. BHAT, H. WEISSIG, I. N. SHINDYALOV, and P. E. BOURNE (2000). "The Protein Data Bank," *Nucleic Acids Res.*, Vol. 28, pp. 235–242.

BÖHM, H. J. (1994). "The development of a simple empirical scoring function to estimate the binding constant for a protein-ligand complex of known three-dimensional structure," *J. Comput. Aid. Mol. Des.*, Vol. 8, pp. 243–256.

BRANDEN, C. and J. TOOZE (1998). *Introduction to Protein Structure*, 2nd ed. Garland, New York.

BROOKS, B. R., R. E. BRUCCOLERI, B. D. OLAFSON, D. J. STATES, S. SWAMINATHAN, and M. KARPLUS (1983). "CHARMM: A program for macromolecular energy, minimization, and dynamics calculations," *J. Comput. Chem.*, Vol. 4, pp. 187–217.

CAVASOTTO, C. N and R. A. ABAGYAN (2004). "Protein flexibility in ligand docking and virtual screening to protein kinases," *J. Mol. Biol.*, Vol. 337, pp. 209–225.

CLARK, K. P. and A. N. JAIN (1995). "Flexible ligand docking without parameter adjustment across four ligand-receptor complexes," *J. Comput. Chem.*, Vol. 16, pp. 1210–1226.

CLAUSSEN, H., C. BUNING, M. RAREY, and T. LENGAUER (2001). "FlexE: Efficient molecular docking considering protein structure variations," *J. Mol. Biol.*, Vol. 308, pp. 377–395.

COLE, C. J., C. W. MURRAY, J. W. M. NISSINK, R. D. TAYLOR, and R. TAYLOR (2005). "Comparing protein-ligand docking programs is difficult," *Proteins*, Vol. 60, pp. 325–332.

COTTERILL, R. (2002). *Biophysics: An Introduction*. Wiley Ltd, West Sussex.

CREIGHTON, T. E. (1997). *Proteins: Structures and Molecular Properties*, 2nd ed. W. H. Freeman, New York.

DILLER, D. J. and C. L. M. J. VERLINDE (1999). "A critical evaluation of several global optimization algorithms for the purpose of molecular docking," *J. Comput. Chem.*, Vol. 20, pp. 1740–1751.

DIXON, J. S. (1993). "Flexible docking of ligands to receptor sites using genetic algorithms," Proceedings of the 9th European Symposium on Structure-Activity Relationships: QSAR and Molecular Modelling, pp. 412–413.

ELDRIDGE, M. D., C. W. MURRAY, T. R. AUTON, G. V. PAOLINI, and R. P. MEE (1997). "Empirical scoring functions. I. The development of a fast, fully empirical scoring function to estimate the binding affinity of ligands in receptor complexes," *J. Comput. Aid. Mol. Des.*, Vol. 11, pp. 425–445.

FERSHT, A. (2000). *Structure and Mechanism in Protein Science: A Guide to Enzyme Catalysis and Protein Folding*, 3rd ed. W. H. Freeman, New York.

FOGEL, L. J. (1962). "Autonomous automata," *Indus. Res.*, Vol. 4, pp. 14–19.

FOGEL, L. J., A. J. OWENS, and M. J. WALSH (1966). *Artificial Intelligence through Simulated Evolution*. Wiley, New York.

FRIESNER, A. R., J. L. BANKS, R. B. MURPHY, T. A. HALGREN, J. J. KLICIC, D. T. MAINZ, M. P. REPASKY, E. H. KNOLL, M. SHELLEY, J. K. PERRY, D. E. SHAW, P. FRANCIS, and P. S. SHENKIN (2004). "Glide: A new approach for rapid, accurate docking and scoring. 1. Method and assessment of docking accuracy," *J. Med. Chem.*, Vol. 47, pp. 1739–1749.

GEHLHAAR, D. K., D. BOUZIDA, and P. A. REJTO (1998). "Fully automated and rapid flexible docking of inhibitors covalently bound to serine proteases," Proceedings of the Seventh International Conference on Evolutionary Programming, pp. 449–461.

GEHLHAAR, D. K. and D. B. FOGEL (1996). "Tuning evolutionary programming for conformationally flexible molecular docking," Proceedings of the Fifth Annual Conference on Evolutionary Programming, pp. 419–429.

GEHLHAAR, D. K., G. VERKHIVKER, P. A. REJTO, D. B. FOGEL, L. J. FOGEL, and S. T. FREER (1995a). "Docking conformationally flexible small molecules into a protein binding site through evolutionary programming," Proceedings of the Fourth International Conference on Evolutionary Programming, pp. 615–627.

GEHLHAAR, D. K., G. M. VERKHIVKER, P. A. REJTO, C. J. SHERMAN, D. B. FOGEL, L. J. FOGEL and S. T. FREER (1995b). "Molecular recognition of the inhibitor AG-1343 by HIV-1 protease: Conformationally flexible docking by evolutionary programming," *Chem. Biol.*, Vol. 2, pp. 317–324.

GIVEN, J. A. and M. K. GILSON (1998). "A hierarchical method for generating low-energy conformers of a protein-ligand complex," *Proteins*, Vol. 33, pp. 475–495.

GOHLKE, H., M. HENDLICH, and G. KLEBE (2000). "Knowledge based scoring function to predict protein-ligand interactions," *J. Mol. Biol.*, Vol. 295, pp. 337–356.

Goodsell, D. S. and A. J. Olson (1990). "Automated docking of substrates to proteins by simulated annealing," *Proteins*, Vol. 8, pp. 195–202.

Griffiths, A. J. F., J. H. Miller, D. T. Suzuki, R. C. Lewontin, and W. M. Gelbart (1999). *An Introduction to Genetic Analysis*, 7th ed. W. H. Freeman and Company, New York.

Halperin, I., B. Ma, H. Wolfson, and R. Nussinov (2002). "Principles of docking: An overview of search algorithms and a guide to scoring functions," *Proteins*, Vol. 47, pp. 409–443.

Hart, W. E., C. Rosin, R. K. Belew, and G. M. Morris (2000). "Improved evolutionary hybrids for flexible ligand docking in AutoDock," *Optim. Comput. Chem. Mol. Biol.*, pp. 209–230.

Hermans, J., H. J. C. Berendsen, W. F. van Gunsteren, and J. P. M. Postma (1984). "A consistent empirical potential for water-protein interactions," *Biopolymers*, Vol. 23, pp. 1513–1518.

Hindle, S., M. Rarey, C. Buning, and T. Lengauer (2002). "Flexible docking under pharmacophore type constraints," *J. Comput. Aid. Mol. Des.*, Vol. 16, pp. 129–149.

Höltje, H.-D., W. Sippl, D. Rognan, and G. Folkers (2003). "Protein-based virtual screening," in *Molecular Modeling: Basic Principles and Applications*, 2nd ed. Wiley-VCH, Weinheim, pp. 145–168.

Hou, T., J. Wang, L. Chen, and X. Xu (1999). "Automated docking of peptides and proteins by using a genetic algorithm combined with a tabu search," *Protein Eng.*, Vol. 12, pp. 639–647.

Jain, A. N. (2003). "Surflex: Fully automatic molecular docking using a molecular similarity based search engine," *J. Med. Chem.*, Vol. 46, pp. 499–511.

Jones, G., P. Willett, and R. C. Glen (1995). "Molecular recognition of receptor sites using a genetic algorithm with a description of desolvation," *J. Mol. Biol.*, Vol. 245, pp. 43–53.

Jones, G., P. Willett, R. C. Glen, A. R. Leach, and R. Taylor (1997). "Development and validation of a genetic algorithm for flexible docking," *J. Mol. Biol.*, Vol. 267, pp. 727–748.

Jones, G., P. Willett, R. C. Glen, A. R. Leach, and R. Taylor (1999). "Further development of a genetic algorithm for ligand docking and its application to screening combinatorial libraries," in *Rational Drug Design: Novel Methodology and Practical Applications*, Parrill and Reddy, Eds. *J. Am. Chem. Soc.*, pp. 271–291.

Jorgensen, W. L., D. S. Maxwell, and J. Tirado-Rives (1996). "Development and testing of the OPLS all-atom force field on conformational energetics and properties of organic liquids," *J. Am. Chem. Soc.*, Vol. 118, pp. 11225–11236.

Judson, R. S., E. P. Jaeger, and A. M. Treasurywala (1994). "A genetic algorithm based method for docking flexible molecules," *J. Mol. Struct.*, Vol. 308, pp. 191–306.

Judson, R. S., Y. T. Tan, E. Mori, C. Melius, E. P. Jaeger, A. M. Treasurywala, and A. Mathiowetz (1995). "Docking flexible molecules: A case study of three proteins," *J. Comput. Chem.*, Vol. 16, pp. 1405–1419.

Kirkpatrick, S., C. D. Gelatt Jr., and M. P. Vecchi (1983). "Optimization by simulated annealing," *Science*, Vol. 220, pp. 671–680.

Kontoyianni, M., L. M. McClellan, and G. S. Sokol (2004). "Evaluation of docking performance: Comparative data on docking algorithms," *J. Med. Chem.*, Vol. 47, pp. 558–565.

Kramer, B., M. Rarey, and T. Lengauer (1999). "Evaluation of the FlexX incremental construction algorithm for protein-ligand docking," *Proteins*, Vol. 37, pp. 228–241.

Kuntz, I. D., J. M. Blaney, S. J. Oatley, R. Langridge, and T. E. Ferrin (1982). "A geometric approach to macromolecule-ligand interactions," *J. Mol. Biol.*, Vol. 161, pp. 269–288.

Leach, A. R. (2001). *Molecular Modelling: Principles and Applications*, 2nd ed. Pearson Education, Essex, UK.

Liu, M. and S. Wang (1999). "MCDOCK: A Monte Carlo simulation approach to the molecular docking problem," *J. Comput. Aid. Mol. Des.*, Vol. 13, pp. 435–451.

London, F. (1930). "Zur Theori und Systematik der Molekularkräfte," *Z. Phys.*, Vol. 63, pp. 245–279.

Mangoni, R., D. Roccatano, and A. D. Nola (1999). "Docking of flexible ligands to flexible receptors in solution by molecular dynamics simulation," *Proteins*, Vol. 35, pp. 153–162.

Martin, W. N., J. Lienig, and J. P. Cohoon (1997). "Island (migration) models: Evolutionary algorithms based on punctuated equilibria," in T. Bäck, D. B. Fogel, and Z. Michalewicz, Eds., *Handbook on Evolutionary Computation*. IOP Publishing and Oxford University Press, Oxford.

McMartin, C. and R. S. Bohacek (1997). "QXP: Powerful, rapid computer algorithms for structure-based drug design," *J. Comput. Aid. Mol. Des.*, Vol. 11, pp. 333–344.

Metropolis, N., A. Rosenbluth, M. Rosenbluth, A. Teller, and E. Teller (1953). "Equation of state calculations by fast computing machines," *J. Chem. Phys.*, Vol. 21, pp. 1087–1092.

Morris, G. M., D. S. Goodsell, R. S. Halliday, R. Huey, W. E. Hart, R. K. Belew, and A. J. Olson, (1998). "Automated docking using a Lamarckian genetic algorithm and an empirical binding free energy function," *J. Comput. Chem.*, Vol. 19, pp. 1639–1662.

Morris, G. M., A. J. Olson, and D. S. Goodsell (2000). "Protein-ligand docking," in *Evolutionary Algorithms in Molecular Design*, Vol. 8. Clark Ed. Wiley-VCH, Weinheim, chapter 3.

Muegge, I. and Y. C. Martin (1999). "A general and fast scoring function for protein-ligand interactions: A simplified potential approach," *J. Med. Chem.*, Vol. 42, pp. 791–804.

Muegge, I. and M. Rarey (2001). "Small molecule docking and scoring," in *Reviews in Computational Chemistry*, Vol. 17. Lipkowitz and Boyd, Eds. Wiley-VCH, New York, pp. 1–60.

Najmanovich, R., J. Kuttner, V. Sobolev, and M. Edelman (2000). "Side-chain flexibility in proteins upon ligand binding," *Proteins*, Vol. 39, pp. 261–268.

Nissink, J. W. M., C. Murray, M. Hartshorn, M. L. Verdonk, J. C. Cole, and R. Taylor (2002). "A new test set for validating predictions of protein-ligand interaction," *Proteins*, Vol. 49, pp. 457–471.

Oshiro, C. M., I. D. Kuntz, and J. S. Dixon (1995). "Flexible ligand docking using a genetic algorithm," *J. Comput. Aid. Mol. Des.*, Vol. 9, pp. 113–130.

Österberg, F., G. M. Morris, M. F. Sanner, A. J. Olson, and D. S. Goodsell (2002). "Automated docking to multiple target structures: Incorporation of protein mobility and structural water heterogeneity in AutoDock," *Proteins*, Vol. 46, pp. 34–40.

Paterlini, S. and T. Krink (2005). "Differential evolution and particle swarm optimization in partitional clustering," *Comput. Stat. Data. An.* Vol. 50, pp. 1220–1247.

Rarey, M., B. Kramer, and T. Lengauer (1995). "Time-efficient docking of flexible ligands into active sites of proteins," in *Proceedings of the Third International Conference on Intelligent Systems for Molecular Biology*, Rawlings et al., Eds. AAAI Press, Menlo Park, CA, pp. 300–308.

Rarey, M., B. Kramer, and T. Lengauer (1999). "Docking of hydrophobic ligands with interaction-based matching algorithms," *Bioinformatics*, Vol. 15, pp. 243–250.

Rarey, M., B. Kramer, T. Lengauer, and G. Klebe (1996). "A fast flexible docking method using an incremental construction algorithm," *J. Mol. Biol.*, Vol. 261, pp. 470–489.

Rosin, C. D., R. S. Halliday, W. E. Hart, and R. K. Belew (1997). "A comparison of global and local search methods in drug docking," in *Proceedings of the Seventh International Conference on Genetic Algorithms*, Bäck, Ed. Morgan Kaufmann, San Francisco, pp. 221–228.

Schnecke, V. and L. A. Kuhn (2000). "Virtual screening with solvation and ligand-induced complementarity," *Perspect. Drug Discov.*, Vol. 20, pp. 171–190.

Shoichet, B. K., A. R. Leach, and I. D. Kuntz (1999). "Ligand solvation in molecular docking," *Proteins*, Vol. 34, pp. 4–16.

Storn, R. and K. Price (1995). "Differential evolution—A simple and efficient adaptive scheme for global optimization over continuous spaces," Technical Report, International Computer Science Institute, Berkley, CA.

Taylor, R. D., P. J. Jewsbury, and J. W. Essex (2002). "A review of protein-small molecule docking methods," *J. Comput. Aid. Mol. Des.*, Vol. 16, pp. 151–166.

Thomsen, R. (2003a). "Flexible ligand docking using evolutionary algorithms: Investigating the effects of variation operators and local search hybrids," *BioSystems*, Vol. 72, pp. 57–73.

Thomsen, R. (2003b). "Flexible ligand docking using differential evolution," in *Proceedings of the 2003 Congress on Evolutionary Computation*. IEEE Press, Piscataway, NJ, pp. 2354–2361.

Thomsen, R. and M. H. Christensen (2006). "MolDock: A new technique for high-accuracy molecular docking," *J. Med. Chem.*, Vol. 49, pp. 3315–3321.

Totrov, M. and R. Abagyan (1997). "Flexible protein-ligand docking by global energy optimization in internal coordinates," *Proteins*, Vol. 1, pp. 215–220.

Ursem, R. K. and P. Vadstrup (2003). "Parameter identification of induction motors using differential evolution," *Proceedings of the 2003 Congress on Evolutionary Computation*. IEEE Press, Piscataway, NJ, pp. 790–796.

VESTERSTRØM, J. and R. THOMSEN (2004). "A comparative study of differential evolution, particle swarm optimization, and evolutionary algorithms on numerical benchmark problems," in *Proceedings of the 2004 Congress on Evolutionary Computation*. IEEE Press, Piscataway, NJ, pp. 1980–1987.

VIETH, M., J. D. HIRST, B. N. DOMINY, H. DAIGLER, and C. L. BROOKS III (1998). "Assessing search strategies for flexible docking," *J. Comput. Chem.*, Vol. 19, pp. 1623–1631.

WANG, R., L. LAI, and S. WANG (2002). "Further development and validation of empirical scoring functions for structure-based binding affinity prediction," *J. Comput. Aid. Mol. Des.*, Vol. 16, pp. 11–26.

WANG, R., Y. LU, and S. WANG (2003). "Comparative evaluation of 11 scoring functions for molecular docking," *J. Med. Chem.*, Vol. 46, pp. 2287–2303.

WEINER, P. K. and P. A. KOLLMAN (1981). "AMBER: Assisted model building with energy refinement. A general program for modeling molecules and their interactions," *J. Comput. Chem.*, Vol. 2, pp. 287–303.

WELCH, W., J. RUPPERT, and A. N. JAIN (1996). "Hammerhead: Fast, fully automated docking of flexible ligands to protein binding sites," *J. Chem. Biol.*, Vol. 3, pp. 449–462.

WESTHEAD, D. R., D. E. CLARK, and C. W. MURRAY (1997). "A comparison of heuristic search algorithms for molecular docking," *J. Comput. Aid. Mol. Des.*, Vol. 11, pp. 209–228.

XIAO, Y. L. and D. E. WILLIAMS (1994a). "A comparison of GA and RSNR docking," in *Proceedings of the First IEEE Conference on Evolutionary Computation*. IEEE Press, Piscataway, NJ, pp. 802–806.

XIAO, Y. L. and D. E. WILLIAMS (1994b). "Genetic algorithms for docking of actinomycin D and deoxyguanosine molecules with comparison to the crystal structure of actinomycin D-deoxyguanosine complex," *J. Phys. Chem.*, Vol. 98, pp. 7191–7200.

YANG, J-M. and C.-C. CHEN (2004). "GEMDOCK: A generic evolutionary method for molecular docking," *Proteins*, Vol. 55, pp. 288–304.

YANG, J.-M., J.-T. HORNG, and C.-Y. KAO (2001). "Integrating adaptive mutations and family competition with differential evolution for flexible ligand docking," in *Proceedings of the 2001 Congress on Evolutionary Computation*. IEEE Press, Piscataway, NJ, pp. 473–480.

YANG, J.-M. and C.-Y. KAO (2000). "Flexible ligand docking using a robust evolutionary algorithm," *J. Comput. Chem.*, Vol. 21, pp. 988–998.

Chapter 9

RNA Secondary Structure Prediction Employing Evolutionary Algorithms

Kay C. Wiese, Alain A. Deschênes, and Andrew G. Hendriks

9.1 INTRODUCTION

Ribonucleic acid (RNA) is an important biological molecule. It plays key roles in the synthesis of protein from deoxyribonucleic acid (DNA). It is also known for its structural and catalytic roles in the cell (Higgs, 2000). For the purpose of structure prediction, it can be simply described as a flexible single-stranded biopolymer. The biopolymer is made from a sequence of four different nucleotides: adenine (A), cytosine (C), guanine (G), and uracil (U). Intramolecular base pairs can form between different nucleotides, folding the sequence onto itself. The most stable and common of these base pairs are GC, AU, and GU, and their mirrors, CG, UA, and UG. These pairs are called canonical base pairs. A more elaborate description of RNA can be found in our previous research (Wiese and Glen, 2002, 2003; Wiese et al., 2003).

Briefly, the secondary structure of RNA is described by a list of base pairs formed from the primary sequence. When pairs form, the RNA strand folds onto itself forming the secondary structure. Figure 9.1 illustrates an example of a secondary structure. Secondary structure elements include *hairpin loops*, which contain one base pair, *internal loops*, which contain two base pairs, and *bulges*, which contain two base pairs with one base from each of its pairs adjacent in the backbone of the molecule. There are also *multibranched* loops, which contain more than two base pairs, and *external bases*, which are not contained in any loop.

RnaPredict (Wiese and Glen, 2002), the evolutionary algorithm (EA) discussed in this chapter, considers stacked pairs, also called a helix, only when three or more adjacent pairs form. In addition, the loop connecting a set of stacked pairs must be

Computational Intelligence in Bioinformatics. Edited by G. B. Fogel, D. Corne, and Y. Pan

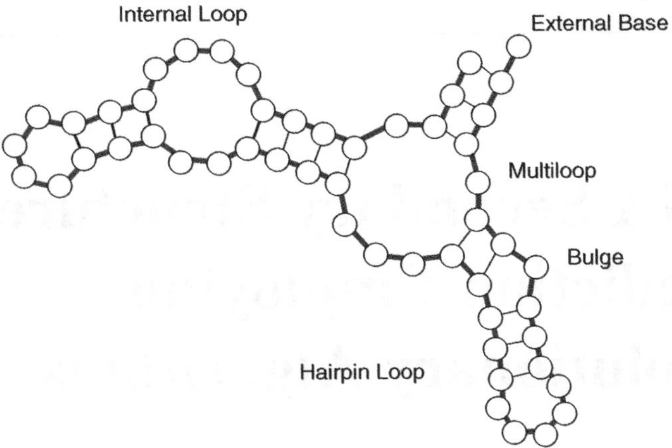

Figure 9.1 Example of RNA secondary structure.

no shorter than three nucleotides in length. Given this model, it is possible to enumerate all the potential helices that can form in a structure. The challenge is in predicting which ones will actually form in the native structure. Determining the secondary structure of RNA through laboratory methods, such as nuclear magnetic resonance (NMR) and X-ray crystallography, is challenging and expensive. Hence, computational prediction methods have been proposed as an alternate approach to this problem.

Evolutionary algorithms (Bäck, 1996) were applied to the field of RNA secondary structure prediction starting in the early 1990s (Shapiro and Navetta, 1994). Since then, there have been many advances in RNA secondary structure prediction using EAs. This type of algorithm was seen as being a very good candidate for deployment on massively parallel supercomputers. EAs can easily be parallelized by dividing the population into *n* subpopulations, one subpopulation for each processor. Shapiro and Navetta (1994) implemented simple operators to test the viability of the EA technique. The parallel computer could compute as many as 16,384 RNA secondary structures at each generation on as many processors.

Shapiro and Wu (1996) modified the algorithm by introducing an annealing mutation operator. This operator controls the probability of mutation by decreasing it linearly at each generation. This new annealing mutation operator decreases the time needed for the EA to converge and produce better results. Van Batenburg et al. (1995; Gultyaev et al., 1995) proposed a modification that allowed for simulation of the folding pathway. The premise was that RNA secondary structure was influenced by kinetic processes. A method used to simulate kinetic folding was to restrict folding to a small part of the strand where this part's size was increased after each iteration. By using this method, they simulated folding of the RNA strand during synthesis. This also helps probe local energy minima. The implementation was done in APL using bit strings to represent the genotype (van Batenburg et al.,

1995). Results showed structures that were more consistent with phylogenetic data than with previous minimum energy solutions.

Around the same time, Benedetti and Morosetti (1995) also compared the accuracy of an EA against known RNA structures with the objective of finding optimal and suboptimal (free energy) structures that were similar. They noted that the shortcomings using the EA were not due to the algorithm itself but rather to the inadequate understanding of the thermodynamic models that influence folding.

Recently, Shapiro et al. (2001a, 2001b) also modified their EA to study folding pathways using a massively parallel genetic algorithm. A completely different approach was taken by Chen et al. (2000). Their method involves using an EA with a thermodynamic fitness function on each sequence of a related set until a certain level of stability is reached. Then, for each structure, a measure reflecting the conservation of structural features among sequences is calculated. Next, an EA is run on the structures where the fitness criterion is the measure of conservation of structural features. The resulting structures are ranked according to a measure of conservation.

Additional related research on EAs and RNA structure includes Chen and Dill (2000), Currey and Shapiro (1997), Fogel et al. (2002), and Titov et al. (2002). The enhancements of secondary structure prediction of the poliovirus 5′ noncoding region by a genetic algorithm were reviewed in Currey and Shapiro (1997). An interesting study on RNA folding energy landscapes was presented in Chen and Dill (2000); it indicated that folding of RNA secondary structures may involve complex intermediate states and rugged energy landscapes. A fast genetic algorithm for RNA secondary structure prediction and subsequent structure analysis, GArna, was introduced in Titov et al. (2002). This algorithm was used for fast calculations of RNA secondary structure; its results were used to interpret how secondary RNA structure influences translation initiation and expression regulation. An evolutionary computation-based method to search across multiple sequences for common elements in possible RNA secondary structures was described in Fogel et al. (2002).

Wiese and Glen (2002) designed a serial EA, RnaPredict, which encodes RNA secondary structures in permutations. A study was performed in Wiese and Glen (2003) that demonstrated the superiority of permutation-based over binary-based representation and studied the effects of various crossover operators and selection strategies such as standard roulette wheel selection (STDS) and keep-best reproduction (KBR) (Wiese and Goodwin, 2001).

A study with improved thermodynamic models was conducted (Deschênes and Wiese, 2004); testing with 3 RNA sequences demonstrated the improved prediction accuracy with the new thermodynamic models. The quality of the predictions was compared to the quality of the predictions from the Nussinov dynamic programming algorithm (DPA) in (Deschênes et al. (2004). RnaPredict was parallelized via a coarse-grained distributed EA for RNA secondary structure prediction (Wiese and Hendriks, 2006); this parallel version was named *P*-RnaPredict.

The focus of this chapter is a comprehensive analysis of the stacking-energy-based thermodynamic models in RnaPredict, and RnaPredict's performance in comparison to known structures and mfold.

The first application of dynamic programming (DP) to RNA secondary structure prediction was developed by Nussinov et al. (1978) and functioned by maximizing the number of base pairs in a predicted structure. Later, Zuker (2003) developed an alternate approach, called mfold. mfold is a dynamic programming algorithm (DPA) applied to the prediction of secondary structures of RNA (Zuker, 1989, 1994, 2000, 2003; Zuker and Stiegler, 1981; Zuker et al., 1999). The software uses a complex thermodynamic model for free energy evaluation of structures. The DPA, along with this model, attempts to find the minimum energy structure. Advances in the algorithm also allow the prediction of suboptimal structures that are important since the native fold does not always correspond to the global minimum (Jaeger et al., 1989; Mathews et al., 1998; Zuker et al., 1991).

The current most complete nearest-neighbor thermodynamic model is incorporated into mfold. The mfold package consists of a group of programs that are written in FORTRAN, C, and C++ tied together with BASH and Perl scripts. The original version was designed to run in the Unix environment. The software was also ported to C++ targeted for the Microsoft Windows platform under the name RNAStructure (Mathews et al., 2004). The latter implementation offers a point-and-click interface.

This chapter will discuss the following topics:

- It will present an investigation of the performance of the stacking-energy-based thermodynamic models, individual nearest-neighbor model (INN) (Freier et al., 1986; Serra and Turner, 1995) and individual nearest-neighbor hydrogen bond model (INN-HB) (Xia et al. 1998) by evaluating the statistical correlation between the free energy in a structure and the number of true positive base pairs.
- It will discuss the prediction performance of RnaPredict, an evolutionary algorithm for RNA secondary structure prediction, by measuring the accuracy of the prediction of nine structures from three RNA classes compared to known structures.
- It will compare the accuracy of predicted structures from RnaPredict with those predicted by the mfold DPA. The prediction accuracy is evaluated via sensitivity, specificity, and *F*-measure.

First, we describe the thermodynamic models in Section 9.2. Section 9.3 describes the evolutionary and dynamic programming algorithms. The experiment designs and their results are reviewed in Section 9.4. Finally, Section 9.5 comments on the results, summarizes the chapter, and offers a brief summary of future work.

9.2 THERMODYNAMIC MODELS

Most prediction methods for RNA secondary structure use Gibbs free energy as their metric. Simply, Gibbs free energy (ΔG) is energy available to do useful work. Differences in ΔG, in a reaction or a conformation change, provide information on process spontaneity. A negative free energy difference in a reaction favors the products and is spontaneous in that direction while a positive free energy difference favors the reactants. Free energy is often represented as a function of enthalpy ΔH (the amount of energy possessed by a thermodynamic system for transfer between itself and the environment), temperature T (a measure of the average kinetic energy in a system), and entropy ΔS (the quantitative measure of the relative disorder of a system):

$$\Delta G = \Delta H - T\,\Delta S. \tag{9.1}$$

It is important to note that differences in free energy between two structures will also dictate the relative amounts at equilibrium.

The term K is the system's equilibrium constant:

$$K = \frac{[C_1]}{[C_2]} = e^{-\Delta G/(RT)}, \tag{9.2}$$

where C_1 and C_2 represent the concentration of two different structures in equilibrium in a system, R is the gas constant, and T is the absolute temperature of the system. The equation simply shows that the concentration ratio in an equilibrium varies exponentially with free energy. The ratio follows an exponential curve where small differences in free energy have large effects on the relative concentration between two conformations.

This is important for RNA because its energy surface is not simple or smooth. The resulting surface looks very rough with many local extrema. Changing a few base pairs can greatly impact the overall secondary structure elements. Therefore, special optimization techniques using good thermodynamic parameters are needed for secondary structure prediction.

Secondary structural elements, and the sequences that form them, account for the bulk of the structure's free energy. Because RNA secondary structure is a list of base pairs, evaluating structures can be done by using different thermodynamic rules on these base pairs or sets of adjacent base pairs. More advanced thermodynamic models also account for destabilizing effects from loops. Each structural element is independent of each other in the way the parameters are described, giving them an additive property. The summation of the individual contributions gives the total free energy of a structure (Borer et al., 1974). The current models are not perfect since there is uncertainty in thermodynamic models. The models are incomplete and are built on noisy data from wet lab experiments. Some models are very simplistic and do not capture all of the free energy contributions. Other problems with current models is that they lack the parameters to correctly model some substructures. The uncertainty in the thermodynamic models translates to uncertainty in the free energy evaluation of the structures. This is why it is believed that the real structure is often a suboptimal one (Jaeger et al., 1989; Mathews et al., 1998; Zuker et al., 1991).

9.2.1 Stacking-Energy Models

Free energy of formation has been determined for many duplexes. Duplexes are two strands of RNA containing some stacked base pairs. One process by which the energy can be determined is described below (Freier et al., 1983). First, the duplex of interest is synthesized in the lab (Kierzek et al., 1986). This is usually done using standard solid-state chemistry techniques. Simply put, each strand is built on a polymer support, such as a small bead. Once the first nucleotide is attached, each subsequent nucleotide can only attach on the uncovered extremity. This way, the sequence identity can easily be controlled. When the desired sequence is synthesized, the strand is removed from the polymer–nucleotide junction. Its identity is then confirmed by NMR and its purity is confirmed by high-pressure liquid chromatography (HPLC). HPLC is a standard chemistry technique used to separate mixtures. A mixture is passed in a stream of solvent (mobile phase) through some material. The components of the mixture interact with the surface of the material through adsorption. The components of the mixture have different adsorptions and therefore different separation rates. Usually, the separation is large enough to detect the number of components in the mixture and their amounts.

Once the component is isolated and purified, the thermodynamic parameter can be determined. Chemical substances have specific frequencies where they absorb electromagnetic radiation. The amount of radiation is related to concentration of a chemical substance. Using a spectrophotometer, by changing the temperature and monitoring the absorbance, the curve of absorbance versus temperature can be plotted. From this curve, the energy needed to break the base pairs (often termed melting) can be determined. From this energy, the free energy loss of base pair formation can be calculated.

More specifically, the measured absorption–temperature profile, with the double and single strand extremes, leads to a measure of the fraction of bases paired. From this, we can determine the equilibrium constant K:

$$K = \frac{f}{2(1-f)^2 c} \tag{9.3}$$

for identical self-complementary strands and

$$K = \frac{2f}{(1-f)^2 c} \tag{9.4}$$

for nonidentical strands. In these equations, c is the total oligonucleotide strand concentration.

From this result, we can determine the free energy change, $\Delta G°$:

$$\Delta G° = -RT \ln K. \tag{9.5}$$

Using data from many duplexes, it is possible to derive thermodynamic parameters for adjacent base pairs by solving sets of equations. Thermodynamic parameters have been determined in this way (and other more sophisticated ways) for almost all possible adjacent bases over the years (Mathews et al., 1999). The next sections will describe two thermodynamic models that make use of these parameters.

9.2.2 Individual Nearest-Neighbor Model (INN)

In 1974, it was hypothesized that the contribution of each base pair in a helix contributes to the stability of that helix and depends on its nearest neighbors (Borer et al., 1974). The study states that the enthalpy of a GC base pair will be different if it is next to an AU base pair instead of being next to a UA base pair. The initial study only considered Watson–Crick base pairs, that is, GC and AU. With only these 2 pairs, there are 16 possible base pair adjacencies, or doublets. Due to rotational symmetry, only 10 of these are unique and are listed in Figure 9.2. The free energy of these 10 nearest-neighbors were determined at 25°C.

The formation of a double-stranded helix can be thought of as a concentration-dependent formation of the first base pair (initiation), followed by a closing of subsequent base pairs (propagation). The first pair involves hydrogen bonding only. The subsequent pairs add to this their stacking interactions. The free energy of each subsequent pair involves free energy changes that depend on sequence. Propagation is independent of concentration since it is a local intramolecular reaction.

Borer et al. (1974) caution that this data only contains parameters for Watson–Crick pairs and does not discuss GU base pairs. Another limitation is that this data was generated from very similar strands (Freier et al., 1986). Further studies would most likely modify the parameter values.

In 1986, the 10 Watson–Crick nearest-neighbor thermodynamic parameters were remeasured at 37°C (Freier et al., 1986). This temperature should more closely model physiological conditions. Calculating the free energy of a strand using the INN model is straightforward. Here is an example for the predicted free energy change of helix formation for 5′GGCC3′–3′CCGG5′:

$$\Delta G^{\circ}_{37}(\text{pred}) = 2\Delta G^{\circ}_{37}(5'\text{GG}3'-3'\text{CC}5') + \Delta G^{\circ}_{37}(5'\text{GC}3'-3'\text{CG}5') + \Delta G^{\circ}_{37\,\text{init}} + \Delta G^{\circ}_{37\,\text{sym}}. \tag{9.6}$$

Using the values from the table provided in Freier et al., 1986, this becomes

$$\Delta G^{\circ}_{37}(\text{pred}) = 2(-2.9) + (-3.4) + 3.4 + 0.4 = -5.4\,\text{kcal/mol}. \tag{9.7}$$

The nearest-neighbor terms are generated by looking at the duplex through a window that is two base pairs wide from left to right. In this example, there is a term for 5′GG3′–3′CC5′ followed by a term for 5′GC3′–3′CG5′ followed by a term for 5′CC3′–3′GG5′. The last term, 5′CC3′–3′GG5′ is the same as 5′GG3′–3′CC5′

5′GG3′ 3′CC5′	5′GC3′ 3′CG5′	5′GA3′ 3′CU5′	5′GU3′ 3′CA5′	5′AG3′ 3′UC5′
5′AA3′ 3′UU5′	5′AU3′ 3′UA5′	5′CG3′ 3′GC5′	5′CA3′ 3′GU5′	5′UA3′ 3′AU5′

Figure 9.2 The 10 Watson–Crick doublets.

except for being rotated by 180°. The initiation term is a constant used to account for the loss of entropy, ΔS, during initial pairing between the two first bases. Entropy is lost because the reaction goes from two strands to a single duplex creating a more ordered system. Equation (9.1) reminds us that this is unfavorable with respect to stability; the entropy change is negative, adding a positive value to the free energy. The last term corrects for symmetry. This self-complementary strand shows twofold rotational symmetry. Again, for reasons of entropy, this destabilizes the strand. The symmetry term has a lesser effect than initiation but must be counted nonetheless.

The parameters described in Freier et al. (1986) do not represent a complete set for use in structure prediction. Canonical base pairs also include GU pairs. Numerous examples of terminal GU pairs are found at the end of helical regions and its stacking energy was found to be approximately equal to a terminal AU pair. Other parameters determined around the same time were those of unpaired terminal nucleotides, terminal mismatches, and parameters for internal GU pairs (Freier et al., 1986; Sugimoto et al., 1986).

In 1991, 11 nearest-neighbor interactions involving GU mismatches were derived from new and existing thermodynamic data (He et al., 1991). The sequences included both isolated and adjacent GU mismatches. An anomaly was discovered where the thermodynamics of 5′UG3′–3′GU5′ sequences are different from those of 5′GU3′–3′UG5′ sequences. However, the most surprising result showed that the nearest-neighbor 5′GU3′–3′UG5′ in the middle of a helix in the contexts of 5′CGUG3′–3′GUGC5′, 5′UGUA3′–3′AUGU5′, and 5′AGUU3′–3′UUGA5′ destabilize the helix. However, addition of the same 5′GU3′–3′UG5′ in the middle of 5′GGUC3′–3′CUGG5′ increases the stability.

He et al. (1991) describe this as a non-nearest-neighbor effect. Whereas the nearest-neighbor 5′UG3′–3′GU5′ is always stabilizing independent of context, the 5′GU3′–3′UG5′ is dependent on context and adds corrections to previously determined parameters (Freier et al., 1986). In this particular situation special parameters are needed to account for this anomaly (Mathews et al., 1999).

It was also found that there is a very weak stabilizing effect from 5′UU3′–3′UU5′ mismatches (Wu et al., 1995). However, since this nearest-neighbor does not contain canonical base pairs, its effects will not be considered here.

A survey on stacking energies was presented in Serra and Turner (1995). Calculations for free energy determination of different strands are presented in a straightforward fashion using nearest-neighbor parameters at 37°C. The following can be used as a guide for an implementation of the INN model.

Serra and Turner (1995) describe how to calculate the free energy of duplexes, and terminal mismatches. For the non-self-complementary duplex 5′CGGC3′–3′GCCG5′, the calculation is described by

$$\begin{aligned}\Delta G^\circ_{37} = \Delta G^\circ_{37}(5'\text{CG}3'\text{–}3'\text{GC}5') + \Delta G^\circ_{37}(5'\text{GG}3'\text{–}3'\text{CC}5') + \\ \Delta G^\circ_{37}(5'\text{GC}3'\text{–}3'\text{CG}5') + \Delta G^\circ_{37\,\text{init}}.\end{aligned} \tag{9.8}$$

For a self-complementary duplex such as 5′GGAUCC3′–3′CCUAGG5′, ΔG°_{37} is calculated the same way but requires the addition of $\Delta G^\circ_{37\,\text{sym}}$ for symmetry:

$$\Delta G^{\circ}_{37} = \Delta G^{\circ}_{37}(5'\text{GG}3'\text{–}3'\text{CC}5') + \Delta G^{\circ}_{37}(5'\text{GA}3'\text{–}3'\text{CU}5') + \\ \Delta G^{\circ}_{37}(5'\text{AU}3'\text{–}3'\text{UA}5') + \Delta G^{\circ}_{37}(5'\text{UC}3'\text{–}3'\text{AG}5') + \\ \Delta G^{\circ}_{37}(5'\text{CC}3'\text{–}3'\text{GG}5') + \Delta G^{\circ}_{37\,\text{init}} + \Delta G^{\circ}_{37\,\text{sym}}. \tag{9.9}$$

These last two equations follow the same rules as Eq. (9.6).

For 3′-terminal unpaired nucleotides such as 5′*GGAUCCA3′–3′ACCUAGG*5′ a mismatch term is added. In this case,

$$\Delta G^{\circ}_{37} = \Delta G^{\circ}_{37}(\text{Core duplex}) + 2\Delta G^{\circ}_{37}(5'\text{CA}3'\text{–}5'\text{G*}5). \tag{9.10}$$

For a helix containing a 5′-terminal unpaired nucleotide such as 5′AGGAUCC*3′–3′*CCUAGGA5′:

$$\Delta G^{\circ}_{37} = \Delta G^{\circ}_{37}(\text{Core duplex}) + 2\Delta G^{\circ}_{37}(5'\text{AG}3'\text{–}3'{*}\text{C}5'). \tag{9.11}$$

Terminal mismatches are handled by using parameters for the mismatches. The method is again the same for the example of 5′AGGAUCCA3′–3′ACCUAGGA5′:

$$\Delta G^{\circ}_{37} = \Delta G^{\circ}_{37}(\text{Core duplex}) + 2\Delta G^{\circ}_{37}(5'\text{CA}3'\text{–}3'\text{GA}5'). \tag{9.12}$$

In 1997, a fairly large study of internal mismatches (2 × 2 internal loops) was performed. Stabilities were confirmed for GU mismatches (Xia et al., 1997). However, some stability has been found in other less common mismatches, UU, CC, GA, AC, and so on. Since these are not canonical base pairs and do not increase stability significantly, they are not considered here.

9.2.3 Individual Nearest-Neighbor Hydrogen Bond Model (INN-HB)

It was noticed that duplexes with the same nearest neighbors but different terminal ends consistently have different stabilities. The duplex with one more terminal GC pair and one less terminal AU pair is always more stable. The reason is that switching a GC pair to an AU pair in base composition decreases the number of hydrogen bonds in the duplex by one.

To account for this difference, the INN-HB model (Xia et al., 1998) also includes a term for terminal AU pairs and therefore for the base composition of the sequence.

The improved thermodynamic parameters were derived from a study of 90 duplexes of short RNA strands containing only Watson–Crick base pairs. In the INN model, the initiation parameter for duplexes with at least one GC base pair were determined, but the initiation parameter for duplexes with only AU base pairs was not determined. It has been shown that the initiation term is dependent on the identities of the two terminal base pairs. Xia et al. (1998) provide not only more accurate parameters but also a penalty term for each terminal AU pair.

The general equation used to calculate the free energy change of duplex formation can be written in INN-HB as the following:

$$\Delta G^{\circ}(\text{duplex}) = \Delta G^{\circ}_{\text{init}} + \sum_j n_j \Delta G^{\circ}_j(\text{NN}) + m_{\text{term-AU}} \Delta G^{\circ}_{\text{term-AU}} + \Delta G^{\circ}_{\text{sym}}. \quad (9.13)$$

Each $\Delta G^{\circ}_j(\text{NN})$ term is the free energy contribution of the jth nearest neighbor with n_j occurrences in the sequence. The $m_{\text{term-AU}}$ and $\Delta G^{\circ}_{\text{term-AU}}$ terms are the number of terminal AU pairs and the associated free energy parameter, respectively. The $\Delta G^{\circ}_{\text{init}}$ term is the free energy of initiation.

The only change from the INN model is the addition of an $m_{\text{term-AU}} \Delta G^{\circ}_{\text{term-AU}}$ penalty when terminal AU pairs exist in a helix. Using the data tables provided by the thermodynamic model of nearest-neighbor parameters, calculating the stability of a helix is straightforward.

For a non-self-complementary duplex such as 5′ACGAGC3′–3′UGCUCG5′:

$$\begin{aligned}\Delta G^{\circ}(\text{duplex}) = \Delta G^{\circ}_{\text{init}} &+ \Delta G^{\circ}_{37}(5'\text{GU}3'-3'\text{CA}5') + \Delta G^{\circ}_{37}(5'\text{CG}3'-3'\text{GC}5') + \\ &\Delta G^{\circ}_{37}(5'\text{GA}3'-3'\text{CU}5') + \Delta G^{\circ}_{37}(5'\text{CU}3'-3'\text{GA}5') + \\ &\Delta G^{\circ}_{37}(5'\text{GC}3'-3'\text{CG}5') + 1 \times \Delta G^{\circ}_{37}(5'\text{A}3'-3'\text{U}5').\end{aligned} \quad (9.14)$$

For a self-complementary duplex such as 5′UGGCCA3′–3′ACCGGU5′:

$$\begin{aligned}\Delta G^{\circ}(\text{duplex}) = \Delta G^{\circ}_{\text{init}} &+ 2 \times \Delta G^{\circ}_{37}(5'\text{CA}3'-3'\text{GU}5') + 2 \times \\ &\Delta G^{\circ}_{37}(5'\text{GG}3'-3'\text{CC}5') + \Delta G^{\circ}_{37}(5'\text{GC}3'-3'\text{CG}5') + \\ &2 \times \Delta G^{\circ}_{37}(5'\text{A}3'-3'\text{U}5') + \Delta G^{\circ}_{\text{sym}}.\end{aligned} \quad (9.15)$$

Terminal GU pairs are treated the same way as terminal AU pairs in the INN-HB model because they also have two hydrogen bonds.

9.3 METHODS

9.3.1 Evolutionary Algorithms

An EA is a stepwise nondeterministic algorithm that follows an evolutionary model mimicking natural evolution (Bäck, 1996). It returns a number of probable solutions at each generation. In the RNA domain, an EA has the goal of finding a set of low ΔG structures. At every generation in the algorithm, it is hoped that the population will contain lower energy structures than during the previous generation. It is expected that the population converges to low ΔG structures. The pseudocode for a standard generational EA is given in Figure 9.3.

Each generation has three key steps.

1. A combination of the parts that make up two parent solutions are chosen to make new offspring solutions. This is called crossover. While many EAs can function well without the use of crossover (e.g., evolution strategies), our

```
Initialize random population of chromosomes;
Evaluate the chromosomes in the population;
while stopping criterion is not reached
        for half of the members of a population
                select 2 parent chromosomes;
                apply crossover operator (Pc);
                apply mutation operator (Pm);
                evaluate the new chromosomes;
                selection strategy;
                elitism;
                insert them into next generation;
        end for
        update stopping criterion;
end while
```

Figure 9.3 Algorithm is based on a standard generational EA. The stopping criterion is the number of generations (Deschênes et al., 2004).

previous research (Deschênes, 2005) has demonstrated that in the RNA folding domain the EA only using mutation did not perform as well as when crossover was included. This suggests that in the RNA folding domain, crossover is an important part of the EA. Since all solutions (members of the population) have parts that are favorable and others that are unfavorable, it is possible that all favorable parts are incorporated into one solution and all unfavorable parts go into the other solution after crossover is applied. Different types of crossover operators exist. Each one of them exhibits its own properties and heuristics.

2. Random changes in the population are introduced via mutations. This step is used to avoid premature genetic convergence in the population. Randomly mutating a part of a solution tends to maintain genetic diversity within the population. Energy minimization problems can be represented by an N-dimensional hypersurface. Using mutation helps probing different parts of the energy hypersurface and avoids converging in local minima.
3. The algorithm selects a new set of solutions from the old solutions. The choice is made from scoring each solution against a fitness function. This criterion selects highly fit solutions with a bias to improve the overall population. All members of the population are evaluated against a fitness function and are ranked. It is the task of the EA to choose good solutions and reject others based on their scores. This way, the EA's solutions converge. Selection can act on parents, the old population, and the new population. It can be local (within a subpopulation) or global (within the entire population). More details on mutation, crossover, and selection can be found in Bäck (1996).

These steps can be repeated for a predetermined number of generations, a predetermined amount of time, or until the population converges, that is, until the population's average diversity reaches a threshold value.

In summary, EAs are stochastic and nondeterministic. The initial population of solutions is generated randomly before the algorithm begins. At the mutation stage, random parts of the solution are changed. When two parents are combined during the crossover, random parts of two solutions are exchanged. Lastly, to generate the new population for the next generation, solutions are chosen randomly where more favorable solutions are given more probability of being chosen.

In RNA secondary structure prediction, the algorithm tries to find low-energy, stable structures. These are the structures that are most likely to be close to the native fold. Thermodynamic models associate changes in free energy to the formation of RNA substructures. In order to calculate the difference in free energy, the free energy contribution (loss) from each substructure is summed. It is expected that the lowest energy structure is close to the native fold. Very often, external interactions such as solvent effects affect the resulting structure. Furthermore, the observed structure may not be the one with minimum free energy (Jaeger et al., 1989; Mathews et al., 1998; Zuker et al., 1991). DPAs are at a disadvantage since they traditionally yield only one optimal (minimum ΔG) structure. Since EA results yield a population of candidate solutions, it is possible to investigate not only the minimum free energy structure found but also other low-energy structures that may be closer to the native fold.

The specifics of our helix generation algorithm are presented in Wiese and Hendriks (2006). In our model, a valid helix is considered to possess a minimum of 3 adjacent canonical base pairs and a minimum hairpin size of 3. The objective in this case is to generate the set of all possible helices under these constraints. First, all valid base pairs are found for the given RNA sequence. Next, the algorithm iterates through each base pair and attempts to build a helix by stacking valid base pairs on it. If the resulting helix meets or exceeds the above requirements, it is added to the set H of possible helices.

Once the set of all potential helices H has been generated, the structure prediction problem becomes one of combinatorial optimization. To ensure chemically feasible structures, no predicted structure may contain helices that share bases. Since these helices may be present in H, RnaPredict must ensure that they are mutually exclusive. To resolve this problem, RnaPredict employs a decoder that eliminates helix conflicts. This decoder is described in more detail below.

RnaPredict attempts to optimize its predicted structures such that structures produced are both chemically feasible and have a free energy close to the minimum ΔG. As discussed before, the minimum ΔG structure itself is usually not identical (or even close) to the native fold. Better structures can be found in the sampling space around the minimum ΔG structure.

Each helix in H is indexed with an integer ranging from 0 to $n - 1$, n being the total number of generated helices. An RNA secondary structure is encoded in RnaPredict via a permutation of these integers. For example, assuming set H contains 5 helices, {0,1,2,3,4} and {3,1,4,0,2} are two possible structures. Depending on how the individual helices conflict, both permutations could result in vastly different structures. Helix conflicts are eliminated by decoding the permutation from left to right. The helix specified at each point in the permutation is checked for conflicts with helices to its left. If no conflicts are found, the helix is retained; otherwise it is discarded. Consequently, only chemically feasible structures are predicted through the use of this permutation-based representation.

9.3.2 Dynamic Programming

Dynamic programming is generally applied to optimization problems possessing two basic properties: optimal substructure and overlapping subproblems. In optimal substructure, the optimal solution to a given problem has optimal solutions to its subproblems. Overlapping subproblems refers to the fact that a recursive algorithm for the problem solves the same subproblems repeatedly. Thus, the DP algorithm divides the prediction problem into subproblems, solves the subproblems and tabulates their solutions, and computes the final structure from the tabulated subproblem solutions.

mfold uses standard INN-HB parameters (Xia et al., 1998) but adds modeling for common RNA substructures. These include stacking energies, terminal mismatch stacking energies (hairpin loops), terminal mismatch stacking energies (interior loops), single mismatch energies, 1 × 2 interior loop energies, tandem mismatch energies, single base stacking energies, loop destabilizing energies, tetra-loops, tri-loops, and other miscellaneous energies.

After generating structures, mfold reevaluates structures with a more complete thermodynamic model using its efn2 helper application. This software adds a more accurate model of multibranch loops among other improvements.

9.4 RESULTS

The sequences in Table 9.1 are used in our experiments. These sequences were chosen as they represent different sequence lengths and come from various genomes of organisms that are exposed to a range of physiological conditions. They represent 3 RNA classes: 5S rRNA, Group I intron 16S rRNA, and 16S rRNA. Due to space

Table 9.1 RNA Sequence Details

Organism (Accession Number)	RNA Class	Length (nt)	Known Base Pairs
Saccharomyces cerevisiae (X67579)	5S rRNA	118	37
Haloarcula marismortui (AF034620)	5S rRNA	122	38
Aureoumbra lagunensis (U40258)	Group I intron, 16S rRNA	468	113
Hildenbrandia rubra (L19345)	Group I intron, 16S rRNA	543	138
Acanthamoeba griffini (U02540)	Group I intron, 16S rRNA	556	131
Caenorhabditis elegans (X54252)	16S rRNA	697	189
Drosophila virilis (X05914)	16S rRNA	784	233
Xenopus laevis (M27605)	16S rRNA	945	251
Homo sapiens (J01415)	16S rRNA	954	266

Source: Comparative RNA website (Cannone et al., 2002).

constraints in some tables we refer to these specific RNA sequences by the name of the organism from which they originated.

9.4.1 Correlation between Free Energy and Correct Base Pairs

The starting premise in this research is that there is a strong relationship between free energy of a structure and the accuracy of the prediction. It is expected that the lower the free energy of a predicted structure, the more correct base pairs will be present. To establish the correlation between free energy and accuracy in predicting base pairs, an experiment was set up. Two different thermodynamic models were tested with RnaPredict. For each sequence, 7010 structures were generated by running RnaPredict with the parameters shown in Table 9.2. For each of the 701 generations (0–700), the 10 lowest energy structures are examined. These parameters were chosen to maximize diversity, yet making as much progress toward low-energy structures as possible.

Figures 9.4, 9.5, and 9.6 show the INN-HB correlation graphs for *X. laevis* 16S rRNA (M27605), *S. cerevisiae* 5S rRNA (X67579), and *C. elegans* 16S rRNA (X54252), respectively. The graphs plot the free energy of 10 structures per generation for 701 generations for a total of 7010 structures. Figures 9.4 and 9.5 show a high correlation where a change in energy corresponds to a change in the number of correctly predicted base pairs.

Figure 9.6 demonstrates imperfections within the INN-HB thermodynamic model. This graph shows very little correlation between free energy and the number of correctly predicted base pairs making INN-HB inadequate for predicting *C. elegans* 16S rRNA (X54252) structures.

Table 9.3 shows the results of the correlation between free energy of the 7010 generated structures for each parameter set and their prediction accuracy. The table shows each sequence along with the correlation coefficient for each thermodynamic

Table 9.2 EA Parameters Used to Generate the Correlation Data

Population size	700
Generations	700
Crossover operators	CX[a]
Crossover probability (P_c)	0.8
Mutation probability (P_m)	0.8
Selection strategy	STDS
Structures per generation	10
Elitism	0
Thermodynamic models	INN, INN-HB

[a]From Oliver et al. (1987).

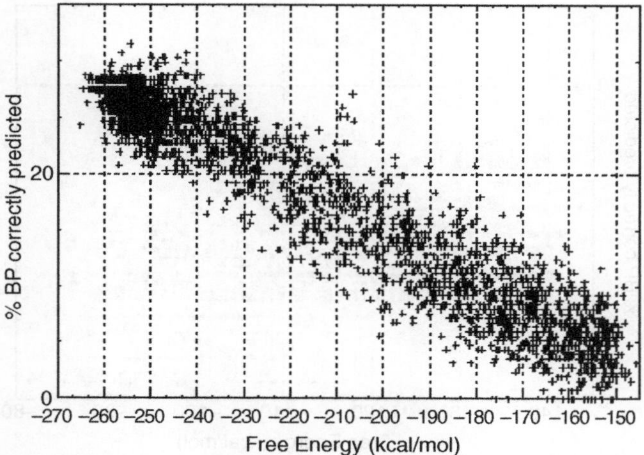

Figure 9.4 Correlation graph for *Xenopus laevis* 16S rRNA (M27605) using INN-HB. The graph plots the free energy of 10 structures per generation for 701 generations for a total of 7010 structures. The correlation for this sequence was evaluated at ρ = −0.96.

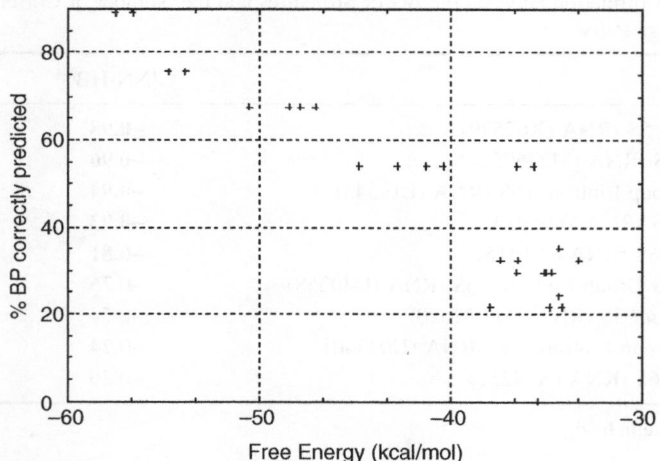

Figure 9.5 Correlation graph for *S. cerevisiae* 5S rRNA (X67579) using INN-HB. The graph plots the free energy of 10 structures per generation for 701 generations for a total of 7010 structures. The correlation for this sequence was evaluated at ρ = −0.98. Note: There are numerous duplicate structures in the population.

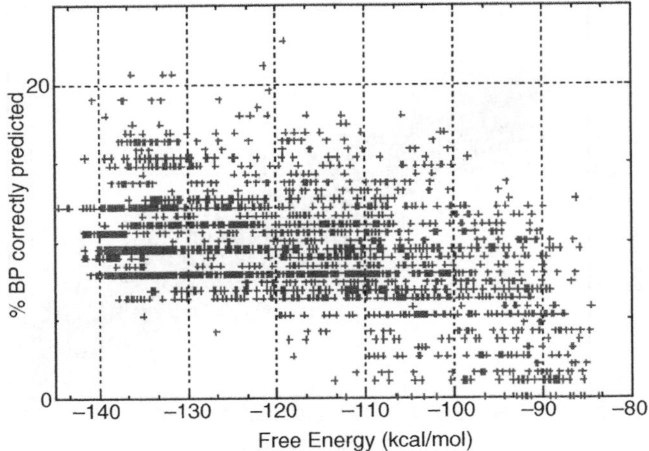

Figure 9.6 Correlation graph for *C. elegans* 16S rRNA (X54252) using INN-HB. The graph plots the free energy of 10 structures per generation for 701 generations for a total of 7010 structures. The correlation for this sequence was evaluated at ρ = −0.26.

Table 9.3 Correlation between the Δ*G* of Structures and the Number of Correctly Predicted Base Pairs[a]

Sequence	INN-HB	INN
S. cerevisiae 5S rRNA (X67579)	**−0.98**	−0.96
X. laevis 16S rRNA (M27605)	**−0.96**	−0.9
H. rubra Group I intron, 16S rRNA (L19345)	**−0.94**	−0.87
D. virilis 16S rRNA (X05914)	**−0.93**	−0.5
H. sapiens 16S rRNA (J01415)	**−0.81**	−0.77
A. lagunensis Group I intron, 16S rRNA (U40258)	−0.76	**−0.77**
H. marismortui 5S rRNA (AF034620)	−0.74	**−0.86**
A. griffini Group I intron, 16S rRNA (U02540)	−0.74	**−0.84**
C. elegans 16S rRNA (X54252)	−0.26	**−0.74**

[a]Best results are in bold.

model. The correlation coefficient is defined as a quantity that gives the quality of a least squares fitting to the original data (Weiss, 1999). For instance, in the first row, *S. cerevisiae* 5S rRNA (X67579) shows a correlation coefficient close to −1 with INN-HB. This value shows that the lower the free energy of a structure of this sequence is, the higher the number of correctly predicted base pairs in the structure.

On initial inspection, the data in Table 9.3 shows a nearly even split, with INN-HB yielding the best correlation for 5 sequences, and INN for 4. There are two instances of notable differences. INN-HB demonstrates a superior correlation over INN on the *D. virilis* 16s rRNA (X05914) sequence, by a margin of 0.43. In contrast, INN demonstrates a better correlation than INN-HB on the *C. elegans* 16S rRNA (X54252) sequence by a margin of 0.48. While both models demonstrate a clear correlation between ΔG and the number of true positive base pairs, there is still considerable variance between sequences.

In summary, it seems to be a general trend that correlation decreases with increasing sequence length. However, this is not the only contributing factor; others may include the number of noncanonical base pairs and general sequence composition.

9.4.2 Comparison with Known Structures and mfold

The quality of structure prediction is related to the similarity of the predicted structures to the known structures. The more "similar" the predicted structure is to the native fold, the higher the accuracy. One particular quantitative metric to measure similarity counts the number of correctly predicted base pairs. The larger the number of base pairs correctly predicted, the higher the quality of the structure.

The results are presented here for nine sequences. Tests were done using various parameters. The discussion is focused on the lowest energy structures found using the cycle (CX) (Oliver et al., 1987) order #2 (OX2) (Syswerda, 1991), and partially matched (PMX) (Goldberg and Lingle, 1985) crossover operators. The structures with the highest number of correctly predicted base pairs found with these parameter sets are also discussed. The generated structures will be compared to known structures and to those predicted by mfold. Assessing the quality of RnaPredict's results is done through several measures described below.

For Tables 9.4 and 9.5, the measures are defined as follows: TP is the total true positive base pairs predicted, FP is the number of false positive base pairs predicted, and FN is the total of false negative base pairs predicted for a given sequence.

A common performance measure for a predictor is *sensitivity*. Sensitivity is computed as TP/(TP + FN) and is the proportion of correctly predicted base pairs to the total number of base pairs found in the native structure. In general, predicting more base pairs will increase the number of true positive base pairs. However, this also will likely increase the number of false positives. While the sensitivity increases in this situation, a higher number of false positives is not desirable.

To better quantify the effectiveness of a prediction method, *specificity* is often used. Specificity is defined as TP/(TP + FP) in Baldi et al. (2000); it is also known as *positive predictive value*. This definition of specificity is different from that utilized in medical statistics. In this instance, specificity is the proportion of all predicted true positive base pairs with respect to all base pairs predicted. For example, a given predictor A with an identical TP count to a predictor B but a higher FP count will have a lower specificity than B and is considered to have a lower performance.

Table 9.4 Comparison of Highest Matching Base Pair RnaPredict Structures (EA) with Lowest ΔG mfold Structures in Terms of TP, FP, and FN[a]

Sequence	Length (nt)	TP		FP		FN		EA Cross.-Sel.-Model
		EA	mfold	EA	mfold	EA	mfold	
S. cerevisiae	118	**33**	**33**	**6**	8	**4**	**4**	OX2-STDS-INNHB
H. marismortui	122	27	**29**	**3**	5	11	**9**	PMX-KBR-INNHB
A. lagunensis	468	**68**	60	**63**	68	**45**	53	CX-STDS-INNHB
H. rubra	543	**79**	49	**82**	127	**59**	89	OX2-STDS-INN
A. griffini	556	**81**	67	**80**	105	**50**	64	OX2-STDS-INNHB
C. elegans	697	**55**	40	**147**	177	**134**	149	OX2-STDS-INN
D. virilis	784	**65**	37	**177**	199	**168**	196	OX2-STDS-INNHB
X. laevis	945	**93**	92	**147**	157	**158**	159	CX-STDS-INN
H. sapiens	954	89	**95**	**161**	163	177	**171**	OX2-STDS-INNHB
Averages	576.33	**65.6**	55.8	**96.2**	112.1	**89.6**	99.3	

[a]Best results are in bold.

Table 9.5 Comparison of Highest Matching Base Pair Structures from RnaPredict (EA) and mfold in Terms of TP, FP, and FN[a]

Sequence	Length (nt)	TP		FP		FN		EA Cross.-Sel.-Model
		EA	mfold	EA	mfold	EA	mfold	
S. cerevisiae	118	**33**	**33**	**6**	8	**4**	**4**	OX2-STDS-INNHB
H. marismortui	122	27	**29**	**3**	5	11	**9**	PMX-KBR-INNHB
A. lagunensis	468	68	**74**	63	**59**	45	**39**	CX-STDS-INNHB
H. rubra	543	79	**83**	**82**	84	59	**55**	OX2-STDS-INN
A. griffini	556	81	**95**	80	**79**	50	**36**	OX2-STDS-INNHB
C. elegans	697	**55**	40	**147**	177	**134**	149	OX2-STDS-INN
D. virilis	784	65	**82**	177	**170**	168	**151**	OX2-STDS-INNHB
X. laevis	945	93	**113**	147	**132**	158	**138**	CX-STDS-INN
H. sapiens	954	89	**95**	**161**	163	177	**171**	OX2-STDS-INNHB
Averages	576.33	65.6	**71.6**	**96.2**	97.4	89.6	**83.6**	

[a]Best results are in bold.

A metric that combines both the specificity and sensitivity measures into one is *F-measure*; it is defined as $F = 2 \times$ specificity $\times$ sensitivity/(specificity + sensitivity) and can be used as a single performance measure for a predictor.

The measures of sensitivity, specificity, and F-measure are used in Tables 9.6 and 9.7 to further quantify the performance of RnaPredict and mfold. The parameters used for RnaPredict are listed in Table 9.8. In Tables 9.4–9.7, the specific EA parameters, which varied for each sequence (crossover, selection strategy, and thermodynamic model), are shown under "EA Cross.-Sel.-Model."

Table 9.6 Comparison of Highest Matching Base Pair RnaPredict Structures (EA) with Lowest ΔG mfold structures in Terms of Percent Sensitivity (Sens.), Specificity (Spec.), and F-Measure (F-Meas.)[a]

Sequence	Length (nt)	Sens.		Spec.		F-Meas.		EA Cross.-Sel.-Model
		EA	mfold	EA	mfold	EA	mfold	
S. cerevisiae	118	**89.2**	**89.2**	**84.6**	80.5	**86.8**	84.6	OX2-STDS-INNHB
H. marismortui	122	71.1	**76.3**	**90.0**	85.3	79.4	**80.6**	PMX-KBR-INNHB
A. lagunensis	468	**60.2**	53.1	**51.9**	46.9	**55.7**	49.8	CX-STDS-INNHB
H. rubra	543	**57.2**	35.5	**49.1**	27.8	**52.8**	31.2	OX2-STDS-INN
A. griffini	556	**61.8**	51.1	**50.3**	39.0	**55.5**	44.2	OX2-STDS-INNHB
C. elegans	697	**29.1**	21.2	**27.2**	18.4	**28.1**	19.7	OX2-STDS-INN
D. virilis	784	**27.9**	15.9	**26.9**	15.7	**27.4**	15.8	OX2-STDS-INNHB
X. laevis	945	**37.1**	36.7	**38.8**	36.9	**37.9**	36.8	CX-STDS-INN
H. sapiens	954	33.5	**35.7**	35.6	**36.8**	34.5	**36.3**	OX2-STDS-INNHB
Averages	576.33	**51.9**	46.1	**50.5**	43.0	**50.9**	44.3	

[a]Best results are in bold.

Table 9.7 Comparison of Highest Matching Base Pair Structures from RnaPredict (EA) and mfold in Terms of Percent Sensitivity (Sens.), Specificity (Spec.), and F-Measure (F-Meas.)[a]

Sequence	Length (nt)	Sens.		Spec.		F-Meas.		EA Cross.-Sel.-Model
		EA	mfold	EA	mfold	EA	mfold	
S. cerevisiae	118	**89.2**	**89.2**	**84.6**	80.5	**86.8**	84.6	OX2-STDS-INNHB
H. marismortui	122	71.1	**76.3**	**90.0**	85.3	79.4	**80.6**	PMX-KBR-INNHB
A. lagunensis	468	60.2	**65.5**	51.9	**55.6**	55.7	**60.2**	CX-STDS-INNHB
H. rubra	543	57.2	**60.1**	49.1	**49.7**	52.8	**54.4**	OX2-STDS-INN
A. griffini	556	61.8	**72.5**	50.3	**54.6**	55.5	**62.3**	OX2-STDS-INNHB
C. elegans	697	**29.1**	21.2	**27.2**	18.4	**28.1**	19.7	OX2-STDS-INN
D. virilis	784	27.9	**35.2**	26.9	**32.5**	27.4	**33.8**	OX2-STDS-INNHB
X. laevis	945	37.1	**45.0**	38.8	**46.1**	37.9	**45.6**	CX-STDS-INN
H. sapiens	954	33.5	**35.7**	35.6	**36.8**	34.5	**36.3**	OX2-STDS-INNHB
Averages	576.33	51.9	**55.6**	50.5	**51.1**	50.9	**53.0**	

[a]Best results are in bold.

To generate the mfold results presented here, the mfold web server version 3.1 was used with default settings. One noteworthy setting is the percentage of suboptimality. This percentage allows the user to control the number of suboptimal structures predicted by mfold. In this experiment, the value was set to return the 5% lowest energy structures. This corresponds to approximately 20 structures for a 1000 nt sequence.

Table 9.8 RnaPredict Parameters for Known Structure and mfold Comparison

Population size	700
Generations	700
Crossover operators	CX[a], OX2[b], PMX[c]
Crossover probability (P_c)	0.7
Mutation probability (P_m)	0.8
Selection strategy	STDS, KBR
Structures per generation	10
Elitism	1
Thermodynamic models	INN, INN-HB
Random seeds	30

[a]Oliver et al. (1987).
[b]Syswerda (1991).
[c]Goldberg and Lingle (1985).

In Table 9.4, the lowest ΔG structure predicted by mfold is compared to the best structure predicted by RnaPredict in terms of TP, FP, and FN base pairs. In six out of nine sequences, RnaPredict found a greater number of TP base pairs than mfold, tying on the seventh sequence. For FP, RnaPredict predicted fewer FP base pairs than mfold for all sequences. Finally, the structures predicted by RnaPredict determined fewer FN base pairs than mfold in six out of nine cases, again tying on the *S. cerevisiae* 5S rRNA (X67579) sequence.

In Table 9.6, the lowest ΔG structure predicted by mfold is compared to the best structure predicted by RnaPredict in terms of sensitivity, specificity, and F-measure. Reviewing sensitivity, we can see that RnaPredict has a higher sensitivity than mfold in six out of nine sequences, tying on the *S. cerevisiae* 5S rRNA (X67579) sequence. For specificity, RnaPredict outperformed mfold in eight out of nine cases, closely approaching mfold on the *H. sapiens* 16S rRNA (J01415) sequence. Finally, examining F-measure, we see that RnaPredict surpassed mfold in seven out of nine cases; it is worth noting that RnaPredict is also quite close to mfold for the *H. marismortui* 5S rRNA (AF034620) sequence. Overall, the average sensitivity, the average specificity, and the average F-measure of RnaPredict are higher than for mfold.

This demonstrates that RnaPredict is capable of predicting structures, for this set of nine sequences, which in most cases are closer to the native fold than those predicted by mfold. This clearly demonstrates that the minimum ΔG structure produced by mfold cannot be considered the best solution and that it is not identical to the native structure.

However, mfold also offers suboptimal solutions. These are structures that have a higher free energy than the minimum ΔG structure but may be closer to the native fold. In the next experiment we have considered the 5% suboptimal structures predicted by mfold. These are covered in Tables 9.5 and 9.7.

Table 9.5 compares the structures with the highest count of true positive base pairs predicted by RnaPredict and mfold in terms of TP, FP, and FN base pairs.

Overall, mfold surpassed RnaPredict in terms of TP base pairs in all but one out of the nine sequences, tying on the *S. cerevisiae* 5S rRNA (X67579) sequence. However, it is worth considering that in four cases RnaPredict is within six or less TP base pairs of the structure predicted by mfold. RnaPredict predicted fewer FP base pairs for five out of nine sequences and closely approaches mfold's result on the *A. griffini* Group I intron, 16S rRNA (U02540) sequence. Finally, considering FN base pairs, mfold again surpassed RnaPredict in all but one case, tying for the *S. cerevisiae* 5S rRNA (X67579) sequence. Here also, RnaPredict was within six base pairs or less of mfold in four out of nine cases.

Table 9.7 compares the sensitivity, specificity, and *F* measure of the best structures predicted by RnaPredict (EA) and mfold. Comparing sensitivity, RnaPredict is exceeded by mfold for all but the *C. elegans* 16S rRNA (X54252) sequence, tying again for the *S. cerevisiae* 5S rRNA (X67579) sequence. In four out of the nine cases, RnaPredict is within 6% of the mfold results. For specificity, RnaPredict improved upon the mfold result in three cases. In five cases, RnaPredict is within 6% of the mfold results. Finally, for *F* measure RnaPredict outperforms the mfold result in two cases; in four cases RnaPredict is within 6% of the mfold results.

These results are very promising as they demonstrate RnaPredict is predicting structures with comparable quality to mfold's considering average sensitivity, specificity, and *F* measure. mfold also employs a much more sophisticated thermodynamic model, giving it a slight advantage.

9.4.3 Secondary Structure Comparison

Figure 9.7 shows the known secondary structure for the *H. marismortui* 5S rRNA (AF034620) sequence. There are a total of 38 base pairs. For this sequence, the highest number of correctly predicted base pairs RnaPredict found was 27 out of 38, or 71.1%.

Figure 9.8 shows the structure predicted by RnaPredict. It is interesting to note that RnaPredict's current thermodynamic models do not account for noncanonical base pairs. However, they do exist in naturally occurring structures including *H. marismortui*. There are two UU pairs at the base of the first helix on the right, which shortens the predicted helix from 7 to 5 base pairs. The left branch of the structure contains another noncanonical base pair, a CU pair. The result is that RnaPredict predicts a shifted helix. The last noncanonical base pair is a GA pair in the right branch of the structure, which again shortens the predicted helix by 2 base pairs. Finally, there is a 2-base-pair-long helix at the end of the right branch, that RnaPredict does not predict as it is constrained to helices that are 3 base pairs in length. Based on the constraints of helix length and canonical base pairs, RnaPredict has predicted all the base pairs it could possibly find.

Figure 9.9 shows mfold's predicted structure. Here, mfold was able to correctly predict two additional base pairs, for a total of 29 out of 38, or 76.3% sensitivity.

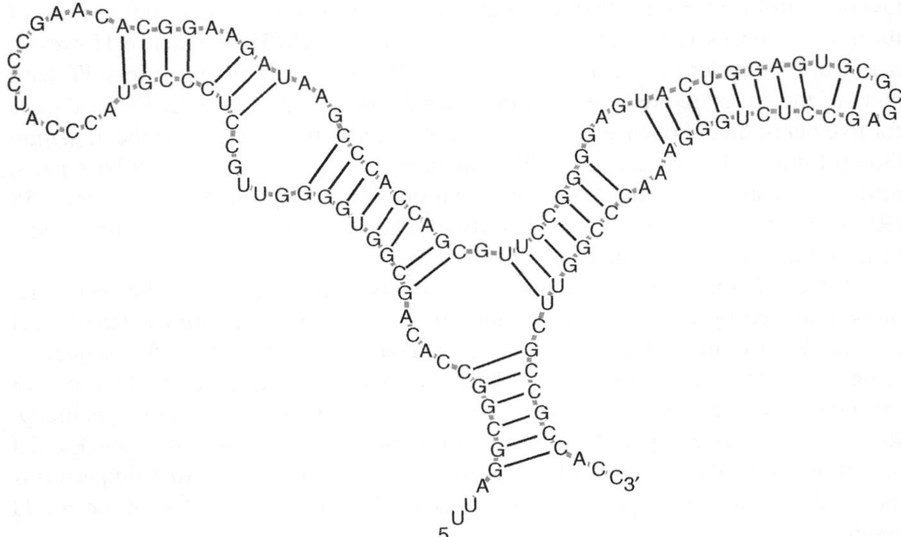

Figure 9.7 Known structure for the *H. marismortui* 5S rRNA (AF034620) sequence.

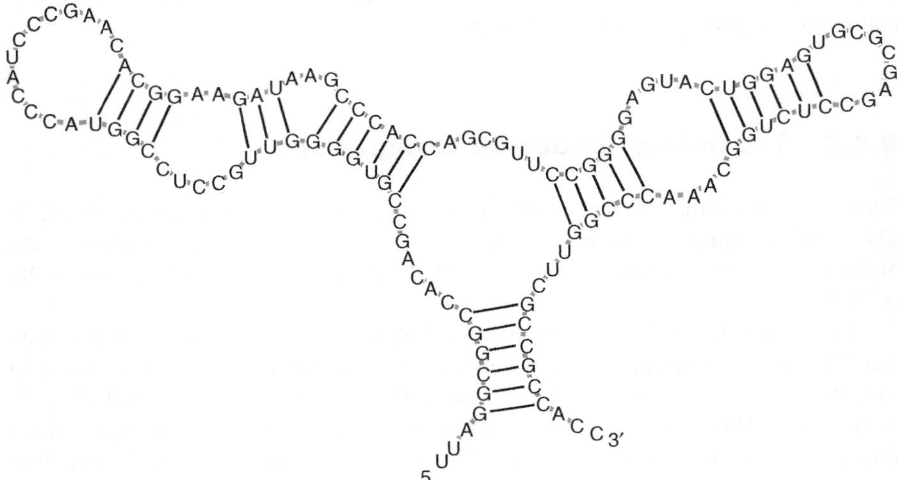

Figure 9.8 Structure for the *H. marismortui* 5S rRNA (AF034620) sequence predicted by RnaPredict.

These pairs can clearly be seen in the 2-base-pair-long helix at the end of the right branch of the structure.

Another 2-base-pair helix at the base of the left branch was also correctly predicted. However, mfold suffers from the same noncanonical base pair issue as RnaPredict. This can be seen in the shifting of the helix in the middle of the left

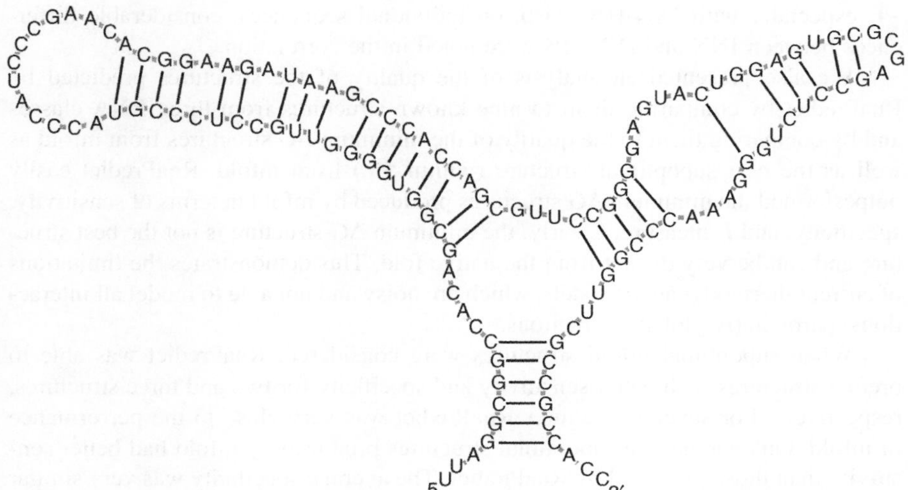

Figure 9.9 Structure for the *H. marismortui* 5S rRNA (AF034620) sequence predicted by mfold.

branch. There is also a single GC base pair correctly predicted in the right branch. Finally, mfold incorrectly predicts a 3-base-pair helix in the center of a bulge within the known *H. marismortui* structure. This results in a significantly lower specificity of 85.3% for mfold, compared with RnaPredict's specificity of 90.0%.

While the *F*-measure for both predicted structures is very similar, the structures predicted by RnaPredict and mfold clearly look quite different. Overall, the RnaPredict structure strongly resembles the native fold. There is one internal loop with three branches; the leftmost branch contains three helices, similar to the native fold with the exception of a bulge. The rightmost branch contains two helices, but is missing the 2-base-pair helix found in the native structure. On the other hand, the mfold structure looks quite different from the native fold in terms of helix content in the left and right branches. We particularly note that mfold has a larger number of helices based on its tendency to predict smaller helices containing 1 or 2 base pairs. While the mfold structure exhibits a higher sensitivity, the overall structural motif does not overlap with the native structure as much as for the RnaPredict structure.

9.5 CONCLUSION

We have presented a detailed description of the two thermodynamic models employed by RnaPredict, namely INN and INN-HB, and we have presented a study of the correlation between ΔG and the sensitivity of the structures predicted.

The results show that while sampling structures in the search space, a strong correlation was found between the structures' free energy and the number of correct base pairs for shorter sequences, but the correlation was not as pronounced for longer sequences. For some of the sequences, the correlation coefficient was very close to

−1, especially with INN-HB. Also, on individual sequences, considerable differences between INN and INN-HB were noted in the correlation.

We also presented an analysis of the quality of the structures predicted by RnaPredict by comparing them to nine known structures from three RNA classes and by comparing them to the quality of the minimum ΔG structures from mfold as well as the best suboptimal structure (within 5%) from mfold. RnaPredict easily outperformed the minimum ΔG structures produced by mfold in terms of sensitivity, specificity, and F-measure. Clearly, the minimum ΔG structure is not the best structure and can be very distant from the native fold. This demonstrates the limitations of current thermodynamic models, which are noisy and not able to model all interactions, particularly global interactions.

When suboptimal mfold structures were considered, RnaPredict was able to predict structures with better sensitivity and specificity for two and three structures, respectively. For several structures RnaPredict was very close to the performance of mfold. On average, the suboptimal structures produced by mfold had better sensitivity than those predicted by RnaPredict. The average specificity was very similar for both approaches. The slightly better average performance of mfold is likely due to a more sophisticated thermodynamic model than the ones used in RnaPredict. We also noted that both approaches performed better on shorter sequences and as sequence length increased, the prediction performance decreased. This is consistent with the correlation findings of the first experiment.

In future work we will incorporate more sophisticated thermodynamic models into RnaPredict. We will also aim at modeling noncanonical base pairs, which occur in native structures, but are not modeled in INN, INN-HB, or the thermodynamic models used by mfold. Both of these measures should lead to a further increase in the prediction accuracy of RnaPredict.

Acknowledgment

The first author would like to acknowledge the support of the Natural Sciences and Engineering Research Council (NSERC) for this research under Research Grant number RG-PIN 238298 and Equipment Grant number EQPEQ 240868. The second author would like to acknowledge support from Simon Fraser University and an NSERC Postgraduate Scholarship (PGS-A). The third author would like to acknowledge support from Simon Fraser University, the Advanced Systems Institute of British Columbia, and a NSERC Canada Graduate Scholarship (CGS-D). All authors would like to acknowledge the support of the InfoNet Media Centre funded by the Canada Foundation for Innovation (CFI) under grant number CFI-3648. All authors would also like to acknowledge that an extended version of this book chapter has been published in the IEEE/ACM *Transactions on Computational Biology and Bioinformatics* (TCBB). The TCBB paper presents a much larger body of data and results, more in-depth technical details, and a larger body of references.

REFERENCES

Bäck, T. (1996). *Evolutionary Algorithms in Theory and Practice: Evolution Strategies, Evolutionary Programming, Genetic Algorithms*. Oxford University Press, New York.

Baldi, P., S. Brunak, Y. Chauvin, C. A. F. Andersen, and H. Nielsen (2000). "Assessing the accuracy of prediction algorithms for classification: An overview," *Bioinformatics*, Vol. 16, No. 5, pp. 412–424.

Benedetti, G. and S. Morosetti (1995). "A genetic algorithm to search for optimal and suboptimal RNA secondary structures," *Biophys. Chem.*, Vol. 55, pp. 253–259.

Borer, P. N., B. Dengler, I. Tinoco Jr., and O. C. Uhlenbeck (1974). "Stability of ribonucleic acid double-stranded helices," *J. Mol. Bio.*, Vol. 86, pp. 843–853.

Cannone, J. J., S. Subramanian, M. N. Schnare, J. R. Collett, L. M. D'Souza, Y. Du, B. Feng, N. Lin, L. V. Madabusi, K. M. Müller, N. Pande, Z. Shang, N. Yu, and R. R. Gutell (2002). "The comparative RNA web (CRW) site: An online database of comparative sequence and structure information for ribosomal, intron, and other RNAs," *BMC Bioinformatics*, Vol. 3:2.

Chen, J. H., S. Y. Le, and J. V. Maizel (2000). "Prediction of common secondary structures of RNAs: A genetic algorithm approach," *Nucleic Acids Res.*, Vol. 28, pp. 991–999.

Chen, S. J. and K. A. Dill (2000). "RNA folding energy landscapes," *Proc. Natl. Acad. Sci. USA*, Vol. 97, pp. 646–651.

Currey, K. M. and B. A. Shapiro (1997). "Secondary structure computer prediction of the poliovirus 5′ non-coding region is improved by a genetic algorithm," *Comput. Appl. Biosc.*, Vol. 13, No. 1, pp. 1–12.

Deschênes, A. (2005). "A genetic algorithm for RNA secondary structure prediction using stacking energy thermodynamic models. Master's thesis, Simon Fraser University, Burnaby, British Columbia, Canada.

Deschênes, A. and K. C. Wiese (2004). "Using stacking-energies (INN and INN-HB) for improving the accuracy of RNA secondary structure prediction with an evolutionary algorithm—A comparison to known structures," in *Proceedings of the 2004 IEEE Congress on Evolutionary Computation*. IEEE Press, New York, Vol. 1, pp. 598–606.

Deschênes, A., K. C. Wiese, and J. Poonian (2004). "Comparison of dynamic programming and evolutionary algorithms for RNA secondary structure prediction," in *Proceedings of the 2004 IEEE Symposium on Computational Intelligence in Bioinformatics and Computational Biology (CIBCB'04)*. IEEE Press, New York, pp. 214–222.

Fogel, G. B., V. W. Porto, D. G. Weekes, D. B. Fogel, R. H. Griffey, J. A. McNeil, E. Lesnik, D. J. Ecker, and R. Sampath (2002). "Discovery of RNA structural elements using evolutionary computation," *Nucleic Acids Res.*, Vol. 30, No. 23, pp. 5310–5317.

Freier, S. M., B. J. Burger, D. Alkema, T. Neilson, and D. H. Turner (1983). "Effects of 3′ dangling end stacking on the stability of GGCC and CCGG double helixes," *Biochemistry*, Vol. 22, No. 26, pp. 6198–6206.

Freier, S. M., R. Kierzek, J. A. Jaeger, N. Sugimoto, M. H. Caruthers, T. Neilson, and D. H. Turner (1986). "Improved free-energy parameters for predictions of RNA duplex stability," *Proc. Nat. Acad. Sci. USA*, Vol. 83, pp. 9373–9377.

Goldberg, D. E. and R. Lingle, Jr. (1985). "Alleles, loci and the travelling salesman problem," in *Proceedings of the First International Conference on Genetic Algorithms*. J. J. Grefenstette, Ed., Lawrence Erlbaum Associates, Mahwah, NJ, pp. 154–159.

Gultyaev, A. P., F. H. D. van Batenburg, and C. W. A. Pleij (1995). "The computer-simulation of RNA folding pathways using a genetic algorithm," *J. Mol. Biol.*, Vol. 250, pp. 37–51.

He, L., R. Kierzek, J. SantaLucia, Jr., A. E. Walter, and D. H. Runer (1991). "Nearest-neighbor parameters for GU mismatches: GU/UG is destabilizing in the contexts CGUG/GUGC, UGUA/AUGU but stabillizing in GGUC/CUGG," *Biochemistry*, Vol. 30, pp. 11124–11132.

Higgs, P. G. (2000). "RNA secondary structure: physical and computational aspects," *Quart. Rev. Biophys.*, Vol. 33, pp. 199–253.

JAEGER, J. A., D. H. TURNER, and M. ZUKER (1989). "Improved predictions of secondary structures for RNA," *Biochemistry*, Vol. 86, pp. 7706–7710.

KIERZEK, R., M. H. CARUTHERS, C. E. LONGFELLOW, D. SWINTON, D. H. TURNER, and S. M. FREIER (1986). "Polymer-supported RNA synthesis and its application to test the nearest neighbor model for duplex stability," *Biochemistry*, Vol. 25, pp. 7840–7846.

MATHEWS, D. H., T. C. ANDRE, J. KIM, D. H. TURNER, and M. ZUKER (1998). "An updated recursive algorithm for RNA secondary structure prediction with improved free energy parameters," in *American Chemical Society*, Vol. 682, N. B. Leontis and J. SantaLucia Jr., Eds. American Chemical Society, Washington, DC, pp. 246–257.

MATHEWS, D. H., M. D. DISNEY, J. L. CHILDS, S. J. SCHROEDER, M. ZUKER, and D. H. TURNER (2004). "Incorporating chemical modification constraints into a dynamic programming algorithm for prediction of RNA secondary structure," *Proc. Nat. Acad. Sci. USA*, Vol. 101, pp. 7287–7292.

MATHEWS, D. H., J. SABINA, M. ZUKER, and D. H. TURNER (1999). "Expanded sequence dependence of thermodynamic parameters improves prediction of RNA secondary structure," *J. Mol. Biol.*, Vol. 288, pp. 911–940.

NUSSINOV, R., G. PIECZENIK, J. R. GRIGGS, and D. J. KLEITMAN (1978). "Algorithms for loop matchings," *SIAM J. Appl. Math.*, Vol. 35, pp. 68–82.

OLIVER, I. M., D. J. SMITH, and J. R. C. HOLLAND (1987). "A study of permutation crossover operators on the traveling salesman problem," in *Proceedings of the Second International Conference on Genetic Algorithms (ICGA-87)*. Lawrence Erlbaum Associates, Mahwah, NJ, pp. 224–230.

SERRA, M. J. and D. H. TURNER (1995). "Predicting thermodynamic properties of RNA," *Meth. Enzymol.*, Vol. 259, pp. 242–261.

SHAPIRO, B. A., D. BENGALI, and W. KASPRZAK (2001a). "Determination of RNA folding pathway functional intermediates using a massively parallel genetic algorithm," in *Proceedings of the ACM SIGKDD Workshop on Data Mining in Bioinformatics BIOKDD 2001*, p. 1. Abstract and references.

SHAPIRO, B. A., D. BENGALI, W. KASPRZAK, and J. C. WU (2001b). "RNA folding pathway functional intermediates: Their prediction and analysis," *J. Mol. Biol.*, Vol. 312, pp. 27–44.

SHAPIRO, B. A. and J. NAVETTA (1994). "A massively-parallel genetic algorithm for RNA secondary structure prediction," *J. Supercomput.*, Vol. 8, pp. 195–207.

SHAPIRO, B. A. and J. C. WU (1996). "An annealing mutation operator in the genetic algorithms for RNA folding," *Comput. Appl. Biosci.*, Vol. 12, No. 3, pp. 171–180.

SUGIMOTO, N., R. KIERZEK, S. M. FREIER, and D. H. TURNER (1986). "Energetics of internal GU mismatches in ribooligonucleotide helixes," *Biochemistry*, Vol. 25, No. 19, pp. 5755–5759.

SYSWERDA, G. (1991). In "Schedule optimization using genetic algorithms," *Handbook of Genetic Algorithms*, L. Davis, Ed., Van Nostrand Reinhold, New York.

TITOV, I. I., D. G. VOROBIEV, V. A. IVANISENKO, and N. A. KOLCHANOV (2002). "A fast genetic algorithm for RNA secondary structure analysis," *Russ. Chem. Bull.*, Vol. 51, No. 7, pp. 1135–1144.

VAN BATENBURG, F. H. D., A. P. GULTYAEV, and C. W. A. PLEIJ (1995). "An APL-programmed genetic algorithm for the prediction of RNA secondary structure," *J. Theor. Biol.*, Vol. 174, pp. 269–280.

WEISS, N. A. (1999). *Elementary Statistics*. Addison-Wesley, Boston.

WIESE, K. C., A. DESCHÊNES, and E. GLEN (2003). "Permutation based RNA secondary structure prediction via a genetic algorithm," in *Proceedings of the 2003 Congress on Evolutionary Computation (CEC2003)*. R. Sarker, R. Reynolds, H. Abbass, K. C. Tan, B. McKay, D. Essam, and T. Gedeon, Eds., IEEE Press, New York, pp. 335–342.

WIESE, K. C. and E. GLEN (2002). "A permutation based genetic algorithm for RNA secondary structure prediction," in *Soft Computing Systems*, Vol. 87 of *Frontiers in Artificial Intelligence and Applications*. A. Abraham, J. Ruiz del Solar, and M. Koppen, Eds., chapter 4, IOS Press, Amsterdam, pp. 173–182.

WIESE, K. C. and E. GLEN (2003). "A permutation-based genetic algorithm for the RNA folding problem: A critical look at selection strategies, crossover operators, and representation issues," *BioSyst.—Special Issue on Computational Intelligence in Bioinformatics*, Vol. 72, pp. 29–41.

WIESE, K. and S. D. GOODWIN (2001). "Keep-best reproduction: A local family competition selection strategy and the environment it flourishes in," *Constraints*, Vol. 6, No. 4, pp. 399–422.

WIESE, K. C. and A. HENDRIKS (2006). "Comparison of P-RnaPredict and mfold—Algorithms for RNA secondary structure prediction," *Bioinformatics*, Vol. 22, No. 8, pp. 934–942.

WU, M., J. A. MCDOWELL, and D. H. TURNER (1995). "A periodic table of symmetric tandem mismatches in RNA," *Biochemistry*, Vol. 34, pp. 3204–3211.

XIA, T., Jr., J. SANTALUCIA, M. E. BURKARD, R. KIERZEK, S. J. SCHROEDER, X. JIAO, C. COX, and D. H. TURNER (1998). "Thermodynamic parameters for an expanded nearest-neighbor model for formation of RNA duplexes with Watson-Crick base pairs," *Biochemistry*, Vol. 37, pp. 14719–14735.

XIA, T., J. A. MCDOWELL, and D. H. TURNER (1997). "Thermodynamics of nonsymmetric tandem mismatches adjacent to GC base pairs in RNA," *Biochemistry*, Vol. 36, pp. 12486–12497.

ZUKER, M. (1989). "On finding all suboptimal foldings of an RNA molecule," *Science*, Vol. 244, pp. 48–52.

ZUKER, M. (1994). "Prediction of RNA secondary structure by energy minimization," in *Computer Analysis of Sequence Data*. A. M. Griffin and H. G. Griffin, Eds., Humana Press, pp. 267–294.

ZUKER, M. (2000). "Calculating nucleic acid secondary structure," *Curr. Opin. Struct. Biol.*, Vol. 10, pp. 303–310.

ZUKER, M. (2003). "Mfold web server for nucleic acid folding and hybridization prediction," *Nucleic Acids Res.*, Vol. 31, No. 13, pp. 3406–3415.

ZUKER, M., J. A. JAEGER, and D. H. TURNER (1991). "A comparison of optimal and suboptimal RNA secondary structures predicted by free energy minimization with structures determined by phylogenetic comparison," *Nucleic Acids Res.*, Vol. 19, No. 10, pp. 2707–2714.

ZUKER, M., D. H. MATHEWS, and D. H. TURNER (1999). "Algorithms and thermodynamics for RNA secondary structure prediction: A practical guide," in *RNA Biochemistry and Biotechnology*. J. Barciszewski and B. F. C. Clark, Eds., NATO ASI Series. Kluwer Academic, Boston.

ZUKER, M. and P. STIEGLER (1981). "Optimal computer folding of large RNA sequences using thermodynamics and auxiliary information," *Nucleic Acids Res.*, Vol. 9, pp. 133–148.

Chapter 10

Machine Learning Approach for Prediction of Human Mitochondrial Proteins

Zhong Huang, Xuheng Xu, and Xiaohua Hu

10.1 INTRODUCTION

10.1.1 Background on Mitochondria and Mitochondria Import Complexes

Mitochondria are important subcellular organelles where many fundamental biological processes take place, including energy production, apoptosis, and biochemical metabolism. The dysfunction of mitochondria has been linked, therefore, to a variety of human diseases including Alzheimer's disease (Chagnon et al., 1999), Parkinson's disease (Langston et al., 1983), type-2 diabetes (Poulton et al., 2002), cardiac disease (Marin-Garcia and Goldenthal, 1997), a variety of neuromuscular diseases (Kamieniecka and Schmalbruch, 1980; DiMauro et al., 1998), and cancer (Polyak et al., 1998). Mitochondria are composed of an outer and inner membrane that divide the organelle into an intermembrane space and matrix. The two-membrane structure separates the mitochondria's environment completely from the cytoplasm.

Human mitochondria contain their own unique circular genome that encodes 13 highly hydrophobic proteins. This circular genome is the result of an ancient endosymbiosis event of an α-proteobacteria by ancient eukaryotes roughly 2 billion years ago (Andersson and Kurland, 1999). According to this theory of evolutionary biology, over time the human mitochondrial genome lost many of its coding genes, partly through integration of these genes into the human genome residing in the nucleus. The mitochondrial genome now encodes only 13 proteins, 22 tRNAs (transfer RNA), and 2 rRNAs (ribosomal RNAs). The 13 proteins encoded by the

Computational Intelligence in Bioinformatics. Edited by G. B. Fogel, D. Corne, and Y. Pan

mitochondrial genome participate in the formation of 5 respiratory chain complexes, along with other proteins imported from cytoplasm.

The majority of proteins localized in the mitochondria are encoded by genes in the nuclear genome whose protein products are then targeted to the mitochondria. The process by which proteins pass from the cytoplasm to the mitochondria (e.g., the mitochondria membrane importing system) still remains largely unknown for humans. Most of the mitochondrial protein precursors are synthesized in cytoplasm with a mitochondria targeting signal peptide at the amino terminus of the sequence (Dolezal et al., 2006). Followed by translation in the cytoplasm, newly synthesized protein precursors with the mitochondrial targeting signal sequence are escorted to surface of mitochondria by chaperone proteins such as heat shock protein 70 (HSP70) and mitochondria import stimulation factor (MSF). The mitochondria targeting proteins associated with chaperones then interact with several large complexes of mitochondria membrane import machinery, which are mainly assembled by mitochondrial protein translocases to function as a protein assembly platform, docking scaffold, and translocation channel. The *translocation of outer membrane* (TOM) complex of the mitochondria is composed of several membrane protein subunits that are responsible for recognizing proteins targeted to the surface of the mitochondria and subsequently transferring them into the mitochondria via a translocation channel. Once imported into the intermembrane space, the imported protein is then bound by tiny TIM (*translocase of inner membrane*) protein and shuttled to the TIM complex, where it is released into TIM channel and into the mitochondria matrix (Stojanovski et al., 2003; Hood and Joseph, 2004; Wiedemann et al., 2004; Dolezal et al., 2006). Although significant progress has been made in characterizing the mitochondria import systems of the yeast *Saccharomyces cerevisiae* and fungus *Neurospora crassa*, the mechanism of this mitochondrial import complex remains largely unknown for mammals.

Because of its fundamental importance for normal cell function, mitochondria have long been a focus of research. To gain better insight into the complete function of the human mitochondria, the full set of mitochondrial proteins must be determined to reveal the complex mitochondrial protein interaction network. A combination of approaches has been applied in this regard. For example, one might consider an empirical approach toward obtaining the complete mitochondria proteome. In such an approach, mitochondria can be purified from yeast, mouse, rat, and human. Purified mitochondrial proteins can then solubilized and separated by polyacrylamide gel electrophoresis (PAGE), such as through isoelectric focusing two-dimensional (IEF 2DE) sodium dodecyl sulfate (SDS) PAGE. Separated mitochondrial proteins can be extracted from the gel, digested with proteases, further purified by liquid-phase chromatography, and finally individually identified using mass spectrometry. This approach is currently the most reliable and accurate way of characterizing mitochondrial proteins. However, the approach suffers from several technical limitations, including the possibility for false positives resulting from minor contamination of nonmitochondrial proteins during purification, and/or false negatives caused by low abundance and insolubility of highly hydrophobic, membrane-associated proteins. A bioinformatics approach to assist in the identification of mitochondrial proteins is desired.

10.1.2 Bioinformatics Approaches for Protein Localization Prediction

Advancements in high-throughput proteomics, rapid DNA (deoxyribonucleic acid) sequencing, and resulting protein prediction, have opened the possibility for the application of machine learning techniques using large quantities of biological data. With the rapid accumulation of experimentally annotated protein sequences deposited in protein databases, along with the genome-wide analysis of mRNA (messenger RNA) transcripts, bioinformatics approaches can be used to assist with protein subcellular location prediction. It is commonly accepted that the total number of human genes in the human genome is roughly 30,000 (Jensen, 2004). Using in silico prediction methods, it is estimated that 5% of the human proteome is composed of mitochondrial proteins, a total of roughly 1500 proteins (Taylor et al., 2003a). In general, methods for prediction of protein subcellular localization are based on protein functional domain structure, sequence similarity searches using BLAST or PSI-BLAST, specific known organelle-targeting signal peptide sequences, amino acid compositions of proteins, or any combination of the above. Currently, several high-performance computational classification methods are available via the Internet to predict the localization of proteins to the mitochondria for eukaryotes. A list of these programs is provided in Table 10.1. For the development of these algorithms, a training set of known examples of mitochondrial-targeted proteins was generated via the SWISSPROT database (Bairoch and Apweiler, 1999; Gasteiger et al., 2001). Candidate mitochondrial proteins with subcellular localization annotations of low confidence, such as "probably" or "predicted," were excluded from the set of known positives. Most of the methods listed in Table 10.1 utilized protein sequences from animals, yeast, fungi, and bacteria for the prediction of multiple subcellular localizations (nuclear, cytoplasm, mitochondria, peroxisome, Golgi apparatus, lysosomes, endoplasmic reticulum, and extracellular/cytoskeleton). In Garg et al. (2005), human proteins were selected from SWISSPROT to train support vector machines (SVMs) for the prediction of multiple subcellular localizations including mitochondrial, cytoplasm, nuclear, and plasma membrane. To our knowledge, there exists no approach specific for the prediction of human mitochondria localization using only experimentally determined human mitochondrial proteins as training data sets. Such an approach is the focus of this chapter.

Human proteome studies by Gaucher et al. (2004) and Taylor et al. (2003b) revealed 615 distinct mitochondrial proteins with highly purified mitochondria extracted from human heart tissue using one-dimensional SDS PAGE, multidimensional liquid chromatography, and mass spectrometry analysis (LC-MS/MS). In this chapter, we utilized the human mitochondrial proteins experimentally determined by Taylor et al. (2003b) to classify mitochondrial protein location using machine learning approaches.

Table 10.1 Selected Methods and Algorithms for Prediction of Mitochondria Subcellular Localization in Eukaryotes[a]

Algorithm Name	Algorithm Approach	Predicted Subcellular Localization	Origin of Protein Sequence	MCC for Overall Performance	URL	References
MitoPred	Combined scoring of Pfam domain pattern, amino acid composition and pI value	Mito	Annotated eukaryotic protein sequences	0.73	http://mitopred.sdsc.edu	Guda et al. (2004a,b)
pTarget	Combined scoring of Pfam domains and the amino acid composition	Cyt, ER, EXC, Golgi, Lyso, Mito, Nuc, Plasma, Perox	Annotated eukaryotic protein sequences	0.74	http://bioinformatics.albany.edu/~ptarget	Guda et al. (2005, 2006)
MitPred	SVMs, amino acid composition, BLAST homology, HMMs	Mito	Annotated eukaryotic protein sequences	0.81	http://www.imtech.res.in/raghava/mitpred	Kumar et al. (2006)
TargetP	ANNs, mitochondrial targeting sequence	Mito	Annotated eukaryotic protein sequences	0.73	http://www.cbs.dtu.dk/services/TargetP/	Emanuelsson et al. (2000)
WoLF PSORT	*k*-nearest neighbors, sequence features, amino acid composition, and sequence length	Mito, Cytosk, ER, Plasma, Golgi, Nuc, Perox	Annotated eukaryotic protein sequences	N/A	http://wolfpsort.org	Horton et al. (2006)
LOCSVMPSI	SVMs, position-specific scoring matrix from PSI-BLAST	Mito, Nuc, Cyto, EXC	Annotated eukaryotic protein sequences	0.81	http://Bioinformatics.ustc.edu.cn/LOCSVMPSI/LOCSVMPSI.php	Xie et al. (2005)
ESL-Pred	SVMs, amino acid composition, amino acid physical-chemical properties, PSI-BLAST	Mito, Nuc, Cyto, EXC	Annotated eukaryotic protein sequences	0.69	http://www.imtech.res.in/raghava/eslpred/	Bhasin et al. (2004)

[a]Cyt, cytoplasm; ER, endoplasmic reticulum; EXC, extracellular/secretory; Golgi, golgi; Lyso, lysosomes; Mito, mitochondria; Nuc, nucleus; Plasma, plasma membrane; Perox, peroxisomes; Cytsk, cytoskeleton; MCC, Matthews correlation coefficient.

10.1.3 Machine Learning Approaches for Protein Subcellular Location Prediction

Artificial neural networks (ANNs) and SVMs are two important machine learning algorithms frequently used in protein structure classification. ANNs simulate the neural network of the brain by adopting the concept of interconnection and parallel distributed processing of neurons. The multiple perceptron is one of the most widely used neural network architectures. It is composed of one input layer, multiple hidden layers, and one output layer. Each layer consists of a number of nodes or "neurons" with weighted directional links to other nodes/neurons in the next layer. Functioning as a processing unit, each node takes input and generates an output based on a non-linear activation function using a modifiable parameter. Through the interconnection of neurons via a hidden layer (or layers), the ANN can learn from the input data and generate parameters to fit this data in the form of a new model. The learning algorithm embedded in each neuron uses an error correction function to adjust its input weight for a desired output. ANNs have proved to be excellent classifiers for data sets where little information is known about the relationship between the input and output.

Support vector machines, on the other hand, are another important classification method that is supported strongly by mathematical theory. Given a set of data labeled as two different classes, SVMs separate the two classes using a hyperplane that has maximum margin with vectors from both classes. SVMs can handle high-dimensional feature spaces efficiently, compared to that of ANN models. Similar to the architecture of ANN models, SVMs use different kernel functions including linear, polynomial, and radial basis function (RBF) Gaussians (Pierre Baldi, 2001). Kernel functions are used to convert the training vector into high-dimensional feature spaces for hyperplane separation. In this chapter, we compare the performance of ANNs and SVMs for the problem of predicting protein localization to the mitochondrial based on protein sequence information.

10.2 METHODS AND SYSTEMS

10.2.1 Data Sets

The sequences of 615 human mitochondrial proteins characterized by one-dimensional electrophoresis and LC/MS mass spectrometry by Taylor et al. (2003b) were downloaded from GenBank using batch retrieval. The mitochondrial genome encodes proteins that have very different physical-chemical properties relative to their nuclear-encoded cytoplasm-imported protein counterparts, which may cause bias when including them in a training set. Therefore proteins encoded by the mitochondrial genome (ND1, ND2, ND4, ND5, ND6, cytochrome b, COX I, COX II, ATPase 6, ATPase 8) were excluded from the data sets. For the inclusion of human non-mitochondrial proteins as a negative training data set, we searched SWISSPROT for proteins with known subcellular localization in the nucleus, cytoplasm, Golgi apparatus, or extracellular matrix. Table 10.2 shows the composition of our training data. For comparison of the classification performance using independent data, we downloaded 879 MitoPred predicted human mitochondrial proteins with confidence >99% from the SWISSPROT database (Guda et al., 2004a). The same negative training data was used with the MitoPred positive data as an independent performance analysis.

10.2.2 Classification Algorithms

We used the Weka data mining package implemented in Java to conduct all classification tasks (Witten and Frank, 2005). For SVM classification, we used both a polynomial and RBF kernel. The parameters used for polynomial kernel are: -C 1.0 -E 1.0 -G 0.01 -A 250007 -L 0.0010 -P 1.0E-12 -N 0 -V -1 -W 1. The parameters used for RBF kernel are: -C 1.0 -E 1.0 -G 0.1 -A 250007 -T 0.0010 -P 1.0E-12 -N 0 -R -V -1 -W 1. It is noted that parameter C is the complexity constant, E the exponent for the polynomial kernel, G gamma for the RBF kernel, A the size of the kernel cache, T the tolerance parameter, P the epsilon for round-off error, N whether to 0 = normalize/1 = standardize/2 = neither, R indicates the use of the RBF kernel, V the number of folds for the internal cross validation, and W the random number seed, respectively. For ANN classification, we used a multiple perceptron with the following parameters: -L 0.3 -M 0.2 -N 500 -V 0 -S 0 -E 20 -H a, where the L option is the learning rate for the backpropagation algorithm, M the momentum rate for the backpropagation algorithm, N the number of epochs to train through, V the percentage size of the validation set used to terminate training, S the value used to seed the random number generator, E the threshold for the number of consequetive errors allowed during validation testing, and H the number of nodes to be used on each layer and the value of wildcard a = (attributes + classes)/2. We also compared the performance of ANN and SVM with naïve Bayes and Bayes network algorithms (Colin Howson, 1993). Due to the relatively small amount of data collected, we performed fivefold cross validation for all classification algorithms. The classification performance was tested by accuracy, sensitivity, specificity, and with the Mathews correlation coefficient (MCC):

Table 10.2 Composition of Training Data and Number of Proteins Used in Mitochondria Subcellular Localization Classification

Subcellular Localization	Number of Proteins
Mitochondria	606
Nuclear	483
Cytoplasm	67
Golgi apparatus	103
Extracellular matrix	36

$$\text{Accuracy} = \frac{\text{TP} + \text{TN}}{\text{TP} + \text{TN} + \text{FP} + \text{FN}} \times 100, \tag{10.1}$$

$$\text{Specificity} = \frac{\text{TN}}{\text{TN} + \text{FP}} \times 100, \tag{10.2}$$

$$\text{Sensitivity} = \frac{\text{TP}}{\text{TP} + \text{FN}} \times 100, \tag{10.3}$$

$$\text{MCC} = \frac{\text{TP} \times \text{TN} - \text{FP} \times \text{FN}}{\sqrt{(\text{TP} + \text{FP}) \times (\text{TP} + \text{FN}) \times (\text{TN} + \text{FP}) \times (\text{TN} + \text{FN})}}, \tag{10.4}$$

where TN is true negatives, TP is true positives, FP is false positives, and FN is false negatives.

10.2.3 Feature Selection

In this study we extracted three feature types from each protein sequence. The first was the simple amino acid composition which follows the IUB/IUPAC amino acid coding standard. The second feature was the combination of amino acid composition and the seven physical-chemical properties of amino acids, namely aliphatic (ILV), aromatic (FHWY), nonpolar (ACFGILMPVWY), polar (DEHKNQRSTZ), charged (BDEHKRZ), basic (HKR), and acidic (BDEZ). The third feature was the traditional (i+1) dipeptide composi-tion, which encapsulates global information on simple amino acid composition and local order of any two adjacent amino acids in a fixed length of 400. Perl scripts were written to extract the above features from protein sequences in FASTA format.

10.3 RESULTS AND DISCUSSION

To explore different ways of capturing protein sequence information, we used simple amino acid composition, a combination of amino acid composition, seven chemico-physical properties, as well as traditional dipeptide composition in following classifications. Our results indicate that dipeptide composition performs better using ANN and SVM classifiers. As expected, the combination of amino acid composition and chemicophysical properties gives better performance compared with simple amino acid composition alone (Tables 10.3 and 10.4). To our surprise, however, multiple perceptron ANNs outperform SVMs in all three training models. As shown in Table 10.4, when a 400-feature dipeptide composition model was used in multiple perceptron ANN training, we achieved 84.69% accuracy, 91.70% sensitivity, and 77.21% specificity, with MCC at 0.7. Using sequence information with 400 dipeptide composition features, we compared classification performances using four popular classification methods including ANN, SVM, naïve Bayes, and the Bayes network.

Table 10.3 SVM Classification Using Polynomial and RBF Kernel on Protein Amino Acid Sequence with Different Feature Spaces

Algorithm (Kernel)	Features	Sensitivity (%)	Specificity (%)	Accuracy (%)	MCC
SVM (polynomial)	20	88.05	68.52	78.59	0.58
SVM (RBF)	20	89.59	64.92	78.59	0.58
SVM (polynomial)	27	88.05	69.12	78.88	0.58
SVM (RBF)	27	89.03	67.47	78.59	0.58
SVM (polynomial)	400	84.08	80.93	82.56	0.65
SVM (RBF)	400	88.17	73.57	81.10	0.63

Table 10.4 Multiple Perceptron Neural Network Classification of Mitochondrial Protein

Algorithm (Kernel)	Number of Features	Sensitivity (%)	Specificity (%)	Accuracy (%)	MCC
ANN (multiple perceptron)	20	82.98	76.16	79.68	0.59
ANN (multiple perceptron)	27	81.29	80.06	80.70	0.61
ANN (multiple perceptron)	400	91.70	77.21	84.69	0.70

Table 10.5 Performance Comparison Using Multiple Perceptron Neural Network, Polynorminal SVM, Naïve Bayes and Bayes Network Classification Methods

Algorithm (Kernel)	Features	Sensitivity (%)	Specificity (%)	Accuracy (%)	MCC
ANN (multiple perceptron)	400	91.70	77.21	84.69	0.70
SVM (polynomial)	400	84.08	80.93	82.56	0.65
Naïve Bayes	400	77.18	73.87	75.58	0.51
Bayes network	400	76.76	69.22	73.11	0.46

Table 10.5 shows that multiple perceptron ANN and SVM outperform naïve Bayes and the Bayes network in classifying mitochondrial proteins.

The high-dimensional feature space of dipeptide composition is very computationally expensive, especially for multiple perceptron neural network classification. Several approaches including information gain based algorithm, rough sets, and decision trees have been utilized to reduce the feature space for better scalability and accuracy. In this study we used information gain with ranker algorithm implemented in Weka package to find the subset of reduced features. Table 10.6 shows that the number of features can be reduced to 250, as opposed to a full set of 400 dipeptide features, without affecting the classification performance. However, in the experiments we have done so far, the feature reduction algorithm fails to improve

Table 10.6 Information Gain-Based Feature Selection Slightly Improves SVM Classification Performance at 250 Feature Spaces

Algorithm (Kernel)	Number of Features	Sensitivity (%)	Specificity (%)	Accuracy (%)	MCC
SVM (polynomial)	200	85.07	80.48	82.85	0.66
	250	85.92	81.38	83.72	0.67
	300	85.35	81.83	83.65	0.67
	350	84.93	81.38	83.21	0.66
	400	84.08	80.93	82.56	0.65

Table 10.7 Comparison of SVM Classification using Two Different Training Datasets[a]

	Experimentally Determined Positive Training Data Sets	MitoPred Predicted Positive Training Data Sets
	SVM (polynomial)	SVM (polynomial)
Features	400	400
Sensitivity (%)	84.08	82.94
Specificity (%)	80.93	75.19
Accuracy (%)	82.56	79.60
MCC	0.65	0.58

[a]One dataset is taken from experimentally determined mitochondrial proteins by Taylor et al. (2003b), and one from computationally predicted human mitochondrial proteins by MitoPred from Guda et al. (2004a). Both include the same negative training datasets selected from the SWISSPROT database.

the ANN performance (data not shown). It may be that in this case all 400 high-dimensional dipeptide composition features are needed in order to reliably achieve the best reults from the ANN; however, we note that in similar ANN research it is common to find that further improvements are possible from a reduced set of features. The findings so far are possibly due to the reduced feature sets from the chosen Weka algorithm being insufficiently informative for exploitation by the ANN. Future work, for example, may instead use evolved feature subsets (see Chapter 6 for discussion of this topic).

Guda et al. (2004a, b) developed a high-performance algorithm, MitoPred, to predict the eukaryotes mitochondrial proteins. In this study, we compared the SVM performance using dipeptide composition of mitochondrial proteins predicted by MitoPred (with >99% confidence) as positive training data, together with the dipeptide composition of nonmitochondrial proteins used in our study. Table 10.7 shows the slightly improved performance with training data sets from experimentally determined sequences than those that were predicted computationally to be located in the mitochondria.

10.4 CONCLUSIONS

Determination of protein subcellular localization not only can give insights into potential functions, but it can also help build knowledge of the protein–protein interaction networks and signaling pathways within the same cellular compartment. Mitochondria are essential for survival of eukaryotes as an organelle to generate adenosine 5′-triphosphate (ATP) through oxidative phosphorylation and participate in apoptosis signaling pathways. Dysfunction of mitochondrial proteins, which is largely due to the dysfunction of imported nuclear encoded proteins, is a direct cause of mitochondrial disease in humans. To predict proteins targeted to mitochondria based on the primary amino acid sequence is a very important and challenging task. The primary sequence of proteins, represented as an ordered amino acid chain, contains information that determines its higher structure and functions. Machine learning has been proved to be an efficient approach to mine the large deposit of protein sequence information for functional implication. SVM, ANN, *k*-means nearest neighbors, and association rule classification have been successfully applied on mitochondrial protein classification and other protein subcellular localization predictions (Emanuelsson et al., 2000; Bhasin and Raghava, 2004; Guda et al., 2004a,b; Garg et al., 2005; Guda and Subramaniam, 2005; Xie et al., 2005; Horton, 2006). In those studies, the selected eukaryotic mitochondrial protein sequences used in model training were downloaded from SWISSPROT by parsing and screening sequence annotations with mitochondria subcellular localization information. Although sequences with low confidence annotations were removed from the data, the overall accuracy of extracted mitochondrial proteins are still subject to the quality of the human-curated annotations. In this study, we utilized sequences of highly purified human mitochondrial proteins identified by one-dimensional SDS-PAGE electrophoresis followed by LC/MS mass spectrometry analysis as sources of positive training data sets. All negative training data sets consisted of human proteins with nuclear, Golgi, cytoplasm, lysosome, ER, and extracellular localizations. The empirically determined mitochondrial proteins and human origin of all training sequences allowed us to classify human mitochondrial proteins using high-quality data sets as input. In summary, we reported neural network and SVM-based methods to classify human mitochondrial proteins with high confidence by using experimentally characterized human mitochondrial proteins. Our results demonstrate that ANNs and SVMs can be reliably utilized to mine large protein data sets, based on their dipeptide composition, for human protein mitochondria localization predictions.

Acknowledgments

This work is supported in part by NSF Career Grant IIS 0448023, NSF CCF 0514679, PA Dept. of Health Tobacco Settlement Formula Grant (No. 240205 and No. 240196), and PA Dept. of Health Grant (No. 239667).

REFERENCES

ANDERSSON, S. G. and C. G. KURLAND (1999). "Origins of mitochondria and hydrogenosomes," *Curr. Opin. Microbiol.* Vol. 2, pp. 535–541.

BAIROCH, A. and R. APWEILER (1999). "The SWISS-PROT protein sequence data bank and its supplement TrEMBL in 1999," *Nucleic Acids Res.*, Vol. 27, pp. 49–54.

BALDI P. and S. BURNAK (2001). *Bioinformatics: The Machine Learning Approach*, The MIT Press, Boston.

BHASIN, M. and G. P. RAGHAVA (2004). "ESLpred: SVM-based method for subcellular localization of eukaryotic proteins using dipeptide composition and PSI-BLAST," *Nucleic Acids Res.*, Vol. 32, pp. W414–419.

CHAGNON, P., M. GEE, M. FILION, Y. ROBITAILLE, M. BELOUCHI, and D. GAUVREAU (1999). "Phylogenetic analysis of the mitochondrial genome indicates significant differences between patients with Alzheimer disease and controls in a French-Canadian founder population," *Am. J. Med. Genet.*, Vol. 85, pp. 20–30.

COLIN HOWSON, P. U. (1993). *Scientific Reasoning: The Bayesian Approach*, Open Court Publishing Company, Chicago.

DIMAURO, S., E. BONILLA, M. DAVIDSON, M. HIRANO, and E. A. SCHON (1998). "Mitochondria in neuromuscular disorders," *Biochim. Biophys. Acta*, Vol. 1366, pp. 199–210.

DOLEZAL, P., V. LIKIC, J. TACHEZY, and T. LITHGOW (2006). "Evolution of the molecular machines for protein import into mitochondria," *Science*, Vol. 313, pp. 314–318.

EMANUELSSON, O., H. NIELSEN, S. BRUNAK, and G. VON HEIJNE (2000). "Predicting subcellular localization of proteins based on their N-terminal amino acid sequence," *J. Mol. Biol.*, Vol. 300, pp. 1005–1016.

GARG, A., M. BHASIN, and G. P. RAGHAVA (2005). "Support vector machine-based method for subcellular localization of human proteins using amino acid compositions, their order, and similarity search," *J. Biol. Chem.*, Vol. 280, pp. 14427–14432.

GASTEIGER, E., E. JUNG, and A. BAIROCH (2001). "SWISS-PROT: Connecting biomolecular knowledge via a protein database," *Curr. Issues Mol. Biol.*, Vol. 3, pp. 47–55.

GAUCHER, S. P., S. W. TAYLOR, E. FAHY, B. ZHANG, D. E. WARNOCK, S. S. GHOSH, and B. W. GIBSON (2004). "Expanded coverage of the human heart mitochondrial proteome using multidimensional liquid chromatography coupled with tandem mass spectrometry," *J. Proteome Res.*, Vol. 3, pp. 495–505.

GUDA, C., E. FAHY, and S. SUBRAMANIAM (2004a). "MITOPRED: A genome-scale method for prediction of nucleus-encoded mitochondrial proteins," *Bioinformatics*, Vol. 20, pp. 1785–1794.

GUDA, C., P. GUDA, E. FAHY, and S. SUBRAMANIAM (2004b). "MITOPRED: A web server for the prediction of mitochondrial proteins," *Nucleic Acids Res.*, Vol. 32, pp. W372–374.

GUDA, C. and S. SUBRAMANIAM (2005). "pTARGET [corrected] a new method for predicting protein subcellular localization in eukaryotes," *Bioinformatics*, Vol. 21, pp. 3963–3969.

GUDA, C. (2006). "pTARGET: a web server for predicting protein subcellular localigation," *Nucleic Acids Res.*, Vol. 34, pp. W210–213.

HOOD, D. A. and A. M. JOSEPH (2004). "Mitochondrial assembly: Protein import," *Proc. Nutr. Soc.*, Vol. 63, pp. 293–300.

HORTON, P., P. K.-J., T. OBAYASHI, and K. NAKAI (2006). "Protein subcellular localization prediction with WoLF PSORT," Proceedings of the 4th Annual Asia Pacific Bioinformatics Conference APBC06, pp. 39–48.

JENSEN, O. N. (2004). "Modification-specific proteomics: Characterization of post-translational modifications by mass spectrometry," *Curr. Opin. Chem. Biol.*, Vol. 8, pp. 33–41.

KAMIENIECKA, Z. and H. SCHMALBRUCH (1980). "Neuromuscular disorders with abnormal muscle mitochondria," *Int. Rev. Cytol.*, Vol. 65, pp. 321–357.

KUMAR, M., R. VERMA, and G. P. RAGHAVA (2006). "Prediction of mitochondrial proteins using support vector machine and hidden markov model," *J. Biol. Chem.*, Vol. 281, pp. 5357–5363.

LANGSTON, J. W., P. BALLARD, J. W. TETRUD, and I. IRWIN (1983). "Chronic Parkinsonism in humans due to a product of meperidine-analog synthesis," *Science*, Vol. 219, pp. 979–980.

MARIN-GARCIA, J. and M. J. GOLDENTHAL (1997). "Mitochondrial cardiomyopathy: Molecular and biochemical analysis," *Pediatr. Cardiol.*, Vol. 18, pp. 251–260.

POLYAK, K., Y. LI, H. ZHU, C. LENGAUER, J. K. WILLSON, S. D. MARKOWITZ, M. A. TRUSH, K. W. KINZLER, and B. VOGELSTEIN (1998). "Somatic mutations of the mitochondrial genome in human colorectal tumours," *Nat. Genet.*, Vol. 20, pp. 291–293.

POULTON, J., J. LUAN, V. MACAULAY, S. HENNINGS, J. MITCHELL, and N. J. WAREHAM (2002). "Type 2 diabetes is associated with a common mitochondrial variant: evidence from a population-based case-control study," *Hum. Mol. Genet.*, Vol. 11, pp. 1581–1583.

STOJANOVSKI, D., A. J. JOHNSTON, I. STREIMANN, N. J. HOOGENRAAD, and M. T. RYAN (2003). "Import of nuclear-encoded proteins into mitochondria," *Exp. Physiol.*, Vol. 88, pp. 57–64.

TAYLOR, S. W., E. FAHY, and S. S. GHOSH (2003a). "Global organellar proteomics," *Trends Biotechnol.*, Vol. 21, pp. 82–88.

TAYLOR, S. W., E. FAHY, B. ZHANG, G. M. GLENN, D. E. WARNOCK, S. WILEY, A. N. MURPHY, S. P. GAUCHER, R. A. CAPALDI, B. W. GIBSON, and S. S. GHOSH (2003b). "Characterization of the human heart mitochondrial proteome," *Nat Biotechnol.*, Vol. 21, pp. 281–286.

WIEDEMANN, N., A. E. FRAZIER, and N. PFANNER (2004). "The protein import machinery of mitochondria," *J. Biol. Chem.*, Vol. 279, pp. 14473–14476.

WITTEN, I. H. and E. F. FRANK (2005). *Data Mining: Practical Machine Learning Tools and Techniques*, Elsevier, Burlington, UK.

XIE, D., A. LI, M. WANG, Z. FAN, and H. FENG (2005). "LOCSVMPSI: A web server for subcellular localization of eukaryotic proteins using SVM and profile of PSI-BLAST," *Nucleic Acids Res.*, Vol. 33, pp. W105–110.

Chapter 11

Phylogenetic Inference Using Evolutionary Algorithms

Clare Bates Congdon

11.1 INTRODUCTION

Phylogenetic inference, often called simply *phylogenetics*, is the process of constructing a model of the hypothesized evolutionary relationships among the species or individuals in a data set. The grand goal of the endeavor is usually considered to be the reconstruction of the "tree of life," a model of how all species currently or formerly inhabiting the Earth are related. More commonly, phylogenetics is used to study the relationships among a set of closely related species. With organisms that mutate rapidly, such as HIV (human immunodeficiency virus), or when longitudinal data is available, such as via the fossil record, phylogenetics can also be used to study the relationships among individuals.

This task of inferring a putative evolutionary tree, or phylogeny, is computationally daunting for modest data sets and intractable for realistically large data sets, and thus has traditionally proceeded as a heuristic search. As with many domains, well-designed evolutionary computation techniques can be a boon to the search for phylogenies. Furthermore, as noted in Fogel (2005), which offers a brief survey of evolutionary computation approaches to phylogenetic analysis, using evolutionary computation in the study of natural evolution itself is particularly satisfying.

Phylogenetic analysis has long been used in paleontology, for example, to study the relationships between samples in the fossil record. Originally, the data used for phylogenetics (also called cladistics in this form) consisted of morphological traits. However, with the increasing availability of high-quality nucleotide and amino acid sequence data, interest in phylogenetic analysis has witnessed a resurgence. Furthermore, the technological advances that enable the acquisition of nucleotide and amino acid sequence data has been accompanied by the development of the Internet and of public databases to facilitate ready access to this data as well as

Computational Intelligence in Bioinformatics. Edited by G. B. Fogel, D. Corne, and Y. Pan

significant increase in the available computational power needed to study large data sets.

Whether working with the fossil record or DNA (deoxyribonucleic acid) sequences, phylogenetics is used to study the relationships among human populations; phylogenetics is also increasingly important in epidemiology, and, for example, is used to study the evolution of pathogens and the evolution of genes within and among species. Phylogenetic analysis of human populations has demonstrated the relatively recent radiation of modern humans from a relatively small population in Africa (Ingman et al., 2000). Phylogenetic analysis of pathogens has helped uncover the origin of HIV in humans (Hahn et al., 2000) and was used to confirm the transmission of HIV from a Florida dentist to his patients (Ou et al., 1992). Phylogenetics is used to identify strains of infectious agents used in bioterrorism (Pattnaik and Jana, 2005), to identify which amino acid substitutions are likely to lead to virulent flu strains (Fitch et al., 1991), and to uncover how recombination of influenza coat and core proteins in nonhuman hosts can generate pandemic strains (Webster et al., 1992). Phylogenetic analysis of genes has helped us understand how point mutations and gene duplications have led to changes in function of both developmental and regulatory genes. For example, the central role of the *Hox* genes in development has been greatly facilitated by an understanding of the phylogenetic context (e.g., Erwin and Davidson, 2002).

This chapter describes the general concept of phylogenetic inference, the challenges and opportunities for applying evolutionary computation to the task, some of the contributions made, and some open questions and future directions for work in this area.

11.2 BACKGROUND IN PHYLOGENETICS

This section cannot hope to cover all the details of phylogenetic inference but will survey some of the main themes. For additional information on phylogenetics, the reader is referred to Felsenstein (2004), Swofford et al. (1996), Page and Holmes (1998), or Hall (2004).

A phylogeny amounts to a hypothesis about the evolutionary relationships in a data set and may provide insights into which species are more closely related than others and the length of time since species have diverged. There are a variety of phylogenetic models in use, varying primarily in the representation of evolutionary relationships and the metric used for evaluating the quality of the phylogenies.

11.2.1 Data

Data sets for phylogenetics consist of a number of different samples. Depending on the task at hand, these represent data from related species or strains of an organism. Each sample consists of phenotypic or genetic attribute values, such as physical or behavioral traits, nucleotide sequences, or amino acid sequences that are of interest. In the following discussions, we will focus on the use of phylogenetics for species and nucleotide sequences specifically, but the techniques of phylogenetics must be understood to be applicable to more than species and sequence data.

It must be noted that most molecular data used for phylogenetic studies is necessarily the output from multiple sequence alignment programs. Nucleotide sequences must be aligned to establish the correspondence between base positions used for the phylogeny construction task. However, multiple sequence alignment is itself a computationally intractable task for data sets of interesting sizes and is therefore also undertaken with heuristic approaches. As with phylogenetics, the quality of the alignments is a product of the scoring system used (which embodies hypotheses about events that are considered more or less likely) and the search method. Multiple alignments generally proceed as pairwise alignments to an anchor species rather than a true global alignment and are commonly accepted to be flawed. The notion of using multiple alignment data for phylogenetics is even more complex when one considers that functional regions move on the genome over evolutionary time. Thus, one must confront the reality that the data upon which phylogenies are built is often flawed. However, as mentioned in Section 11.1, when properly used, phylogenetics has contributed considerably to our understanding of relationships among organisms.

11.2.2 Phylogeny Representation

Phylogenies are usually represented as binary trees. The leaves of the tree represent the samples in the data, with exactly one leaf node for each sample. The interior nodes are usually considered to be common ancestors to the subtrees they connect and are typically not present in the data; the root node is considered to be the common ancestor to all the data. Each branching of a node into subtrees represents a divergence in one or more attribute values of the species within the two subtrees. A binary tree with five species (labeled A to E) is illustrated in Figure 11.1.

While most phylogenetic models are binary, representing a split from each common ancestor into exactly two species, in some variations, the trees are not binary, which allows one to represent a simultaneous divergence into more than two species, as illustrated in Figure 11.2. In some variations, a split of a node into subtrees represents a species or strain diverging from another rather than two new species arising from a common ancestor; in this case, the interior node is not considered to be a distinct species. In one of the more important variations, the branches

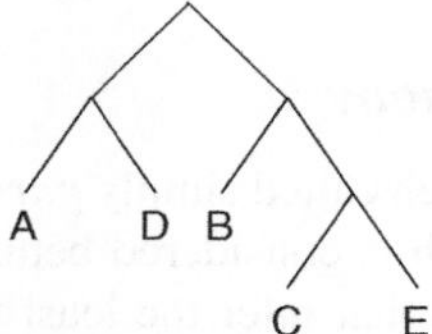

Figure 11.1 Illustration of a binary phylogeny with five species.

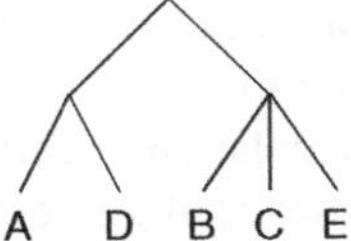

Figure 11.2 Illustration of a nonbinary phylogeny with five species.

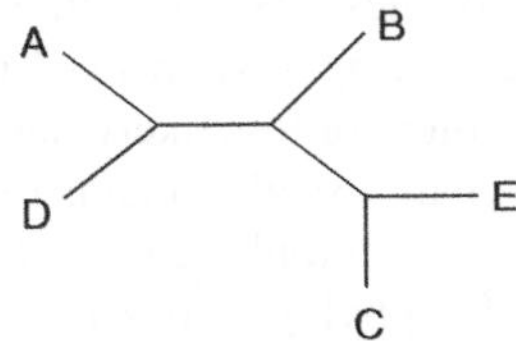

Figure 11.3 Illustration of an unrooted phylogeny with five species.

are said to have lengths representing the amount of divergence in the species they connect.

Finally, for phylogenetic models that do not incorporate assumptions about the directions of evolutionary change (the phylogenies do not model whether species A evolved into species B or vice versa), the "trees" do not have an explicit "root," which would correspond to a hypothetical ancestor. In this case, the trees are considered "unrooted" and are often drawn as networks, although they are typically still called "trees," as illustrated in Figure 11.3. Most phylogenetics software illustrates phylogenies as rooted trees whether or not the direction of evolutionary change is part of the model, and, thus, it is common to see phylogenies illustrated as binary trees regardless of the model in use.

11.2.3 Scoring Metrics for Phylogenies

In general, the metric used to evaluate the phylogenies embodies hypotheses about the likely paths of evolutionary change; for example, some metrics embody the assumption that species will grow only more complex via evolution (that features will be gained, but not lost, in the evolutionary process). Thus, there are several different metrics used within the phylogenetics community. The two most prevalent metrics used to evaluate the quality of a phylogeny are maximum parsimony and maximum likelihood, discussed below.

11.2.3.1 Maximum Parsimony

Using maximum parsimony, often called simply *parsimony*, a phylogeny with fewer changes in the evolutionary path is considered better than one with more changes. In other words, the phylogenies that infer the least amount of evolutionary change are the most parsimonious and thus preferred using this metric. The general idea of maximum parsimony is akin to Occam's razor (Blumer et al., 1987), the notion that the simplest explanation of the data is the preferable one.

A hypothetical data set is shown in Figure 11.4. In this example, there are four six-base DNA sequences (e.g., representing samples from four related species); the task is to find one or more most parsimonious phylogenies that connect these. That is, we are searching for possible descriptions of the evolutionary relationships between these species that minimize the number of changes in the genetic sequence over evolutionary history. In this example, we do not assume to know a common ancestor or direction of evolutionary change (whether A's are more likely to mutate to T's or vice versa), so we are looking for an unrooted phylogeny.

Figure 11.5 shows one possible phylogeny for the data in Figure 11.4. The phylogeny contains four observed sequences (the leaf nodes) and the sequences for two hypothesized ancestors (interior nodes). Hash marks indicate the number of base changes between adjacent species. The phylogeny shown is a "most parsimonious" phylogeny for this data, requiring six base changes to relate the four species. The phylogeny is unrooted and drawn as a network.

Note that it is possible to have multiple "best" solutions that are equally parsimonious. With four species, as in the example above, there are only three distinct phylogenies possible, corresponding to pairing different species in the subtrees. In list form, these phylogenies can be represented as [(1,2),(3,4)], [(1,3),(2,4)], and [(1,4),(2,3)]; these phylogenies require 7, 8, and 10 base changes, respectively, to relate the four species in the phylogeny. (Note that swapping left and right subtrees does not constitute a distinct phylogeny. For example, [(1,2),(3,4)] and [(4,3),(1,2)] are the same.) In this example, there is only one "most parsimonious" solution. However, in general, there may be multiple phylogenies that are "most parsimonious," and one would typically like to find all of these when searching, as they correspond to equally plausible evolutionary hypotheses.

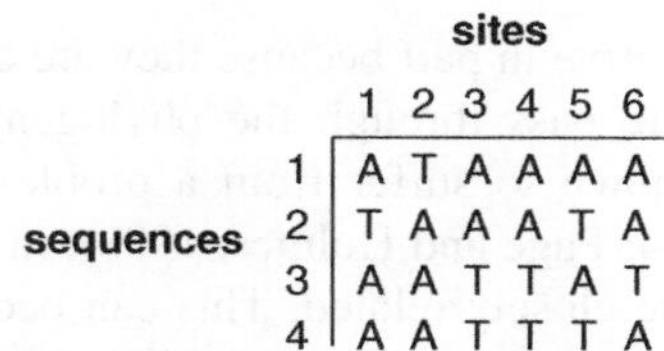

Figure 11.4 Four hypothetical six-base DNA sequences.

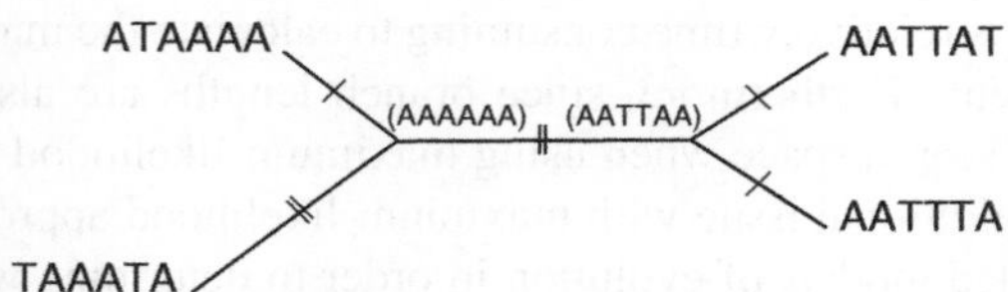

Figure 11.5 One possible phylogenetic tree relating the sequences in Figure 11.4. Sequences in parentheses are hypothesized ancestors; the hash marks indicate the number of base changes connecting adjacent sequences.

11.2.3.2 Maximum Likelihood

The idea of maximum likelihood is to maximize the probability that the evolutionary history represented in the phylogeny produced the data in the data set. A phylogeny that is considered more likely to have given rise to the observed data is considered more probable.

As with parsimony criteria, there are many possible maximum-likelihood criteria, each embodying different assumptions about how likely different forms of evolutionary change are. For example, one might like to embody the assumption that transitions are more likely than transversions. [A transition is a mutation changing a purine (A or G) to the other purine, or a pyrimidine (C or T) to the other pyrimidine, while a transversion refers to a mutation from a purine to a pyrimidine or vice versa. Transversions occur less frequently than transitions.]

Calculating maximum likelihood for a given phylogeny is typically computationally expensive in part because more information must come to bear in doing this calculation. For example, maximum-likelihood criteria are not concerned just with the topology of the phylogenies (how the species are connected to each other) but are also concerned with how long the branches are (a measure of hypothesized distance since divergence events); models of evolution typically include hypotheses about rates of evolutionary change, which is incorporated into the scoring system. Methods for estimating branch lengths for a given topology require optimizing the inferred length of each branch separately (Swofford et al., 1996). Thus, use of a maximum-likelihood criterion requires multiple levels of inference, and approaches must search for and infer both the topology of the tree and the appropriate branch lengths for that tree in order to apply the chosen maximum-likelihood criterion.

11.2.3.3 Pros and Cons of Maximum Parsimony and Maximum Likelihood

Parsimony metrics are attractive in part because they are computationally simple to compute, requiring just one pass through the phylogeny to calculate the score. However, parsimony is known to suffer from a problem known as long-branch attraction (Felsenstein, 2004; Page and Holmes, 1998), in which lineages known to be distant are inferred to be closely related. This can occur due to rapid mutation rates in these lineages, which increases the probability of evolving the same nucleotide at a given site. The nucleotides in common have happened by chance but are attributed to a common ancestor.

Conversely, it is relatively time consuming to calculate the maximum likelihood of a given phylogeny. Furthermore, since branch lengths are also included in the representation, the search space when using maximum-likelihood metrics is considerably larger. An additional issue with maximum-likelihood approaches is that they require more detailed models of evolution in order to determine which phylogenies are “more likely” than others.

The result of the phylogenetic process is one or more best phylogenies; with maximum parsimony, one is more likely to have multiple equally best trees at the end of the search. (One can see the reason for this effect when one considers that parsimony is counting the number of changes as the tree is traversed, which is necessarily an integer, with an upper bound determined by the number of species and the length of the sequences. Likelihood, on the other hand, is a real number.) More equally plausible solutions can be an advantage or a disadvantage depending on what one plans to do with the result of the phylogenetic inference.

11.2.4 Search Task

One might like to choose a representation and a metric and then deterministically compute the best possible phylogeny for a given data set, but there is not an algorithmic means of directly computing the best tree. Search is required, and exhaustive search is computationally intractable for most problems that use a interesting number of species.

11.2.4.1 Size of Search Space

Inferring phylogenetic history is not a trivial task; the number of potential trees quickly becomes unwieldy as the number of sequences increase. The number of unrooted binary trees is (2n – 5)!! or (2n – 5) (2n – 7) … (3) (1), where *n* is the number of sequences. For example, when there are 4 sequences, there are 3 possible unrooted binary trees and when there are 5 sequences, there are 15 possible unrooted binary trees; if only 10 sequences are being compared, there are more than 2 million trees that could potentially describe the evolutionary relationships among those sequences. For a discussion of how many possible trees exist for different tree representations see Felsenstein (2004).

Eleven species results in 34,459,425 different unrooted binary phylogenies, which is thought to be about the upper limit on exhaustive search (Swofford et al., 1996). Realistically, the upper limit on the number of species depends on computational speed; however, phylogeny inference belongs to the class of computer algorithms known as NP-complete (Day, 1987). Thus, as identified by Lewis (1998), "There are no known efficient solutions for this class of problems. . . . It will never be possible to examine all the trees when hundreds of taxa are included in the analysis."

11.2.4.2 Standard Phylogenetics Software

There are several different software packages available for inferring phylogenetic trees, but two are most prominent. PHYLIP [Phylogeny Inference Package (Felsenstein, 1995)] is the software most widely used by molecular biologists for finding phylogenies. PHYLIP is freeware and open source software, making it especially attractive to genomics researchers (Gibas and Jambeck, 2001). Although PHYLIP is most commonly used by molecular biologists, PAUP* [phylogenetic analysis under parsimony (Swofford, 1996)] also has a strong following and uses slightly different search mechanisms. Felsenstein (2004) contains a listing of several different phylogenetic software packages. Both PHYLIP and PAUP* are highly parameterized to allow phylogeny inference with a variety of types of data, tree representations, and scoring metrics.

11.2.4.3 Search Strategies

A typical phylogenetics approach uses a deterministic hill-climbing methodology to find phylogenies for a given data set, saving one or more "best" phylogenies as the result of the process, where "best" is defined by the specific metric used (in this chapter, we are using parsimony). The phylogeny-building approach adds each species into the phylogeny in sequence, searching for the best place to add the new species. The search process is deterministic, but different phylogenies may be found by running the search with different random orderings of the species in the data set. For example, a PHYLIP search for a most parsimonious phylogeny consists of specifying the number of random orderings of the data set (called "jumbles") to search with. Each of these searches proceeds independently, and each will end when the search has added all species into the phylogeny and thereby reached a local optimum. In contrast, an evolutionary algorithm is able to explore multiple phylogenies in parallel and, by recombining these, has the potential to jump from one region of the search space to another. Felsenstein (2004) contains discussions of some of the many different approaches to searching for phylogenies.

11.3 CHALLENGES AND OPPORTUNITIES FOR EVOLUTIONARY COMPUTATION

Evolutionary computation approaches employ a radically different search strategy than the hill-climbing or branch-and-bound searches typically used in phylogenetics software. Evolutionary computation has demonstrated resilience to the problem of getting stuck on local optima in complex problems (Fogel, 1995; Forrest, 1996; Goldberg, 1989; Mitchell, 1996) and offers great promise as an approach for phylogenetic inference.

11.3.1 Interest in Evolutionary Computation for Phylogenetic Inference

As noted by Congdon (2001), there are several factors to consider in evaluating the utility of evolutionary computation for a particular task:

1. Is there a more straightforward means of finding a "best" solution to the problem? (If so, there is no point in using the evolutionary computation approach.)

   In phylogenetic inference, there is a standard approach to forming the phylogenies, but that process also has a stochastic element, so the standard approach is not guaranteed to find "the best" phylogeny for a given data set.

2. Can potential solutions to the problem be represented using simple data structures such as bit strings or trees? (If not, it may be difficult to work with the mechanics of evolutionary computation.)

   In phylogenetics, solutions to the problem are naturally represented as trees.

3. Can a meaningful evaluation metric be identified that will enable one to rate the quality of each potential solution to your problem? (Without such a measure, the mechanisms of evolutionary computation are unable to determine which solutions are more promising to work with.)

   Metrics such as parsimony and maximum likelihood are part of the phylogenetic inference task.

4. Can operators be devised to produce (viable) offspring solutions from "parent" solutions? (If the offspring do not potentially retain some of what made the parents "good", an evolutionary computation approach will not be markedly better than random trial and error.)

   Devising appropriate reproduction operators introduces a slight challenge to the use of evolutionary computation for phylogenetics in that naive crossover and mutation operators do not tend to produce viable offspring (if each species does not appear exactly once in the resulting tree). However, this can be overcome by careful operator design.

Thus, evolutionary computation is a natural approach to apply to the problem of phylogenetic inference. The main issue in this pursuit seems to be how to best design the evolutionary search to enable parent solutions the ability to pass on especially useful parts of their solution to their offspring. This issue may arise in the design of the reproduction operators themselves, in the mechanics of parent selection, elitism, preventing convergence, and in other facets of the evolutionary search.

Assessing the relative contributions of these design decisions is further exacerbated by the variety of data sets used in different projects and the fact that relatively few data sets have been investigated using evolutionary computation approaches at all.

11.4 ONE CONTRIBUTION OF EVOLUTIONARY COMPUTATION: GAPHYL

There are relatively few projects described in the literature that use evolutionary computation approaches for phylogenetic inference. This section will describe one such project in detail to exemplify some of the salient concepts; the following section will describe other evolutionary computation approaches to phylogenetics. Gaphyl (Congdon, 2001; Congdon and Septor, 2003) used evolutionary computation with maximum parsimony to search for phylogenies and has been shown to outperform PHYLIP in limited circumstances, as will be described in this section.

11.4.1 Gaphyl Overview

Gaphyl was constructed to pursue the question of whether evolutionary computation could improve on PHYLIP's abilities to find most parsimonious trees. The system is illustrated in Figure 11.6.

PHYLIP serves two roles in the work described here. On the one hand, it is used as a point of comparison to the evolutionary search approach; on the other hand, PHYLIP code is used to evaluate the phylogenies, serving as the fitness function for the evolutionary search. (Using the PHYLIP source code rather than writing new phylogeny evaluation modules also helps to ensure that Gaphyl phylogenies are properly comparable to the PHYLIP phylogenies.) PHYLIP was selected for the base code because it is the software predominantly used by molecular biologists and its source code is freely available.

Genesis (Grefenstette, 1987) was used as the basis for the evolutionary mechanisms added. These mechanisms were changed considerably, as Gaphyl evolves tree structures not bit strings. However, the basic cycle of fitness evaluation, parent selection, reproduction via crossover and mutation, and generational replacement was retained from Genesis. The code from PHYLIP is used only to evaluate the phylogenies (using parsimony), and the evolutionary computation process is used to construct the phylogenies.

Gaphyl begins with an initial population of trees, which may be generated at random or provided by the user. In each generation, the parsimony of each tree is evaluated. A lower score corresponds to a more parsimonious explanation of the data; thus, the lower the score, the more likely the tree is to be chosen as a parent to the next generation.

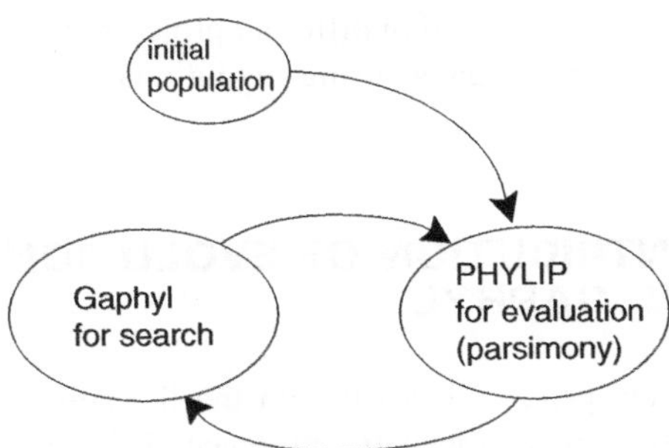

Figure 11.6 Gaphyl uses an evolutionary algorithm for searching through the space of possible phylogenies and PHYLIP to evaluate the phylogenies.

11.4.2 Evolutionary Search Component

The reproduction process in Gaphyl differs slightly from a typical generational evolutionary algorithm applied to tree structures, as will be explained in this section. The processes that warrant clarification are the use of elitism, the reproduction operators, and the use of multiple populations. Elitism was added to Gaphyl to ensure that the best solutions are available as parents each generation and that the final population contains the best solutions discovered during the search. Also, the crossover and mutation operators are specialized, due to the constraints of phylogenetics. A typical evolutionary algorithms approach to doing crossover with two parent solutions with a tree representation is to pick a subtree (an interior or root node) in both parents at random and then swap the subtrees to form the offspring solution. A typical mutation operator would select a subtree and replace it with a randomly generated subtree. However, these approaches do not work with the phylogenies because each species must be represented in the tree exactly once. The use of multiple populations was added after initial experimentation revealed that the population tended to converge on a set of suboptimal solutions. Mechanisms designed specifically for reproduction with phylogenies are described in the following sections.

11.4.2.1 Elitism and Canonical Form

In order to save the best solutions found while searching, both as potential parents for future generations and so that the best solutions are present at the end of the process, Gaphyl uses an elitist replacement strategy. The percentage of the population to save from one generation to the next is specified at runtime as a parameter to the system. These best solutions are retained from one generation into the next, so that there is not full generational replacement.

In order to identify duplicates and to ensure that no equivalent trees are saved among the best ones, trees are put into a canonical form when saving the best trees found in each generation. This is a standardized representation designed for this work, and is necessary because the trees are manipulated as if using a rooted phylogeny although they are more correctly networks (or unrooted trees). For example, the three binary trees illustrated in Figure 11.7 all correspond to the same evolutionary hypothesis.

The process of putting trees into canonical form is illustrated in Figure 11.7. The process picks the first species in the data set to be an offspring of the root, and "rotates" the tree (and switches subtrees, if necessary) to keep the species relationships intact, but to reroot the tree at a given species. (To simplify comparisons, Gaphyl follows the default PHYLIP assumption of making the first species in the data set the direct offspring of the root of the tree.) Second, the subtrees are (recursively) rearranged to make left subtrees smaller (fewer nodes) than right subtrees; also, when left and right subtrees have the same number of nodes, subtrees are swapped so that a preorder traversal of the left subtree is alphabetically before a preorder traversal of the right subtree.

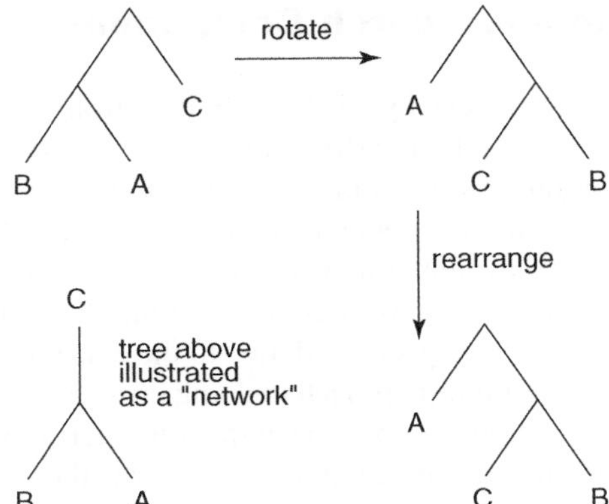

Figure 11.7 Illustration of putting a tree into canonical form. The tree starts as in the top left; an alternate representation of the tree as a "network" is shown at the bottom left. First, the tree was rotated so that the first species in the data set is an offspring of the root. Second, subtrees were rearranged so that smaller trees are on the left and alphabetically lower species are on the left.

Trees were saved in their original form, which is used for reproduction. The canonical form is used primarily for elitism and is also helpful at the end of a run, so that if there are multiple best trees, they are known to be distinct solutions.

11.4.2.2 Crossover Operator

The Gaphyl operator was designed to preserve some of the species relationships from the parents, and combine them in new ways. Because each species in the data must appear exactly once in a leaf node, conceptually, the needs for a crossover operator bear some similarity to traveling salesperson problems (TSPs), where each city is to be visited exactly once on a tour. There are several approaches in the literature for working on this type of problem with an evolutionary algorithm, although the TSP naturally calls for a string representation, not a tree. Thus, while evolutionary algorithms approaches to TSPs are informative, it was necessary to design an approach specifically for crossover with phylogenies.

The Gaphyl crossover operator proceeds as follows:

1. From the first parent tree, a random subtree is selected, excluding leaf nodes and the entire tree. (The exclusions prevent crossovers where no information is gained from the operation.)
2. In the second parent tree, the smallest subtree containing all the species from the first parent's subtree is identified.
3. To form an offspring tree, the subtree from the first parent is replaced with the subtree from the second parent. The offspring is then pruned (from the "older" branches) to remove any duplicate species.
4. This process is repeated using the other parent as the starting point, so that this process results in two offspring trees from two parent trees.

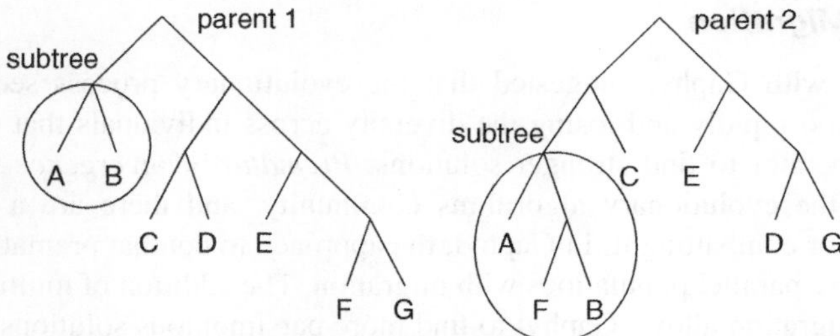

Figure 11.8 Two example parent trees for a phylogenetics problem with seven species. A subtree for crossover has been dentified for each tree.

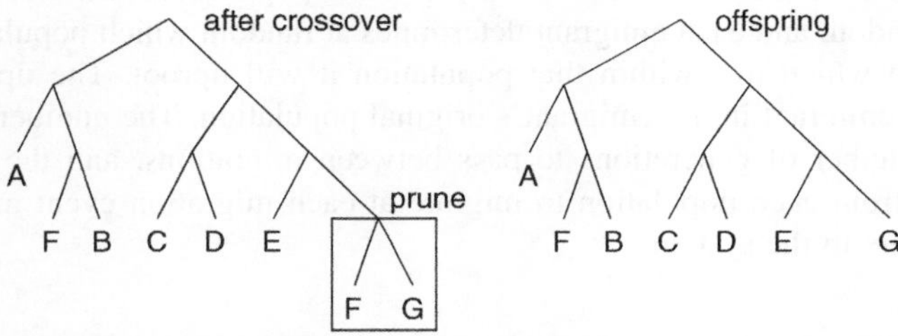

Figure 11.9 At left, the offspring initially formed by replacing the subtree from parent1 with the subtree from parent2; on right, the offspring tree has pruned to remove the duplicate species F.

An example crossover is illustrated in Figures 11.8 and 11.9. In this example, the offspring tree can be considered to retain the [A (F B)] subtree from the second parent and the [(C D) (E G)] subtree from the first parent.

11.4.2.3 Mutation Operators

There are two mutation operators in Gaphyl. The first mutation operator selects two leaf nodes (species) at random and swaps their positions in the tree. This operator may make only a slight change to the tree, when the two leaf nodes swapped are connected by only a few branches. However, if the leaf nodes that are swapped are many branches apart in the phylogeny, this operator may result in a large change to the parsimony score.

A second mutation operator selects a random subtree and a random species within the subtree. The subtree is rotated to have the species as the direct offspring of the root and reconnected to the parent. The idea behind this operator was that within a subtree, the species might be connected to each other in a promising manner, but not well connected to the rest of the tree. The mechanics of rotating the subtree are similar to the rotation for canonical form, illustrated in Figure 11.7.

11.4.2.4 Migration

Initial work with Gaphyl suggested that the evolutionary process seemed to be converging too rapidly and losing the diversity across individuals that enables the crossover operator to find stronger solutions. *Premature convergence* is a known problem in the evolutionary algorithms community, and there are a number of approaches for combatting it. In Gaphyl, the approach to combat premature convergence is to use parallel populations with migration. The addition of multiple populations with migration allows Gaphyl to find more parsimonious solutions than it had been able to find with a single population.

The population is subdivided into multiple subpopulations which, in most generations, are distinct from each other (crossovers happened only within a given subpopulation). After a specified number of generations have passed, each population migrates a number of its individuals into other populations; the emmigrants are chosen at random and each emigrant determines at random which population it will move to and which tree within that population it will uproot. The uprooted tree replaces the emigrant in the emigrant's original population. The number of populations, the number of generations to pass between migrations, and the number of individuals from each population to migrate at each migration event are specified via parameters to the system.

11.4.3 Gaphyl Evaluation: Materials and Methods

This section describes the data used to evaluate Gaphyl and the means used to compare Gaphyl and PHYLIP.

11.4.3.1 Data Sets

The Gaphyl research described here was conducted using published data sets available over the Internet via TreeBase (Donoghue, 2000). Recall that there is little reason to consider evolutionary search for data sets with fewer than a dozen species because a problem of that size is within the range of exhaustive search.

The binary data sets used were:

1. The families of the superorder of Lamiiflorae data set (An-Ming, 1990), consisting of 23 species and 29 attributes. This data set was chosen as being large enough to be interesting but small enough to be manageable.
2. The major clades of the angiosperms (Dahlgren and Bremer, 1985), consisting of 49 species and 61 attributes, was used for further experimentation.

These data sets were selected because the attributes are binary,which simplified the initial development of the system. As a preliminary step in evaluating evolutionary computation as a search mechanism for phylogenetics, "unknown" values for the attributes were replaced with 1's to make the data fully binary. This minor alteration to the data does impact the meaningfulness of the resulting phylogenies as evolutionary hypotheses but does not affect the comparison of Gaphyl and PHYLIP as search mechanisms.

Three nucleotide sequence data sets were chosen because they were available in TreeBase, used DNA sequence data, and had an appropriate number of species in the data set:

1. A study of angiosperms (Donoghue and Mathews, 1998) consisting of 30 species and 3264 nucleic acid characters. (This data set is matrix accession number M184c3x27x98c11c12c31 in TreeBase, abbreviated as M184 in the text.)
2. A study of angiosperms (Mathews and Donoghue, 1999) consisting of 50 species and 1104 nucleic acid characters (matrix accession number M608 in TreeBase).
3. A study of polyporaceae (Hibbett and Donoghue, 1995) consisting of 63 species and 1588 nucleic acid characters. (This data set is matrix accession number M194c3x30x98c09c57c27 in TreeBase, abbreviated as M194 in the text.)

Based on previous work with Gaphyl, it appeared that when the number of species was small, the hill-climbing method in PHYLIP was sufficient to find the best solutions and no gain is achieved through the evolutionary search processes in Gaphyl. The hypothesis was that as the number of species in the data set increases, Gaphyl might be more likely to show a gain over the PHYLIP search process, perhaps by sidestepping local optima that hinder the performance of PHYLIP.

11.4.3.2 Methodology for Comparing Gaphyl and PHYLIP

It is difficult to know how to fairly compare the work done by the two systems in order to understand their relative strengths and weaknesses.

Run time is a primary concern, however, both Gaphyl and PHYLIP are far from optimized, so strong conclusions cannot be drawn from run time alone. Recall that Gaphyl was constructed from existing systems, neither one of which was optimized for speed. In particular, Genesis was designed to simplify evolutionary computation experimentation and modifications (much like the project here).

It is possible to make some comparisons of trees evaluated or operations performed by the two systems in their search, but these are apples and oranges, since the work done to get from one phylogeny to the next varies between the systems. For example, PHYLIP constructs trees incrementally and considers many partial solutions while doing its hill-climbing search. One PHYLIP jumble, for example, may evaluate on the order of 10,000 trees for the angiosperm data. Thus, comparing the number of PHYLIP jumbles to the number of Gaphyl trees evaluated is an erroneous comparison. Similarly, if each "run" corresponds to running the system with a distinct random seed, one run of PHYLIP corresponds to one jumble, while one Gaphyl run corresponds to evaluating a number of trees determined by the population size and number of generations. Thus, the number of restarts with different random seeds does not serve as a comparable measure of computational effort.

It is not clear that either run time or number of phylogenies evaluated are the proper metrics to use in comparing the two systems; neither seems quite appropriate. However, run time is used here as a coarse metric and because it is a practical concern when the searches must run for days. As defaults for the experiments reported below, PHYLIP was run for 10,000 jumbles, while Gaphyl was run with 2 populations of 500 trees each and for 2000 generations. With the angiosperm data, this resulted in comparable run times. However, the run times for the two systems on different specific data sets is quite variable, so additional experiments were done in an effort to determine how much run time is required for a solution with each data set. In practice, the question pursued in these experiments is to determine how long each of these systems need to run to find the best known solution(s) for each data set.

11.4.3.3 System Parameters

PHYLIP experiments were run with the "mix" module for binary data and the "dnapars" module for nucleotide data. Default parameters were used (including the use of Wagner parsimony), save for the number of jumbles to do for each data set.

The default Gaphyl parameters for these experiments are shown in Table 11.1. In the experiments reported here, the number of populations, population size, and number of generations were varied.

Fifty percent elitism means that half of the trees are preserved into the next generation (the best trees) and that half of the trees are replaced through the evolutionary process. Parents are selected (with replacement) from the full population to generate offspring; all of the parents are subject to crossover and the second form of mutation. Of the parents 10% are subject to the first form of mutation. With a population size of 500, this means that 250 parents are selected to produce 250 new

Table 11.1 Default Gaphyl Parameters Used for Experiments Reported Here

Elitism	50%
Crossover	100%
Mutation1	10%
Mutation2	100%
Population size	500
Generations	2000
Number of populations	2
Migration rate	5%
Migration interval	25% of the generations
Number of repetitions	10

solutions each generation in each population. With a population size of 500 and 2000 generations and the migration parameters from the table, 25 trees will migrate to the other population on generations 500, 1000, and 1500. Ten repetitions were done of each experiment (and run time is counted as the time required to complete all 10 experiments), and "the answer" from the run is considered the population from the best of these runs.

These are the default parameters for the experiments reported here. However, the simpler data sets do not require multiple populations or as many generations to find solutions, so there is some variation in the number of populations and the number of generations in the experiments reported here.

11.4.4 Results of Comparison of Gaphyl and PHYLIP on Binary Data

Summaries of the data sets, including best known parsimony and the maximum number of trees at that score, are shown in Table 11.2. Although it is impossible to know for certain without doing an exhaustive search, empirically, these appear to be the best solutions possible for these data sets. Results of running PHYLIP and Gaphyl on these data sets are shown in Table 11.3.

Lamiiflorae and M184 are relatively simple data sets, able to find all known trees at the maximally parsimonious scores in relatively short run times. PHYLIP in particular is able to find these trees quickly, and Gaphyl does not require multiple populations to find the most parsimonious trees.

With the M608 and M194 data sets, PHYLIP is able to find all of the maximally parsimonious trees quickly, while Gaphyl needs more run time.

The angiosperms data set appears to be the most difficult of the five. PHYLIP is able to find one most parsimonious tree more quickly than Gaphyl but has a difficult time finding all 252 known trees. While Gaphyl is able to find them all in roughly half a day of run time, 600,000 PHYLIP jumbles, requiring over 8 days, find only 200 of the 252 known trees.

Table 11.2 Characteristics of Five Data Sets Run for These Experiments[a]

Data	Number of Species	Number of Attributes	Best Known Parsimony	Number of Trees
Lamiiflorae	23	29	72	45
Angiosperms	49	61	279	252
M184	30	3,264	18,676	2
M608	50	1,104	7,497	4
M194	63	1,588	7,671	1

[a]The number of species and attributes, the best known parsimony score, and the maximum number of distinct trees found at this score.

Table 11.3 Results of Running Five Data Sets with PHYLIP and Gaphyl, Including Some Parameter Variations, to Poke at Capabilities of Systems[a]

	PHYLIP		Gaphyl			
Date	Jumbles	Run Time	Parameters	Run Time	Best Score Found?	Max Trees Found?
Lamiiflorae	10,000	10 min	2/500/500	30 min	Y	Y
	1,000	1 min	1/500/500	14 min	Y	Y
Angiosperms	—	—	2/500/2000	4 h	N (280)	—
	—	—	2/1000/2000	13 h	Y	Y
	10,000	3 h	—	—	Y	N (16)
	100,000	30 h	—	—	Y	N (83)
	200,000	3 days	—	—	Y	N (162)
	400,000	5.5 days	—	—	Y	N (181)
	600,000	8.25 days	—	—	Y	N (200)
	1,000,000	14 days	—	—	Y	N (226)
M184	10,000	6.5 h	2/500/500	3.5 h	Y	Y
	1,000	30 min	1/125/500	26 min	Y	Y
M608	10,000	4.5 h	2/500/2000	11.5 h	Y	Y
	1,000	36 m	—	—	Y	Y
M194	10,000	10 h	2/500/2000	18 h	Y	Y
	1,000	1 h	2/250/2000	8.5 h	Y	Y

[a]Gaphyl parameters are abbreviated as number of generations/population size/number of generations.

11.4.5 Discussion

In the binary data sets there were many equally best solutions. In general, PHYLIP seemed able to find one of the best solutions more quickly than Gaphyl was able to find one, but then Gaphyl was able to find more of the multiple bests than PHYLIP. Since the data sets used here seem to have only a small number of best solutions, they may be more amenable to PHYLIP search. In particular, the M608 data set appears to have only one best solution and it is therefore not anomalous that PHYLIP is able to best Gaphyl in finding this one phylogeny in less run time. Thus, differences from previous observations may reflect the nature of a specific data set and not the difference between binary and genetic data.

However, it is also possible that there is an important difference between the binary data and the genetic data. For example, multiple equally best phylogenies might be more common in data with binary character states, which tends to have fewer attributes and has only the two values for each attribute. The effect of this could lead to fewer differences in the observed values among species, and this would tend to result in subtrees where two species could be swapped without altering the parsimony. When two leaf nodes can be swapped without affecting the parsimony (and these are not sibling nodes), two different equally parsimonious solutions will necessarily result.

The pattern that emerges is that as the problems get more complex, Gaphyl is able to find a more complete set of trees with less work than what PHYLIP is able to find. The work done to date illustrates that Gaphyl is a promising approach for phylogenetics work, as Gaphyl finds a wider variety of trees on the angiosperms problem than PHYLIP does. This further suggests that Gaphyl may be able to find solutions better than those PHYLIP is able to find on data sets with a larger number of species and attributes because it appears to be searching more successful regions of the search space.

11.4.6 Additional Experiments with Gaphyl

A study of the contributions of the operators to the search has also been conducted for Gaphyl (Congdon, 2001), which indicated that, in general, higher crossover rates were better for the data investigated, while low rates for the first mutation operator were preferable. In general, high rates of the second mutation operator were also helpful in terms of finding more equally parsimonious trees. This work also included a study of the number of populations, finding that two populations improved the search but that more than that did not lead to great gains in the search.

Additional work (Congdon and Septor, 2003) also included an investigation of a hybrid system that used PHYLIP runs to create phylogenies as the starting point for the evolutionary computation search. The motivation for this investigation was the observation that PHYLIP is faster than Gaphyl at finding relatively high-scoring trees. These experiments found that while seeding from PHYLIP runs may help the progress of the search, the initial seeds must be sufficiently diverse for this to be helpful. It appears that choosing the seed trees from a single PHYLIP jumble is comparable to starting the genetic algorithm (GA) with a population that has already converged, while selecting seed trees from distinct jumbles may help the search process.

11.4.7 Summary of Gaphyl Contributions

Gaphyl has been shown to be a promising approach for searching for phylogenetic trees using data with binary character states and Wagner parsimony, finding more of the most parsimonious trees than PHYLIP, and finding them in less time. With the molecular data described here, Gaphyl is able to find the same set of maximally parsimonious phylogenies faster than PHYLIP. Additional work is required with additional data to fully evaluate Gaphyl's abilities, but the ability of Gaphyl to outperform PHYLIP with binary data suggests that there will be a niche where it is the preferable search approach to use with molecular data. Furthermore, PHYLIP has been so well explored over the years that even equaling its performance in initial results is extremely promising.

11.5 SOME OTHER CONTRIBUTIONS OF EVOLUTIONARY COMPUTATION

Gaphyl is described above as an example of an application of evolutionary computation to phylogenetics that saw an early modest success. This section will describe other systems using evolutionary computation for phylogenetic inference.

Most of the systems described here are focused on maximum-likelihood phylogenies, though two address parsimony. As mentioned before, the systems differ most in the approaches used to enable parent solutions to pass on especially useful parts of their solution to their offspring.

11.5.1 Matsuda: Maximum Likelihood

Matsuda (1996) appears to be the first application of evolutionary computation to phylogenetic inference. The work addresses using maximum likelihood to construct phylogenies based on amino acid sequences. The topology of the tree and the branch lengths are evolved jointly.

Matsuda (1996) uses a crossover operator that seeks to identify the two species with the greatest difference in distance in the two parent trees. The tree with the greater distance between these two species is modified to incorporate the subtree of the other parent that connects the two species, creating a single offspring. The mutation operator used swapped leaf nodes.

Matsuda (1996) reported that the GA approach did not find the optimum likelihood tree but argued that the performance of the evolutionary computation algorithm was comparable to other phylogenetic approaches.

11.5.2 Lewis: Maximum Likelihood

Lewis (1998) describes GAML, which used maximum likelihood on phylogenies based on nucleotide sequences. In this work, branch lengths are represented as additional genes in the evolutionary computation process and thus evolve distinctly from the toplogy of the tree.

GAML uses a crossover operator that takes a random subtree from one parent and grafts it onto the other parent, removing duplicate species in the process. On the topological level, the GAML mutation operator swaps random subtrees; branch lengths are also subject to random mutations.

This work reports evolutionary search as considerably faster than the likelihood implementation of PAUP* and is able to find the same tree topology. However, the evolutionary search does not optimize the branch lengths as well as PAUP* does.

Later work by Lewis and colleagues (Brauer et al., 2002) did not find best-known solutions with evolutionary computation search.

11.5.3 Moilanen: Maximum Parsimony

Moilanen (1999) used a design similar to that of Gaphyl for finding maximally parsimonious trees. The crossover operator is implemented slightly differently, in effect crossing over two subtrees and then repairing the new trees to remove duplicate species and to insert missing species. Instead of a typical mutation operator, Moilanen uses what is called a "local search" operator, which may reorder a subtree or may move a subtree to another position. This operator is tried on all trees in the population but takes effect only if the modified tree is an improvement.

Moilanen's work focuses on the speed of the search as compared to other software; the system was designed to find only one most parsimonious tree. The system is evaluated using six previously published data sets, ranging from 19 to 101 species and from 27 to 70 characters.

11.5.4 Katoh, Kuma, and Miyata: Maximum Likelihood and Steady-State Reproduction

Katoh, Kuma, and Miyata (2001) describe GA-mt, an evolutionary computation approach to maximum-likelihood phylogenetic inference that outputs multiple trees. These trees are found to be most likely, and alternative trees are not significantly worse than the best tree.

GA-mt uses steady-state reproduction instead of generational reproduction, so only one new tree is produced at a time. When this happens, the new tree is produced via either crossover or mutation (but not both); these two operators proceed similarly to Lewis (1998).

GA-mt is compared to four other inference methods on artificial data from 5 to 24 taxa and is reported to have always predicted more accurate trees than the other methods did. The system was also used to infer the tree of life based on 39 genes from 27 organisms (sequence lengths of 5959 amino acids); the inferred tree supports alternate hypotheses about the eukaryal and two archaeal (Euryarchaeota and Crenarchaeota) groups.

11.5.5 Lemmon and Milinkovitch: Maximum Likelihood and Multiple Populations

METAPIGA, described by Lemmon and Milinkovitch (2002), uses evolutionary computation search with maximum likelihood to search for phylogenies and also uses multiple populations of phylogenies.

METAPIGA uses branch-length mutations and a recombination operator similar to GAML. Subpopulations of phylogenies evolve largely independently; communication between populations identifies parts of the phylogenies that are not protected from mutation events.

METAPIGA was evaluated on artificial data sets as well as the data used in Lewis (1998) and reported a significant speedup as compared to other approaches.

The research investigated different parameter settings, such as varying numbers of populations and communication strategies; some of these variations found the best known maximum-likelihood (ML) tree and some did not.

11.5.6 Poladian: Maximum Likelihood and Genotype–Phenotype Distinction

Poladian (2005) uses different representations for the genotype and phenotype of the phylogenies; evolving the nontree genotype alters the search space and allows the use of qualitatively different reproduction operators. In this work, the genotype is a distance matrix and the phenotype is the phylogeny; crossover is a process of crossing over entire rows of the distance matrix and later repairing the offspring matrix to be symmetric. Because the process introduces novelty into the population, no distinct mutation operator was used.

This system was evaluated on artificial data with 7, 10, and 20 species and nucleotide sequences 500 base pairs (bp) long. As number of species increased, the best maximum-likelihood values found by the system were slightly worst than those found by PHYLIP.

11.5.7 Williams and Smith: Maximum Parsimony, Parallel Computation, and Large-Scale Phylogenies

Williams and Smith (2005, 2006) describe a population-based version of the Rec-I-DCM3 algorithm for large-scale phylogenies (on the order of thousands of species), specifically comparing a population-based version to the standard local-search version and evaluating different communication strategies among the populations. (Rec-I-DCM3 is a recursive and iterative approach to disk-covering methods, which construct a phylogenetic tree as a covering of subtrees and then combine these into the full phylogeny.)

Cooperative Rec-I-DCM3 uses an unusual parent-selection process that divides the population into three ranked sets used to select parents. The details of the recombination operator are not discussed; the mutation operator proceeds as a local search using the Rec-I-DCM3 algorithm on each tree in the population.

Cooperative Rec-I-DCM3 uses a parallel computation approach to utilize multiple processors effectively. It was evaluated on four data sets from 921 to 4114 sequences and from 713 to 1263 sites, finding the best known parsimony score on all 4 data sets while the original Rec-I-DCM3 found the best in only 2 of the 4 data sets.

11.6 OPEN QUESTIONS AND OPPORTUNITIES

Evolutionary computation is an appealing approach to apply to the problem of phylogenetic inference, not only because it is rewarding to apply the principles of evolution to the study of evolution, but also because the complexity of the computational task seems a natural fit for an evolutionary computation approach.

When one looks at the struggles and successes of systems that have applied evolutionary computation to phylogenetic inference, it becomes apparent that a difficulty is in designing the evolutionary mechanisms to best enable parents to share promising parts of their solution with offspring. While this concern is most naturally addressed via the design of the reproduction operators, it is also apparent in the mechanics of parent selection and elitism used and in the mechanisms used to prevent the population from converging too quickly.

The relatively small number of publications reporting on the application of evolutionary computation to phylogenetic inference is likely in part a reflection of the fact that the phylogenetic inference task is difficult and that the previously existing nonevolutionary approaches are doing a reasonably good job with it, thus leaving room for only incremental improvements via evolutionary computation.

Evolutionary computation applied to phylogenetic inference appears to be helpful in inferring more equally parsimonious trees when using a parsimony approach, but only on some data sets. In the case of parsimony, evolutionary computation approaches may be finding trees that standard nonevolutionary approaches would not find given a reasonable amount of time. Several evolutionary computation approaches have been demonstrated that considerably lessen the run time over conventional phylogenetic approaches.

It appears that the specific data sets used for investigation and the specific implementation of operators may have quite a bit to do with the degree of success reported in the application of evolutionary comptuation to phylogenetic inference. Assessing how challenging a data set might be for phylogenetic inference is more complicated than counting the numbers of species and attributes. Relatively few data sets have been investigated using evolutionary computation approaches at all; contrariwise, the use of benchmarking data across systems would assist in understanding the relative contributions of systems.

Further work must be done in understanding the fitness landscape of phylogenetics tasks in general and how a given data set challenges or simplifies the search process. Ideally, one would like to determine ahead of time whether an evolutionary computation approach or a standard system is appropriate to use for a given data set. Working with artificial data may help to pursue these questions.

A study of different sorts of reproduction operators, as well as parent selection and anticonvergence strategies, applied to phylogenies representing different sorts of data sets would help shed light on these questions. Different design decisions lead to different characteristics of the parents being passed on to offspring. A careful analysis and decomposition of the task into subtasks can assist in distinguishing the facets in which evolutionary computation might best assist from the facets in which existing approaches are particularly hard to improve on. This includes the design of reproduction operators, where, for example, local hill climbing as a mutation operator may hasten the search time considerably.

Hybrid approaches, which make use of evolutionary computation for some facets of the task but proceed with more standard phylogenetics approaches in other facets, offer promise. Examples include systems that do "local search" in lieu of (or in addition to) a more standard mutation and Meade et al. (2001), who used an evolutionary computation approach to optimize a vector of transition rates used for maximum-likelihood phylogenies.

Phylogenetic inference is a computationally daunting task that provides multiple opportunities for contributions from evolutionary computation techniques. Furthermore, it is immensely rewarding to use evolutionary computation to assist in the study of natural evolution and to contribute to our understanding of the relationships among species.

Acknowledgments

I would like to thank Judy L. Stone for sharing her knowledge of phylogenetic theory and practice. Thanks also to Emily F. Greenfest, who worked with me in the initial design and implementation of Gaphyl, and to Joshua R. Ladieu, Kevin J. Septor, and Skyler S. Place, who also provided research assistance. This publication was made possible by NIH Grant Number P20 RR-016463 from the INBRE Program of the National Center for Research Resources, as well as the Salisbury Cove Research Fund from the Mount Desert Island Biological Laboratory.

REFERENCES

AN-MING, L. (1990). "A preliminary cladistic study of the families of the superorder lamiiflorae," *Biol. J. Linn. Soc.*, Vol. 103, pp. 39–57.

BLUMER, A., A. EHRENFEUCHT, D. HAUSSLER, and M. K. WARMUTH (1987). "Occam's razor," *Info. Process. Lett.*, Vol. 24, No. 6, pp. 377–380.

BRAUER, M. J., M. T. HOLDER, L. A. Q. DRIES, D. J. ZWICKL, P. O. LEWIS, and D. M. HILLIS (2002). "Genetic algorithms and parallel processing in maximum-likelihood phylogeny inference," *Mol. Biol. Evol.*, Vol. 19, No. 10, pp. 1717–1726.

CONGDON, C. B. (2001). "Gaphyl: A genetic algorithms approach to cladistics," in *Principles of Data Mining and Knowledge Dicovery (PKDD 2001)*, L. De. Raedt and A. Siebes, Eds. Lecture notes in artificial intelligence 2168. Springer, New York, pp. 67–78.

CONGDON, C. B. and K. S. SEPTOR (2003). *"Phylogenetic trees using evolutionary search: Initial progress in extending gaphyl to work with genetic data," in Proceedings of the Congress on Evolutionary Computation (CEC 2003)*, Canberra, Australia, pp. 1717–1726.

DAHLGREN, R. and K. BREMER (1985). "Major clades of the angiosperms," *Cladistics*, Vol. 1, pp. 349–368.

DAY, W. H. (1987) "Computational complexity of inferring phylogenies from dissimilarity matrices," *Bull. Math. Biol.*, Vol. 49, pp. 461–467.

DONOGHUE, M. J. (2000). Treebase: A database of phylogenetic knowledge. web-based data repository. http://phylogeny.harvard.edu/treebase.

DONOGHUE, M. J. and S. MATHEWS (1998), "Duplicate genes and the root of the angiosperms, with an example using phytochrome genes," *Mol. Phyl. Eval.*, Vol. 9, pp. 489–500.

ERWIN, D. H. and E. H. DAVIDSON (2002). "The last common bilaterian ancestor," *Development*, Vol. 129, pp. 3021–3032.

FELSENSTEIN, J. (1995). "Phylip source code and documentation," available via the Web at http://evolution.genetics.washington.edu/phylip.html.

FELSENSTEIN, J. (2004). *Inferring Phylogenies*. Sinauer Associates, Sunderland, MA.

Fitch, W. M., J. M. E. Leiter, X. Li, and P. Palese (1991). "Positive darwinian evolution in influenza A viruses," *Proc. Natl. Acad. Sci.*, Vol. 88, pp. 4270–4274.

Fogel, D. B (1995). *Evolutionary Computation: Toward a New Philosophy of Machine Learning*. IEEE Press, New York.

Fogel, G. B. (2005). "Evolutionary computation for the inference of natural evolutionary histories," *IEEE Connect.*, Vol. 3, No. 1, pp. 11–14.

Forrest, S. (1996). "Genetic algorithms," *ACM Comput. Surv.*, Vol. 28, No. 1, pp. 77–80.

Gibas, C. and P. Jambeck (2001). *Developing Bioinformatics Computer Skills*. O'Reilly & Associates, Sebastopol, CA.

Goldberg, D. E. (1989). *Genetic Algorithms in Search, Optimization and Machine Learning*. Addison-Wesley, Reading, MA.

Grefenstette, J. J (1987). "A user's guide to GENESIS," Technical report, Navy Center for Applied Research in AI, Washington, DC. Source code updated 1990; available at http://www.cs.cmu.edu/afs/cs/project/ai-repository/ai/areas/genetic/ga/systems/genesis/.

Hahn, B. H., G. M. Shaw, K. M Cock, and P. M. Sharp (2000). "AIDS as a zoonosis: Scientific and public health implications." *Science*, Vol. 287, pp. 607–614.

Hall, B. G (2004). *Phylogenetic Trees Made Easy: A How-To Manual*, 2nd ed. Sinauer Associates, Sunderland, MA.

Hibbett, D. S. and M. J. Donoghue (1995). "Progress toward a phylogenetic classification of the polyporaceae through parsimony analysis of mitochondrial ribosomal dna sequences," *Can. J. Bot.*, Vol. 73, pp. S853–S861.

Ingman, M., H. Kaessmann, and U. Gyllensten (2000). "Mitochondrial genome variations and the origin of modern humans," *Nature*, Vol. 408, pp. 708–713.

Katoh, K., K.-I. Kuma, and T. Miyata (2001). "Genetic algorithm-based maximum-likelihood analysis for molecular phylogeny," *J. Mol. Evol.*, Vol. 53, pp. 477–484.

Lemmon, A. R. and M. C. Milinkovitch (2002). "The metapopulation genetic algorithm: An efficient solution for the problem of large phylogeny estimation," *Proc. Natl. Acad. Sci.*, Vol. 99, No. 16, pp. 10516–10521.

Lewis, P. O. (1998). "A genetic algorithm for maximum-likelihood phylogeny inference using nucleotide sequence data." *Mol. Biol. Evol.*, Vol. 15, No. 3, pp. 277–283.

Mathews, S. and M. J. Donoghue (1999). The root of angiosperm phylogeny inferred from duplicate phytochrome genes," *Science*, Vol. 186, No. 5441, pp. 947–950.

Matsuda, H (1996). "Protein phylogenetic inference using maximum likelihood with a genetic algorithm," in *Pacific Symposium on Biocomputing '96*, L. Hunter and T. E. Klein, Eds., World Scientific, London, pp. 512–523.

Meade, A., D. Corne, M. Pagel, and R. Sibly (2001). "Using evolutionary algorithms to estimate transition rates of discrete characteristics in phylogenetic trees," in *Proceedings of the 2001 Congress on Evolutionary Computation*. IEEE, New York, pp. 1170–1176.

Mitchell, M. (1996). *An Introduction to Genetic Algorithms*. MIT Press, Cambridge, MA.

Moilanen, A. (1999). "Searching for most parsimonious trees with simulated evolutionary optimization," *Cladistics*, Vol. 15, pp. 39–50.

Ou, C.-Y., C. A. Ciesielski, G. Myers, C. I. Bandea, C.-C. Luo, B. T. M. Korber, J. I. Mullins, G. Schochetman, R. L. Berkelman, A. Nikki Economou, J. J. Witte, L. J. Furman, G. A. Satten, K. A. Maclnnes, J. W. Curran, H. W. Jaffe, Laboratory Investigation Group, and Epidemiologic Investigation Group (1992). "Molecular epidemiology of HIV transmission in a dental practice," *Science*, Vol. 256, No. 5060, pp. 1165–1171.

Page, R. D. M. and E. C. Holmes (1998). *Molecular Evolution: A Phylogenetic Approach*. Blackwell Science, Malden, MA.

Pattnaik, P. and A. Jana (2005). "Microbial forensics: Applications in bioterrorism," *Environmental Forensics*, Vol. 6, No. 2, pp. 197–204.

Poladian, L. (2005). "A GA for maximum likelihood phylogenetic inference using neighbour-joining as a genotype to phenotype mapping," in *Proceedings of the Genetic and Evolutionary Computation Conference (GECCO 2005)*, pp. 415–422.

Swofford, D. L. (1996). "PAUP*: Phylogenetic analysis using parsimony (and other methods)," Version 4.0.

SWOFFORD, D. L., G. J. OLSEN, P. J. WADDELL, and D. M. HILLIS (1996). "Phylogenetic inference," in *Molecular systematics*, 2nd ed., D. H. Hillis, C. Moritz, and B. K. Mable, Eds. Sinauer Associates, Sunderland, MA, pp. 407–514.

WEBSTER, R. G., W. J. BEAN, O. T. GORMAN, T. M. CHAMBERS, and Y. KAWAOKA (1992). "Evolution and ecology of influenza a viruses," *Microbiol. Rev.*, Vol. 56, pp. 152–179.

WILLIAMS, T. and M. SMITH (2005). "Cooperative rec-i-dcm3: A population-based approach for reconstructing phylogenies," in *Proc. of IEEE Symposium on Computational Intelligence in Broinformatics and Computational Biology*, pp. 127–134.

WILLIAMS, T. L. and M. L. SMITH (2006). "The role of diverse populations in phylogenetic analysis," in *Proceedings of the Genetic and Evolutionary Computation Conference (GECCO 2006)*. ACM Press, New York, pp. 287–294.

Part Four

Medicine

Chapter 12

Evolutionary Algorithms for Cancer Chemotherapy Optimization

John McCall, Andrei Petrovski, and Siddhartha Shakya

12.1 INTRODUCTION

Cancer is a serious and often fatal disease, widespread in the developed world. One of the most common methods of treating cancer is chemotherapy with toxic drugs. These drugs themselves can often have debilitating or life-threatening effects on patients undergoing chemotherapy. Clinicians attempting to treat cancer are therefore faced with the complex problem of how to balance the need for tumor treatment against the undesirable and dangerous side effects that the treatment may cause.

In this chapter we will consider ways in which evolutionary algorithms (EAs) can be used to assist clinicians in developing approaches to chemotherapy that offer improvements over current clinical practice. Section 12.2 provides a general introduction into the nature of cancer in its many forms. In Section 12.3 we explain how chemotherapy is used to control or reverse tumor growth by exposing cells to the action of toxic drugs. The section explores the complexity that clinicians face in managing toxic side effects and the process of clinical optimization that leads to the treatments conventionally used in practice. In order to apply computational power to this problem, it is first necessary to generate in silico models that, to an acceptable degree, can be relied upon to simulate the growth of cancerous tumors and the effect upon them of chemotherapy. Section 12.4 describes the process of modeling tumor growth and its response to chemotherapy while Section 12.5 develops mathematical

Computational Intelligence in Bioinformatics. Edited by G. B. Fogel, D. Corne, and Y. Pan

formulations of the constraints under which chemotherapy can be applied. In Section 12.6, we review a body of chemotherapy optimization work based on mathematical and numerical optimization approaches developed over the past 30 years. This work provides a strong foundation for in silico chemotherapy design, but the techniques used become intractable as more realistic features and constraints are incorporated. In Section 12.7, we argue the general case for applying EAs to chemotherapy optimization. Sections 12.8 and 12.9 contain a review of our work in this area as it has developed over the last 10 years, including some previously unpublished work. Interest in this application area has been growing in the EA community in recent years, and so Section 12.10 contains a review of related work. In Section 12.11 we describe a decision support system, the Oncology Workbench, that we have developed to make models and algorithms accessible to clinicians. Finally, we conclude in Section 12.12, summarizing the achievements in this area to date and identifying challenges that have yet to be overcome.

12.2 NATURE OF CANCER

Wheldon (1988) contains a useful description of the nature of cancer, which we summarize in this section. Cancer is a broad term covering a range of conditions known as neoplastic disease. All forms of cancer have in common the phenomenon of uncontrolled cell proliferation. Normal healthy cells go through a cycle of growth and differentiation. In response to external chemical stimuli, cells undergo a process called mitosis. A mitotic cell subdivides into two copies. Under the right circumstances, child cells may be different from the parent cells. This is a process known as cell differentiation and is the means by which generic *stem* cells can be turned into cells specialized to function as particular components of the body (e.g., skin, liver, brain, and lung cells).

The human body is subjected to a vast range of harmful external influences, such as solar radiation, cigarette smoke, or harmful ingested material. These can cause damage to particular cells impairing their function. Normally, when cells are damaged, a number of repair mechanisms come into operation. Damaged cells self-destruct and are absorbed into the body. Chemical signals cause quiescent cells to become active and proliferate, thus replacing the damaged ones. When the repair has been affected, cells in the surrounding region return to a quiescent state. This behavior is known as the homeostatic control mechanism.

Occasionally, however, these processes malfunction. Damaged cells do not self-destruct but rather begin to proliferate, producing more cells with the same damaged instructions. Moreover, this proliferation ignores homeostasis. Over time, the result is an increase in the number of unhealthy cells. Such an aggregation of cells is known as a tumor. Tumors of sufficient size result in impairment of bodily function and ultimately death. Moreover, large tumors can metastasize (i.e., spread cells through the bloodstream), resulting in the growth of secondary tumors elsewhere in the body. Cancer, then, can be thought of as uncontrolled growth of a distinct cell population in the body.

12.3 NATURE OF CHEMOTHERAPY

A cytotoxic drug (also termed a "cytotoxic") is a chemical that destroys the cells to which it is exposed. Chemotherapy is the use of cytotoxics to control or eliminate cancer. The drugs are administered to the body by a variety of routes, the objective of which is to create a concentration of drug in the bloodstream.

When tumors reach a particular size, further growth becomes restricted by the availability of oxygen to the cells. The tumor releases chemical signals that stimulate the growth of blood vessels connecting the tumor to the body's circulatory system. This process is known as angiogenesis and has been well-studied and modeled. A concentration of a cytotoxic in the bloodstream will act, therefore, to kill cells systemically. This means that both tumor cells and healthy cells in other organs of the body will alike be killed by the drugs. The intention is to eradicate the tumor or at least to control its growth. However, the treatment has toxic side effects on the rest of the body. Most chemotherapies that fail do so because the patient is unable to bear the drug treatment. In some instances, treatment fails because the tumor has acquired resistance to the drugs being delivered. The success of chemotherapy depends crucially on maintaining damage to the tumor while effectively managing the toxic side effects.

This is a complex control problem, and it is instructive to explore how chemotherapies are designed. There are around 30 cytotoxics used regularly in chemotherapy. These have been developed over several decades by pharmaceutical companies. The process of developing an anticancer drug is a complex and expensive one, most of which is beyond the scope of this chapter. Of particular interest here is the clinical trials process through which all drugs must pass on the route to market, a process that significantly influences how drugs are used in practice. In clinical trials, drugs are administered to patients in controlled studies. There are four phases of trials. The early phases determine the activity of the drug and its therapeutic limits. These are the minimum dose at which the drug is observed to be effective against tumors and the maximum dose that patients can tolerate. Later trial phases are aimed at determining a recommended treatment and establishing the drug's efficacy against particular cancer types. The treatment selected here will be strongly influential on how the drug is administered in clinical practice once it has come to market.

Once a drug is on the market, it will be adopted for practical use by clinicians as appropriate. Clinicians use the drug in its recommended form on the particular disease against which it was tested. However, clinical experience has shown that the most effective treatments often involve combinations of drugs. There is a motivation to integrate a new anticancer drug into more sophisticated treatment designs. A major goal of cancer medicine is the optimization of treatment. This is taken to mean the selection of a treatment that provides, for an individual patient, the best outcome as measured against a set of clinical objectives. Considerable effort is expended on the search for alternative treatments that significantly improve outcomes measured against a range of clinical goals in comparison with conventionally accepted treatments. This effort operates at two levels. Small studies are carried out on individual groups of patients and the results shared by research publication. Occasionally, larger, more expensive, studies are commissioned. The costs of this search, both in human and financial terms, are often prohibitive and progress occurs at an incremental pace. It can take several years of empirical trial to develop significantly improved treatments (Cassidy and McLeod, 1995). It is natural to seek approaches to optimization that can significantly improve this search.

12.4 MODELS OF TUMOR GROWTH AND RESPONSE

In this chapter, we are interested in the use of EAs to assist in the design of chemotherapy. In order to achieve this, we must be able to evaluate potential treatments in silico. Therefore, it is essential to have a model of treatment that gives us a sensible expectation of its effectiveness in terms of treating cancer.

Over the last 30 to 40 years, significant attention has been focused on mathematical modeling of tumor growth and how tumors respond to chemotherapy. This is a very broad area, ranging from microscale models at the cellular or even biomolecular level to macroscale models involving large cell populations making up tumors, organs, and other tissues.

A major focus of this work has been to develop explanations for observed behaviors in the growth of tumors. It is not perhaps surprising that, as a general rule of thumb, the more subtle the behavior of the disease, the more sophisticated a model has to be to capture it. Sophisticated models, however, tend to require large numbers of parameters and behave in ways that are quite sensitive to parameter values. A good example of this is pharmacokinetic/pharmacodynamic (PK/PD) models that describe how drugs move around the vascular system and permeate tissues. Typically these represent tissues, organs, and the vascular system as compartments between which drugs can flow. Parameters are used to set rates of drug flow between compartments, thus calibrating a set of coupled differential equations governing the concentration of drug in each compartment.

In an influential book, Wheldon (1988) described a range of different approaches to the mathematical modeling of cancer. In particular he describes the Gompertz growth model, which has been shown in many studies to reliably model the growth of tumors both in vitro and in vivo. The Gompertz Model has the following form:

$$\frac{dN}{dt} = N(t)\lambda \ln \frac{\Theta}{N(t)}. \tag{12.1}$$

The model describes how a population of tumor cells of size N at time t will grow at a rate that depends on both an intrinsic growth rate λ and the closeness of its current size N to an absolute limiting size Θ. As the tumor grows, it needs to compete with other tissues and organs for resources such as blood supply and oxygen. The maximum size is limited by the supply of blood that the body can produce to support it and by the physical volume available to it. The strain on the body to support and accommodate tumor growth ultimately proves fatal as the functioning of other organs is impaired.

When N is very small compared to Θ, Gompertz growth is very close to exponential. The coefficient λ can be thought of as the "exponential" growth rate of the tumor. Thus, Gompertz growth is characterized by two parameters λ and Θ. A significant advantage of this model is that realistic estimates of λ and Θ can be made from clinical observation. Other cell population growth models exist and have been fitted to clinical data. These include the Verhulst (or logistic) and von Bertalanffy models (Marusic et al., 1994). However, Gompertz growth is the most widely used and extensively validated against clinical data and so will be adopted from now on in this chapter.

From the point of view of clinical treatment, tumors are first observed long after the cancer has started. Typically, at the point where it is first observed, a tumor will have a volume of about 1 mL and consist of approximately 10^9 cells. The growth parameter λ can be estimated from the tumor doubling time, the time it takes for the tumor to grow to twice its first observed size. As it is unethical to allow tumors to grow untreated simply to provide realistic estimates of growth rates, experiments have been conducted with cell lines in vitro and with mouse xenografts in order to determine typical values for λ. Θ is typically thought to be in the range of 10^{12} to 10^{14} cells (Wheldon, 1988).

Wheldon (1988) also discussed how to model the effect of chemotherapy on tumor growth. According to the way in which they kill tumor cells, cytotoxics can be to some extent distinguished into two broad classes: cycle specific and cycle nonspecific. Cancer cells have a life-cycle consisting in a number of phases. For much of the time a cell will be in a quiescent state. When it is ready to proliferate, the cell then passes through a process called mitosis, the result of which is that it subdivides into two daughter cells, each bearing the parent cell DNA (deoxyribonucleic acid). This process, unrestrained in cancer cells, leads to the growth of a tumor.

Cycle-specific drugs tend to be more effective during the mitotic phase of the cell cycle. Cycle-non-specific drugs are more or less equally effective at all stages of the cell cycle. In a large tumor, particular cells will exist at all stages of the cell cycle so there is some justification in regarding all drugs as being cycle nonspecific. Models of the effects of cycle-specific drugs on cell populations can be found in Agur (1986).

It is observed that the ability of an anticancer drug to kill cells increases with increasing concentration of drug present. We can express these relationships as a differential equation as follows:

$$\frac{dN}{dt} = -kC(t)N(t). \tag{12.2}$$

Here N is again the number of tumor cells at time t, $C(t)$ is the drug concentration that the cells are exposed to at time t, and $k > 0$ is a constant representing the cytotoxicity of the drug. Combining this with (12.1) we obtain

$$\dot{N}(t) = N(t)\lambda \ln\left(\frac{\Theta}{N(t)}\right) - kC(t)N(t). \qquad (12.3)$$

This differential equation can be solved numerically. A treatment may be thought of as a time-dependent function of drug concentration over time. Given a treatment plan, parameters λ, Θ, and k, and a starting volume N_0, for the tumor, then the response model allows us to simulate what happens to the tumor over a course of chemotherapy. Note that, more generally, (12.3) can be replaced by

$$\dot{N}(t) = F(N(t)) - kC(t)N(t), \qquad (12.4)$$

where $F(N)$ can be replaced by a variety of cell growth models, such as one of those mentioned in Marusic et al. (1994).

It is important to be aware of the approximations and assumptions involved in these models. $C(t)$ is the effective drug dose to which the tumor is exposed. As stated earlier, PK/PD effects mean that distribution of drug around the body will not be uniform. Furthermore, there are many documented barriers to drug delivery that protect a tumor from the effects of drugs (Jain, 1994). Also the choice of drug delivery method will affect how effectively the drug reaches the tumor. Therefore, there is a complex relationship between the dose of drug administered and the concentration of drug to which a tumor is actually exposed.

Typically, a tumor will be treated over a period of several months. Treatment consists of a number of doses with usually some days or a few weeks between successive administrations. When a particular dose is delivered, the drug is excreted from the body in a matter of hours. Therefore, it is not unrealistic to regard a treatment as a series of instantaneous doses with intervening periods of recovery. Consequently, in our work we regard the potency coefficient k as encapsulating the relationship between delivered dose and tumor response in terms of volume reduction.

It needs to be stated that the relationship between tumor volume and number of tumor cells is not an exact one. Measurements of tumor volume are often difficult in patients. Moreover, tumors contain tissues and fluids that are not cancer cells. However, there is a strong positive correlation between tumor volume and number of cells, and we make the assumption that they are roughly proportional (Wheldon, 1988). Typically, the progress of treatment is monitored by measuring tumor volume, and we use data reported from clinical trials to estimate values for k, following a technique developed in Henderson (1997).

12.5 CONSTRAINTS ON CHEMOTHERAPY

The drugs used in chemotherapy are highly toxic and growth of the tumor is life threatening. Also, chemotherapy creates often debilitating side effects that can cause treatment to fail. In this section we describe the main constraints that apply when designing a cancer treatment.

12.5.1 Dose Constraints

There are limits to both the instantaneous dose and the overall amount of drug that can be given to a patient. We formulate two constraints associated with drug dose. The *maximum tolerable dose* is the largest amount of a drug, measured in milligrams/meter squared, that can be delivered in a single dose.

The *maximum cumulative dose* is the total amount, measured in milligrams/meter squared, of a drug that can be delivered to a patient over a period of treatment. Many anticancer drugs, such as epirubicin, cause long-term damage to organs such as the heart or liver. The consequence is that there are limits to the amount of drug that can be given to a patient over the course of a treatment.

12.5.2 Tumor Growth Constraints

If the tumor is allowed to grow unimpaired, it will eventually become a fatal burden on the body. Long before this, it is likely to cause severe pain to the patient or impair the functioning of organs. We therefore postulate an upper limit, N_{max}, on the number of cells the tumor can have at any point during the treatment.

12.5.3 Toxicity Constraints

There are a large number of recognized toxic side effects of chemotherapy. Typical effects are nausea, hair loss, and bone marrow depletion. A categorization of side effects into symptoms and damage to specific organs is given in Dearnaley et al. (1995) along with a simple severity index associating particular side effects with particular drugs. For a treatment we can formulate constraints on toxicity by summing side effects. We impose a maximum limit on this severity index at any point in the treatment.

12.5.4 Delivery Constraints

It is also possible to introduce constraints associated with the delivery schedule. Standard chemotherapies often use fixed dosages and combinations of drugs, at equally spaced intervals throughout the treatment period. One of the reasons for this is that it is easier to organize patients to present for treatment at fixed times. Another is that fixed dosages and combinations are more easily communicated and administered, the consequences of erroneous treatment being severe.

12.6 OPTIMAL CONTROL FORMULATIONS OF CANCER CHEMOTHERAPY

Much of the modeling work discussed so far has been aimed at understanding the progression of cancer and the way it responds to treatment. We now review work done over the last 30 years that applied the mathematics of optimal control to derive optimal treatment schedules.

12.6.1 Basic Description of Optimal Control

An optimal control problem is usually characterized by two types of variables: a state vector N and a control vector C. Both N and C may be discrete or continuous and vary over time. As time varies, the state N will change and the values of control C are assumed to influence the way in which this occurs.

An objective function J is formulated in terms of C and the values N takes when C is applied. If the problem is constrained, then general equality and inequality constraints involving the values of C and N (and, in general, their derivatives where they exist) can also be formulated.

In the case of chemotherapy, we are interested in controlling the size N of a tumor over a treatment period of length T by applying a treatment C. Constraints can be formulated as indicated in Section 12.5. The earliest work in this area was carried out by Swan and Vincent (1977). They developed an objective that was expressed in terms of the total amount of drug delivered over the treatment period:

$$J(c) = \int_0^T c(t)\, dt. \tag{12.5}$$

Here T is the end of the treatment period, a fixed time point.

A constraint was also applied that the tumor should be reduced to a specified size by the end of the therapy. The aim of this was to find treatments that reduced the tumor to an eradication point while minimizing the toxicity of the treatment. A weakness of this approach is that the growth of the tumor between endpoints of the treatment period is unconstrained.

Swan (1990) later refined his approach to include explicit minimization of the tumor in the objective function:

$$J(c) = \int_0^T [(N(t) - N_d)^2 + \rho c^2(t)]\, dt. \tag{12.6}$$

Here N_d is a lower limit on tumor size that is taken to represent eradication of the tumor. It is difficult to determine by clinical observation whether a tumor has been eradicated. Once the tumor has been reduced to fewer that 10^9 cells, it is no longer clinically observable. However, it is postulated that a tumor would need to be reduced to as few as 10^3 cells to be properly eradicated. At that point other mechanisms take over, and the body can naturally eliminate the remaining cells. We can distinguish between the level N_{cure} at which the tumor has been effectively eradicated and N_d at which the tumor is no longer observable. In Swan's approach, N_d is used as an observable clinical target.

Another feature of this model is the coefficient ρ. This is a penalty coefficient that can be varied in order to increase or decrease the penalty attached to applying drug. This provides flexibility to take into account the differing reactions of patients to treatment. Some patients have a high tolerance to side effects and can bear more toxic treatments than can others. It should also be noted here that Swan uses the logistic growth model for the tumor, modified by a cell kill term as in (12.4).

These examples illustrate nicely one of the fundamental conflicts in chemotherapy. To eradicate a tumor, it is necessary to expose it to sufficient toxicity. At the same time, it is essential to maintain toxicity below acceptable bounds because of side effects and potential long-term damage to the patient. Therefore, it is natural to think of chemotherapy in terms of multiobjective optimization.

An earlier approach to multiobjective optimization of chemotherapy, still in terms of optimal control, was explored by Zeitz and Nicolini (1979). They formulated objectives in terms of two cell populations. The first population consists of normal tissues exposed to the treatment. The second consists of the tumor cell population. The objectives are formulated as follows:

$$J_1(c) = \int_0^T N_1(t)\left[-\lambda_1 \ln\left(\frac{N_1(t)}{\Theta_1}\right) - \kappa_1 c(t)\right] dt, \tag{12.7}$$

$$J_2(c) = \int_0^T N_2(t)\left[-\lambda_2 \ln\left(\frac{N_2(t)}{\Theta_2}\right) - \kappa_2 c(t)\right] dt. \tag{12.8}$$

Here N_1 represents the population of normal cells and N_2 the tumor cell population. The objectives integrate the instantaneous sizes of each population over the whole treatment period under Gompertz growth with an exponential cell kill (12.3). The normal population operates under homeostatic control and so the value of Θ_1 will be close to the initial population size. The value of $J_1(c)$ is therefore bounded above by $T\Theta_1$. The fewer normal cells that are killed during treatment the better, and so maximization of $J_1(c)$ is taken to be a surrogate for minimization of toxicity to normal cells. The second objective, $J_2(c)$, is often referred to as the tumor burden. Minimization of $J_2(c)$ is a surrogate for tumor eradication, and it also has the advantage of addressing tumor size throughout the treatment, rather than solely at the endpoint. The overall objective used in the optimal control was to maximize $J(c) = J_1(c) - J_2(c)$.

In 1990, Murray explored a variant of this approach that expresses the objective solely in terms of the tumor population but retaining the normal cell population as a constraint (Murray, 1990a,b). Murray used a weighted version of the tumor burden as his objective:

$$J(c) = \int_0^T f(t, N_2(t))\, dt \tag{12.9}$$

This was governed by normal and tumor cell populations growing according to (12.3) and the following constraints:

$$\text{Maximum dose constraint:} \quad c(t) \in [0, C_{\max}], \tag{12.10}$$

$$\text{Cumulative dose constraint:} \quad \int_0^T c(t)\,dt \le C_{\text{cum}}, \tag{12.11}$$

$$\text{Normal cell toxicity constraint:} \quad N_1(t) \ge N_1^{\min} \forall t \in [0, T]. \tag{12.12}$$

Murray explored a number of different scenarios by changing the weighting of the objective function. A consistent feature of the treatments generated using his approach was a phased approach. An intensive phase used the maximum possible dose for a period and aimed to greatly reduce the number of tumor cells. A maintenance phase used a level of drug to hold the tumor size at a prescribed level. Murray's treatment designs typically also included rest phases where no treatment was applied. These occurred at the beginning and end of treatment and also separating the intensive and maintenance phases. When the objective was strongly weighted to tumor eradication by the end of the treatment period, both the intensive and rest phases were pushed toward the end of the treatment period. In practical chemotherapy, clinicians will not leave a tumor untreated for long periods of time because of the potential of tumor cells to develop drug resistance, a feature not included in Murray's models.

An important feature of Murray's work is that it suggested irregular schedules, which may be more successful than standard chemotherapies in achieving complex objectives. A disadvantage of the approaches we have considered so far is that it is difficult to determine the nature and growth characteristics of the normal cell population exposed to the therapy.

Martin and Teo (1994) developed an approach that focused solely on objectives related to the tumor cell population and its growth–response characteristics. Toxicity constraints were expressed in terms of maximum and cumulative dose constraints, removing the need for modeling a normal cell population. They formulated the following objective:

$$J(c) = N(T). \tag{12.13}$$

Here $N(T)$ is the final size of the tumor at a fixed endpoint T of the treatment interval. The control c consists of a series of drug boluses that are given at discrete time intervals. This resulted in the following system of equations:

$$\dot{N}(t) = N(t)\left\{\lambda \ln\left[\frac{\Theta}{N(t)}\right] - \kappa[c(t) - c_{\text{th}}]H[c(t) - c_{\text{th}}]\right\}, \tag{12.14}$$

$$c(t) = \sum_{i-1}^{n} C_i \exp[-\delta(t - t_i)H(t - t_i)]. \tag{12.15}$$

Martin also applies three constraints, the first limiting tumor size to below a comfort threshold and the remaining two, identical to Eqs. (12.10) and (12.11), limiting the toxicity of the treatment that can be applied:

$$\text{Tumor size constraint:} \quad N(t) \le N_{\max} \forall t \in [0, T]. \tag{12.16}$$

This approach has a number of advantages over the other work we have considered. The chemotherapy is delivered as a series of discrete doses. This is closer to clinical practice, and so the results are more directly relevant to the clinician. Eliminating consideration of a normal cell population substantially reduces the number of parameters required for the model. In particular, it is no longer necessary to estimate the overall size and asymptotic limit of the normal cell population, its growth characteristics, and the cell kill rate of normal cells by the anticancer drugs. Also the approach is extensible to different objective functions, and a range of tumor growth models can be adapted to incorporate acquired drug resistance by tumor cells at advanced stages of the disease (Martin and Teo, 1994).

Martin and Teo (1994) explored a range of objectives including minimization of final tumor size, forced decrease of tumor size over the treatment period, and maximizing patient survival time in cases where the tumor was incurable due to the existence of drug-resistant cancer cells in the tumor. Like Murray, they discovered optimal treatments that differed radically from conventional treatments.

12.6.2 Solution Methods

In all of the studies discussed above, unique theoretical optimal solutions are known to exist as solutions of the Euler–Lagrange equations derived from the optimal control problem. In practice, when tumor size and drug–dose constraints are included, these problems do not have closed-form solutions that can be derived mathematically (Martin and Teo, 1994). Therefore, computational optimization techniques are applied to approximate the optimal control. Martin and Teo (1994) approximated the control by piecewise constant control and apply a sequential quadratic programming technique to their set of problems. Proofs of some useful convergence results for the algorithms they use were contained in their publication.

12.7 EVOLUTIONARY ALGORITHMS FOR CANCER CHEMOTHERAPY OPTIMIZATION

The numerical techniques applied in Martin and Teo (1994) do not guarantee optimality. The approach is liable to become trapped in local optima. The heuristic adopted to avoid this behavior is to run the algorithm repeatedly with randomized starting values for the controls until a solution is found that is sufficiently close to a theoretically derived optimal bound (Martin and Teo, 1994). Moreover, as the number of constraints increases, the computation becomes increasingly intractable to classical numerical methods. In practical chemotherapy, it is desirable to use multiple drugs in combination therapies, perhaps involving three to five drugs (Henderson et al., 1988). Each drug has its own constraints on maximum and cumulative dose and is responsible for its own set of side effects. Also, in order to explore the full range of possibility for phasing drug treatment, dosing times cannot be fixed in advance.

Petrovski (1998) explored a range of nonlinear programming techniques and their applicability to cancer chemotherapy optimization. In particular, a detailed exploration of two pattern-search methods, the complex method (CM) and Hooke and Jeeves (HJ) was conducted. These methods can be efficient in optimizing unimodal objective functions when the location of feasible regions is known. Typically, in chemotherapy optimization, it is difficult to find solutions that satisfy all constraints. Reducing the tumor size usually requires the application of drugs at levels close to maximal allowable doses. Conversely, using smaller drug doses leads to large tumors that are close to tumor size boundaries. Moreover, chemotherapy optimization problems tend to be multimodal: There are many local optima. In cases where multiple drugs are used there may be a number of equally valued global optima representing treatments that achieve the objectives of chemotherapy equally well but using different drug combinations and dosing strategies. It is desirable to locate all of these global optima as this gives clinicians insight into the range of possible treatments. Petrovski (1998) demonstrated that both CM and HJ tended to converge to nonglobal optima or fail to find a feasible solution at all.

For a number of years, we have been interested in developing chemotherapy optimization algorithms that can cope with realistic numbers of drugs and constraints and can be flexibly applied to optimize a range of clinical objectives. In this chapter, we will describe work we have carried out in applying a range of EAs to such problems. The next section will describe how we encode and evaluate chemotherapy treatments.

12.8 ENCODING AND EVALUATION

We now review our work in applying EAs to cancer chemotherapy optimization. There are a number of aims in our research. The most important aim has been to explore as fully as possible the space of cancer treatments. The mathematical optimization work described in Section 12.6 has shown that conventional therapy is not necessarily optimal, suggesting in particular that more intensive therapy should be phased later in the treatment period. The computational complexity of these approaches, however, prohibits wide exploration of the space. EAs are known to be successful in multimodal highly constrained optimization problems. In particular, they scale well as additional constraints are added.

The second aim in our work is to shift emphasis away from precise mathematical modeling toward learning from clinical data. The motivation for this is the desire to eliminate overdependence upon model parameters that are difficult or impossible to measure clinically. We therefore select an approach that can, to some extent, be adapted to patient variability in a clinical setting.

The third aim of our work is to make our algorithms accessible to clinicians, so that they can specify particular treatment design problems. We have developed a problem-solving environment, the Oncology Workbench, to assist with the exploration of possible treatments in silico. An important aim of this environment is that clinicians can solve treatment design problems in a way that is natural to their understanding of the problem. This is discussed in more detail in Section 12.11.

Our final aim is to widely explore different EA approaches to optimizing chemotherapy. By comparing and contrasting different algorithms, over a range of single and multiobjective problems, we hope ultimately to develop a robust and efficient solver that can be used clinically to develop radically improved therapies for different cancer types.

12.8.1 General Response Model Formulation

Anticancer drugs are usually delivered according to a discrete dosage program in which there are n doses given at times $t_1, t_2, \ldots, t_n$. In the case of multidrug chemotherapy each dose is a cocktail of d drugs characterized by the concentration levels C_{ij}, $i = 1, \ldots, n$, $j = 1, \ldots, d$ of anticancer drugs in the blood plasma. Optimization of the treatment is achieved by modification of these variables. Therefore, the solution space of the chemotherapy optimization problem is the set of control vectors $c = C_{ij}$ representing the drug concentration profiles.

The response of the tumor to treatment is governed by a growth–response model. Following Martin and Teo (1994), we use the Gompertz growth model with a linear cell loss effect. The model takes the following form:

$$\frac{dN}{dt} = N(t)\left\{\lambda \ln\left(\frac{\Theta}{N(t)}\right) - \sum_{j=1}^{d} \kappa_j \sum_{i=1}^{n} C_{ij}[H(t-t_i) - H(t-t_{i+1})]\right\}, \qquad (12.17)$$

where $N(t)$ represents the number of tumor cells at time t; λ, Θ are the parameters of tumor growth, $H(t)$ is the Heaviside step function, C_{ij} denote the concentrations of anticancer drugs, and κ_j are the quantities representing the efficacy of anticancer drugs. In order to allow particular named drugs to be used in our therapy designs, it is necessary to estimate coefficients κ_j. Henderson (1997) developed a procedure for calibrating κ_j using published clinical trial results. In Petrovski (1998) we applied this technique to estimate coefficients for 10 drugs commonly used in the treatment of breast cancer. It is important to note that these coefficients are dependent on the particular cancer type chosen. Moreover, Henderson's technique gives an expectation of drug activity for a population of patients with normally distributed variability in response to the drug. In the case of individual patients, factors such as general condition and genetic makeup can play a significant, though as yet not fully understood, role in the effectiveness of particular drugs.

In relation to breast cancer, a typical value of λ gives an initial doubling time of 4 weeks for the tumor cell population. The asymptotic limit Θ of tumor growth is assumed to be 10^{12} cells as observed in Martin and Teo (1994). We use the following canonical formulation for the constraints of chemotherapy:

$$\begin{aligned}
g_1(\mathbf{c}) &= C_{\max j} - C_{ij} \ge 0 && \forall i = \overline{1, n},\ j = \overline{1, d},\\
g_2(\mathbf{c}) &= C_{\text{cum } j} - \sum_{i=1}^{n} C_{ij} \ge 0 && \forall j = \overline{1, d},\\
g_3(\mathbf{c}) &= N_{\max} - N(t_i) \ge 0 && \forall i = \overline{1, n},\\
g_4(\mathbf{c}) &= C_{\text{s-eff } k} - \sum_{j=1}^{d} \eta_{kj} C_{ij} \ge 0 && \forall i = \overline{1, n},\ k = \overline{1, m}.
\end{aligned} \qquad (12.18)$$

Constraints $g_1(\mathbf{c})$ and $g_2(\mathbf{c})$ specify the boundaries for the maximum instantaneous dose $C_{\max,j}$ and the maximum cumulative dose $C_{\text{cum},j}$ for each drug used. Constraints $g_3(\mathbf{c})$ limit the tumor size during treatment. These constraints are reformulations of Eqs. (12.10), (12.11), and (12.16).

In our work we introduce a further set of constraints $g_4(\mathbf{c})$. These represent the toxic side effects of multidrug chemotherapy. Different drugs are associated with different side effects. These have been observed clinically and documented during studies. A widely used guide to toxicity in clinical practice (Dearnaley et al., 1995) documents several different classes of side effects and gives severity indices for each side effect for each drug. The factors η_{kj} are based on these indices and represent the severity of the kth side effect (e.g., damage to the heart, bone marrow, lung) of administering the jth drug. One of the main reasons for failure of chemotherapy is that side effects become too severe for patients to continue with treatment. There is therefore a need to balance severe side effects against the need to reduce tumor volume. It is hoped that EAs can find schedules that exploit differences in drugs to effectively manage side effects while maintaining pressure on the tumor.

12.8.2 Encoding

Multidrug chemotherapy control vectors $\mathbf{c} = (C_{ij})$, $i = \overline{1, n}$, $j = \overline{1, d}$ are encoded as bit strings. The representation space I can be expressed as a Cartesian product $I = A_1^1 \times A_1^2 \times \ldots \times A_1^d \times A_2^1 \times A_2^2 \times \ldots \times A_2^d \times \ldots \times A_n^1 \times A_n^2 \times \ldots \times A_n^d$ of allele sets A_i^j. This represents a series of doses of d drugs delivered at n discrete points in a treatment period. There are a number of possibilities for the allele values. In most of our work, we have used a 4-bit scheme $A_i^j = \{a_1 a_2 a_3 a_4 : a_k \in \{0, 1\} \forall k = \overline{1, 4}\}$, so that each concentration C_{ij} takes an integer value in the range of 0 to 15 concentration units. A concentration unit has a translation in terms of milligrams/meter squared depending on the drug related to that particular section of the encoding. The encoding allows the drug dosage to vary from 0 to the maximum allowable dose for each drug, thus enforcing constraints $g_1(\mathbf{c})$ irrespective of the treatment generated. The use of standard concentration units was developed in order to conveniently add the effects of doses from different drugs given simultaneously.[1] The actual volume of drug administered can vary across a range of scales. For example, 5-fluorouracil has a maximum tolerable dose of around 3000 mg/m^2, whereas epirubicin has a maximum tolerable dose of around 75 mg/m^2.

In general, with n treatment intervals and up to 2^p concentration levels for d drugs there are up to 2^{npd} chromosomes. The assumptions made in McCall and Petrovski (1999), giving a search space of 2^{400} chromosomes, are not unrealistic. Other encodings have been explored: in McCall and Petrovski (1999), a delta-coding was used to improve solutions generated by a binary-coded genetic algorithm (GA);

[1] It should be noted that drug effects will not, in general, add in this simple way. Drugs can potentially reinforce or inhibit the effects of each other because they affect the enzymatic reactions that govern metabolism. So far we have not included drug–drug interactions in our models.

in Petrovski et al. (2005), an integer-coded GA is developed and optimized using statistical parameter tuning; and in Petrovski et al. (2004) an integer encoding was used for a particle swarm optimization (PSO) approach. To date we have not used floating-point encodings, although other researchers have done so—see Section 12.10. Given patient variability and the practical difficulty of delivering particular drug concentrations to a high level of precision, it is arguable whethere treatment schedules defined to fractions of a milligrams/meter squared can be reliably delivered. We have chosen to explore a more manageable search space over any benefits in precision that a floating-point representation might bring. A range of 0 to 15 concentration levels for each drug was considered sensible by the clinical oncologists with whom we have collaborated.

12.8.3 Evaluation

Evolutionary algorithms use evaluation or fitness functions to drive the evolution of optimal solutions. Our general strategy is to use the tumor growth–response model, combined with penalties for constraint violation, to formulate an evaluation function for encoded treatments. A candidate treatment is simulated using the growth–response model. The treatment is then evaluated using objective functions relating to the aims of the treatment, augmented by penalties for each constraint violated.

We now define precisely the approaches to evaluation that we have used in our work. Figure 12.1 captures the essential data required to evaluate a treatment. The diagram shows the tumor volume N graphed against time. The tumor is shown growing from a small size to the point where treatment can commence. For most of this time the tumor has been undetected. At the beginning of the treatment period, time T_0, the tumor has size N_0. One overall objective of therapy might be to eradicate the tumor. This is typically represented in two ways: minimize final tumor size or minimize the tumor burden. The final tumor size, N_{final}, is the size of the tumor at time T_{final}, the end of the treatment period. The tumor burden is the integral of the tumor response curve, represented by the gray shaded area in Figure 12.1. A further

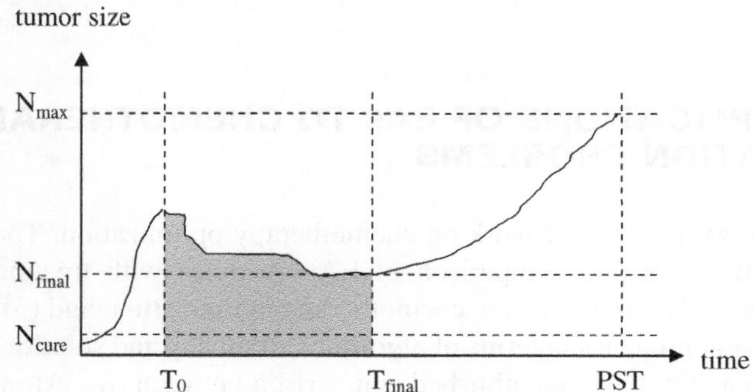

Figure 12.1 Objectives of chemotherapy.

measure that can be introduced is the cure level, N_{cure}. If a treatment drives the tumor below this size, the tumor is deemed to be cured. If the tumor remains above N_{cure} for the entire treatment period, the chemotherapy has failed and the tumor will regrow. Ultimately, if treatment is unsuccessful, the tumor will reach a fatal size, N_{fatal}. The time taken for this to occur, measured from the start of the treatment period, is defined as the patient survival time (PST). An alternative aim of chemotherapy, where cure is not thought to be possible, is to palliate the tumor. Here the aim is to maintain quality of life for the patient for as long as possible. The objective then is to hold the tumor size below a level N_{max} at which the patient is comfortable for as long as possible, that is, maximizing PST.

The objective for a curative treatment to minimize the tumor burden is now formulated mathematically as:

$$\text{Minimize } J_1(c) = \int_{t_1}^{t_n} N(\tau)\, d\tau. \tag{12.19}$$

The objective of prolonging the PST, denoted here by T is formulated as

$$\text{Maximize } J_2(c) = \int_{T_0}^{T} d\tau. \tag{12.20}$$

Treatments **c** that violate constraints are penalized using metrics

$$d_s = \max{}^2\{-g_s(c), 0\}, \quad s = 1, \ldots, 4, \tag{12.21}$$

to measure the distance from the feasible region corresponding to the toxicity constraints $g_s(\mathbf{c})$ formalized in (12.18). Applying penalty coefficients P_s to each metric, we obtain augmented objective functions as follows:

$$\tilde{J}_i(c) = J_i(c) - \sum_{s=1}^{4} P_s d_s. \tag{12.22}$$

Particular evaluation functions can now be derived from these mathematical formulations. This will typically involve numerical solution of the tumor response model, providing terms relating to the tumor size at each time step. The exact form of the fitness function will depend on the particular treatment design objectives chosen.

12.9 APPLICATIONS OF EAs TO CHEMOTHERAPY OPTIMIZATION PROBLEMS

In this section we review our work on chemotherapy optimization. There are three features of this work: (1) comparison of EA schedules with treatments used in clinical practice, (2) multiobjective chemotherapy optimization, and (3) comparison of different EA approaches in terms of algorithm efficiency and solution quality. We conclude with a previously unpublished comparison between two estimation of distribution algorithms (EDA) for chemotherapy optimization.

12.9.1 Comparison with Conventional Treatment

In Petrovski and McCall (2000), we implemented four commonly used models of tumor growth from Marusic et al. (1994):

Exponential model: $\dfrac{dN}{dt} = \lambda N(t),$

von Bertalanffy model: $\dfrac{dN}{dt} = \lambda \sqrt[3]{N(t)^2} - \mu N(t),$

Verhulst model: $\dfrac{dN}{dt} = \lambda N(t) - \mu N(t)^2,$

Gompertz model: $\dfrac{dN}{dt} = \lambda N(t) - \mu N(t) \ln N(t).$

Each of these was developed into a tumor response model by adding a standard cell kill term in the differential equation; see Eq. (12.4). For each model, two treatment optimization scenarios were defined: (1) eradication of the tumor by minimizing tumor burden and (2) tumor palliation by maximization of PST. Tumor burden was measured relative to the initial tumor size and PST is measured in weeks. These were solved as single objective optimization problems using genetic algorithms. The characteristics of the optimal treatments found by the GAs were compared with those of a commonly used breast cancer schedule CAF. CAF uses a combination of three anticancer agents (cyclophosphamide, doxorubicin and 5-fluorouracil) and is fully described in Dearnaley et al. (1995). The GAs were restricted to using the same drugs and total amounts of drug as used in the CAF treatment. Results are summarized in Table 12.1 and show that, irrespective of the tumor response model used, the GA-designed treatments outperform the treatment used in clinical practice. This is consistent with other optimization results in suggesting that more effective chemotherapies are possible by adopting nonstandard treatments.

Table 12.1 Comparison of the GA Optimal Treatment with CAF

Response Model	Treatment Method	Tumor Burden	PST (weeks)
Exponential	GA	6.1002	35
	CAF	10.3616	31
Von Bertalanffy	GA	7.9120	40
	CAF	9.7691	40
Verhulst	GA	7.3572	37
	CAF	10.1039	32
Gompertz	GA	4.6843	38
	CAF	11.2897	24

12.9.2 Sequential Multiobjective Optimization

As discussed earlier, the objectives of curative and palliative treatments conflict with each other in the sense that drug treatments that tend to minimize tumor size are highly toxic and therefore have a negative effect on patient quality of life. Moreover, it has been shown that a severe treatment that fails to cure can result in a shorter patient survival time (PST) than a milder palliative treatment (Martin and Teo, 1994). It is therefore natural to consider multiobjective approaches to treatment optimization.

In McCall and Petrovski (1999), we used a delta-coding GA approach to optimize against each objective in sequence. First, we attempted to eradicate the tumor by minimizing tumor burden. The objective function f_1 uses the negative log of tumor size, so in this case tumor eradication was posed as a maximization problem. A treatment is deemed to be curative if the tumor is reduced to 1000 or fewer cells for three consecutive treatment intervals. The tumor size was then fixed at zero for the remainder of the treatment. If, after a fixed number of runs, the algorithm failed to find a curative treatment, the best treatment found to date was recorded and the objective switches to palliation.

The work flow of this algorithm is best illustrated by considering two typical runs, giving rise to treatments A, B, and C. Figure 12.2 is a graph of tumor size against time for each treatment. The cure threshold of 1000 cells is also marked. Table 12.2 shows the outcomes of each treatment against each objective, f_1 and f_2.

In the first run, cure is impossible so the algorithm outputs its best attempt at eradicating the tumor (treatment A) followed by an optimal palliative treatment (treatment B). In the second run, cure is possible and treatment C is the treatment found to produce the highest value of the eradication objective. As can be seen from

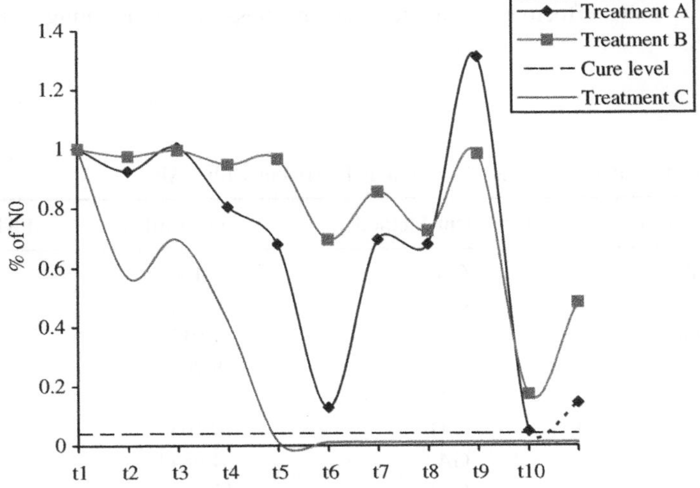

Figure 12.2 Tumor response curves for three multidrug treatments.

Table 12.2 Characteristics of Optimal Treatments[a]

Optimal treatment	f_1	f_2	Avg. Tumor Size as a Fraction of N_0	PST
Treatment A	6.9942	N/A	0.6226	N/A
Treatment B	N/A	0.3435	0.8040	32
Treatment C	115.015	N/A	0.1630	N/A

[a]All constraints are satisfied for each treatment type A, B, or C.

Figure 12.2, under treatment A, the tumor did not reach the cure threshold and grew back to 1.4 times the original size during the treatment. It is likely that, before the end of treatment, this tumor would be composed of cells highly resistant to chemotherapy, and patient survival would not extend much beyond the end of the treatment period. Under treatment B the tumor is maintained at a larger size throughout the treatment and only substantially reduced toward the end. As a result, PST is extended and the drug dosing is lower, giving a better quality of life. Under treatment C, the tumor is eradicated half way through the treatment period. This approach might be useful in assisting a clinician to choose between an eradication or a palliation strategy by indicating how severe treatment needs to be in order to be curative.

12.9.3 Pareto Optimization

As it is often impossible to determine beforehand which objective should be pursued, another approach is to design treatments that balance both objectives as well as possible. In Petrovski and McCall (2001), we formulate a multiobjective optimization problem with a vector-valued objective function $F(c) = [f_1(c)_1, f_2(c)]^{\mathrm{T}}$. Here $f_1(c)$ refers to the tumor eradication objective, formulated as a maximization problem as follows:

$$f_1(c) = \int_{t_1}^{t_n} \ln\left(\frac{\Theta}{N(\tau)}\right) d\tau. \tag{12.23}$$

Simlarly, $f_2(c)$ refers to the tumor palliation objective, formulated as a maximization problem as follows:

$$f_2(c) = \int_{t_1}^{T} d\tau = T - t_1. \tag{12.24}$$

These objectives were augmented by constraint violation penalties shown in Eq. (12.18). We adapted the SPEA algorithm (Zitzler, 1999) to evolve a population of treatments spread out along the Pareto front. The algorithm found treatments that performed well on the single objectives as in earlier work but also discovered new possible treatments that offered a good balance between the objectives. Thus the multiobjective approach offers a powerful tool to assist in chemotherapy design. Clinicians can set model and constraint parameters and then select from the Pareto-optimal set of treatments generated by the algorithm.

12.9.4 Performance Optimization

The evaluation of treatments requires the simulation of tumor response, which can become computationally expensive, depending on the complexity and sophistication of the tumor-response model. Ideally, one would want to evaluate a treatment against a population of simulated tumors, thus building in robustness against patient variability. Therefore, some of our work has been to apply a range of EA approaches to treatment optimization, comparing with the GA in terms of efficiency and solution quality.

In Petrovski et al. (2004), we compared a GA with local-best and global-best PSO algorithms on a single-objective treatment optimization problem with a tumor burden minimization objective. PSO algorithms use a population of particles that swarm around the search space. In our implementation, each particle was located at a treatment, encoded as a string of integers. We measured the efficiency of each algorithm as the number of fitness evaluations necessary to find a treatment satisfying all constraints. The global-best PSO algorithm was the most efficient, closely followed by local-best PSO. Both PSO algorithms outperformed the GA to a very high level of significance. We also compared solution quality at convergence and found that there was no significant difference among the three approaches. Analysis of run time distributions showed that the PSO algorithms had a very predictable performance and always found a feasible solution. The GA, by comparison, had a highly variable performance, failing, in some runs, to find a feasible solution. Martin and Teo (1994) showed mathematically that, for the problems they considered, optimal treatments lay on the boundaries of feasible regions. Recombination and mutation in the GA may often give rise to infeasible solutions causing inefficiencies in the search. The PSO algorithms on the other hand tend to focus exploration around remembered locations in the search space, a particular advantage when good solutions are right on the boundary of feasible regions. We conclude that, on the whole, the ability to retain information on the best solutions found by each particle and pass it round the whole population is a valuable asset in solving chemotherapy optimization problems. Another way of explicitly retaining information learned from the population is in a probabilistic model. In recent years, estimation of distribution algorithms (EDAs) have been the focus of considerable interest in the EA community. EDAs replace the traditional recombination operators in a GA with a process of sampling a probabilistic model. Thus the successor population is generated by sampling a probabilistic model of "good solutions" learned from the current and possibly earlier generations.

In Shakya et al. (2006), we applied population-based incremental learning (PBIL) to the problem of treatment optimization with the objective of tumor eradication and compare it with a GA. PBIL is one of the earliest and most established EDAs. It operates on bit-string solutions and uses a univariate probabilistic model. The model is implemented in the form of a vector, each coordinate of which gives the probability that the corresponding variable in the bit string takes value 1. New solutions are generated from this model by sampling the associated probability distribution.

We now present previously unpublished work applying a different univariate EDA, $DEUM_d$ to the chemotherapy optimization problem. $DEUM_d$, an algorithm developed by Shakya et al. (2005), resembles PBIL in its approach but uses solution fitness to bias the probabilistic model. It has been shown to outperform PBIL in terms of number of fitness evaluations and solution quality on a range of problems (Shakya et al., 2005). It is natural therefore to extend the work in Shakya et al. (2006) to compare with $DEUM_d$.

We compared the performance of a GA, PBIL, and $DEUM_d$ on the problem of multidrug cancer chemotherapy optimization. We performed a single-objective optimization with the tumor eradication objective defined in Eq. (12.19). Two different measures are adopted to evaluate the performance of each algorithm:

1. *Efficiency:* For the algorithms considered, the evaluation of a particular treatment is the most computationally expensive task. Therefore, we measure efficiency in terms of the number of fitness evaluations taken by the algorithm to find a treatment that satisfies all of the tumor size and toxicity constraints.
2. *Solution Quality:* Solution quality is the value of the fitness function (12.25) that indicates how successful a particular chemotherapy schedule is in reducing the overall tumor burden. We compare the best treatments found by each algorithm:

$$F(c) = \sum_{p=1}^{n} \sum_{d=1}^{j} \kappa_j \sum_{i=1}^{p} C_{ij} e^{\lambda(t_{i-1}-t_p)} - \sum_{s=1}^{4} P_s d_s. \tag{12.25}$$

The terms in the triple summation are derived from a transformed version of the Gompertz growth model and relate to the negative natural log of tumor size. Therefore, minimization of tumor burden corresponds to maximization of (12.25). The d in the final summation represents the distance measures specifying how seriously the constraints are violated and the P_s are the corresponding penalty coefficients. Once all constraints are satisfied, the second term in (12.25) will be zero, significantly increasing the value of the fitness function.

12.9.5 Efficiency Comparison

In the efficiency comparison experiment, each algorithm was run 1000 times and for each run the number of fitness evaluations taken to find at least one feasible treatment was recorded. Run length distribution (RLD) curves (Hoos and Stutzle, 1999) were plotted to measure the performance (Fig. 12.3). The RLD curves show, for each algorithm, the cumulative percentage of successful runs that terminated within a certain number of function evaluations. The parameters for each of the algorithms were chosen either on the basis of the previous work (Petrovski et al., 2005) (for the GA) or empirically (for PBIL and $DEUM_d$), taking into consideration the fairness of the comparison.

In accordance with the best GA parameter settings found in Petrovski et al. (2005) for the cancer chemotherapy problem, we used a GA population of 100

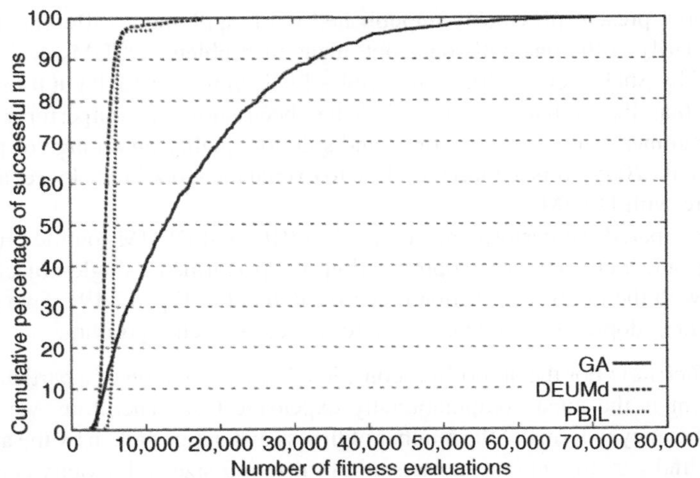

Figure 12.3 Experimental results in the form of RLD curves showing, for each algorithm, the cumulative percentage of successful runs that terminated within a certain number of function evaluations.

chromosomes, one-point crossover with the probability of 0.614, uniform mutation with probability of 0.075, and elitism retaining the two best chromosomes. The algorithm terminated when it discovered a solution that did not break any constraints.

Similarly, the PBIL population consisted of 100 chromosomes, the selection size N was 20, and the learning rate λ was 0.3 (Shakya et al., 2006). For DEUM$_d$, the population consisted of 100 chromosomes, the selection size N was 6, the temperature coefficient τ was 2, and elitism retaining the two best chromosomes was used.

Figure 12.3 shows that the GA had the poorest performance of all three algorithms, whereas DEUM$_d$ had the best performance, marginally but significantly better than PBIL. For DEUM$_d$, 90% of the runs found feasible solutions within 6100 function evaluations in comparison to 6700 function evaluations for PBIL and 32,900 function evaluations for GA. Also, the steepness of the RLD curves for DEUM$_d$ and PBIL show that these algorithms very reliably find feasible solutions within a predictable number of function evaluations, though PBIL proved more precisely predictable than DEUM$_d$. Conversely, the gradual slope of the RLD curve for the GA shows that the number of GA fitness evaluations required to find a feasible solution varies significantly and cannot be predicted accurately.

The mean value and the standard deviation of the number of function evaluations taken by each algorithm over 1000 runs is presented in Table 12.3. Also shown is the success rate of finding a feasible solution over 1000 runs for each algorithm. The means for each pair of algorithms are compared by applying a two-tailed t test as shown in Table 12.4. As indicated by the very low p value for each pair, the differences in the means are strongly statistically significant.

Table 12.3 Mean and Standard Deviation of Number of Fitness Evaluations

	Mean	Std. Dev.	Success (%)
GA	16,208.10	12,045.81	100
PBIL	5,953.45	522.40	97.1
$DEUM_d$	5,015.75	1,426.90	99.7

Table 12.4 *t*-Test Analysis of Differences in Mean Fitness Evaluations

	Diff. in Means	Std. Error	*t* Test	*p* Value
GA vs. PBIL	10,254.65	381.29	26.89	<0.0001
GA vs. $DEUM_d$	11,192.35	383.58	29.18	<0.0001
PBIL vs. $DEUM_d$	937.70	48.05	19.52	<0.0001

Table 12.5 Mean and Standard Deviation of Number of Fitness Evaluations

	Mean	Std. Dev.
GA	0.4520	0.0354
PBIL	0.4917	0.0005
$DEUM_d$	0.4920	0.0006

12.9.6 Comparison of Treatment Quality

The goal of the second experiment was to find the best chemotherapeutic treatment within a specified maximum number of function evaluations, which was set to 200,000. In order to better achieve this goal, the population size for each of the algorithms was increased to 200 chromosomes. Also, to encourage more thorough exploration of the solution space, the mutation probability of the GA was adjusted in accordance with the strategies suggested in Petrovski et al. (2005). Also, the learning rate of PBIL was decreased to 0.05 and the temperature coefficient of $DEUM_d$ was decreased to 0.1. The results of the treatment quality experiment are presented in Table 12.5, which shows the mean and the standard deviation of the fitness values found by the algorithms over 200 runs.

It is evident that the solution found by $DEUM_d$ and PBIL is better than that found by the GA. Also, $DEUM_d$, with a mean fitness of 0.4920, very slightly outperforms PBIL with mean fitness of 0.4917. These conclusions are again supported by a two-tailed *t* test applied to the mean fitness values found by the algorithms and are presented in Table 12.6. The *p* values in Table 12.6 indicate that the differences in the mean best fitness values for all three pairs are highly significant.

Table 12.6 *t*-Test Analysis of Differences in Mean Fitness Evaluations

	Diff. In Means	Std. Error	*t* Test	*p* Value
GA vs. PBIL	−0.03970	0.003	15.87	<0.0001
GA vs. $DEUM_d$	−0.04000	0.003	15.98	<0.0001
PBIL vs. $DEUM_d$	−0.00028	0.000	4.72	<0.0001

In conclusion, $DEUM_d$ and PBIL outperform the GA on the chemotherapy optimization problem not only with respect to the computational speed of finding a feasible treatment, but also with respect to the quality of the treatment found. $DEUM_d$ has a slightly better performance than PBIL on both measures.

12.10 RELATED WORK

In this section, we compare our approach with other research that uses heuristic approaches to the optimization of chemotherapy.

12.10.1 Simulated Annealing

Agur et al. (2006) considered the problem of optimizing chemotherapy where drugs have a cell-cycle-specific drug action. The aim of the optimization was to compute efficient patient-specific chemotherapies. This approach modeled populations of healthy and cancer cells as they moved through phases of the cell cycle, tracking the position in the cell cycle of each cell at each moment. The model incorporated pharmacodynamics of a model patient to determine the response of cells to treatment with cycle-specific drugs. The model was mathematically intractable, and so the authors compared the performance of three local search heuristics: simulated annealing (SA) and two derivative algorithms, threshold acceptance (TA) and old bachelor acceptance (OBA).

The model admits bang-bang control, and so treatments were encoded as a bit string where 1 denoted treatment and 0 denoted no treatment. A given treatment was simulated in the model, and fitness was calculated as a weighted sum of factors based on cancer cell reduction, time of cure (if it occurred), maintenance of the healthy cell population, and time of patient death through toxicity (if it occurred).

The heuristics were compared on a number of problems, defined by choices of model parameters. OBA had the best performance overall in terms of computational expense though it requires more tuning than TA. SA was notably slower that the other two heuristics.

All three heuristics produced optimal treatments of similar quality. The best treatments produced fall into two categories: (1) intensive strategies involving a small number of unbroken sequences of treatment and (2) nonintensive treatments. The nature of the winning strategy of any particular problem seems to depend critically on patient-specific parameters in the problem. However, the research again suggests that a departure from conventional schedules could prove beneficial for patients.

12.10.2 Genetic Algorithm, Evolution Strategies, and Simulated Annealing

In Villasana et al. (2004), the authors formulated a chemotherapy optimization problem based on a model developed in Villasana and Radunskaya (2003). The model simulated the response to chemotherapy of three cell populations: immune cells, tumor cells in mitosis, and tumor cells in interphase. The chemotherapy consisted of doses of a single drug with a cycle-specific action. The optimal control problem admitted bang-bang control, and so a treatment could be expressed as a series of switching times between a maximum drug dose and a resting phase where no drug was administered. The model had an attractor with the tumor cell population at zero. Therefore, the objective of the chemotherapy was to drive the tumor cells below a threshold, at which point the tumor population will be inevitably eliminated.

Switching times were encoded as real numbers, which indicated the length of a drug administration period or a rest period. These value ranges were typically based on clinically realistic considerations. A parameter P (typically 18) encoded the number of switching times in a treatment giving rise to a real vector encoding of a sequence of alternating administration time lengths and resting time lengths. Treatments were evaluated according to a fitness function that minimized the average and final tumor size, modified by a penalty term for damage to the immune cell population.

Villasana and Ochoa (2004) compared the performance of three heuristics on this problem: a real-valued GA, an evolution strategy (CMA-ES), and simulated annealing (SA). The computation time for the model precluded extensive parameter tuning so the algorithms were compared using standard parameters. All algorithms were able to find treatments that drove the tumor size to the cure threshold. CMA-ES proved to have the fastest convergence and also achieved high solution quality. GA converged faster than SA but failed to achieve the same quality of solution as CMA-ES or SA. The treatments produced by these models are more likely to be clinically acceptable because they are encoded to occur in cycles with treatment and resting periods. However, as with many of the approaches discussed in this chapter, the best treatments exhibited in Villasana and Ochoa (2004) departed from conventional treatments in that they delay intensive treatment until near the end of the treatment period.

12.10.3 Multimodal Optimization Using a Genetic Algorithm

Liang et al. (2006a,b) considered a chemotherapy optimization that also built on the models introduced by Martin and Teo (1994). Chemotherapy with a single drug was optimized with the objective of tumor eradication measured in terms of final tumor size. Liang et al. (2006a,b) adapted the model in the way drug is cleared from the body to more explicitly reflect aspects of enzymatic reaction. In contrast with our approach, Liang et al. adopted a highly structured encoding cyclical encoding with an initial period of intensive dosing. Doing so imposes significant constraints on the nature of the treatments that can be represented. The aim is to create treatments that conform to a priori clinical assumptions and is designed explicitly to avoid the delay of intensive scheduling that is a feature of optimal treatments generated by other approaches. There is a real underlying question here as to whether the effect of this is to introduce a realistic and necessary clinical constraint or to eliminate potentially superior treatments unnecessarily. We will return to this question in the conclusion.

Liang et al. (2006a,b) solved their problem using a multimodal optimization genetic algorithm (AEGA). The motivation of seeking many optima is to provide the clinician with a choice of treatment when faced with different patients. This can be contrasted with our multiobjective approach, which produces a set of Pareto-optimal treatments aimed at assisting the clinician to balance conflicting objectives.

In Tse et al. (2005), the same authors experiment with multidrug schedules using the same algorithm but with a different encoding, more like ours. A feature of this study is a compartment-based model of drug resistance. The tumor cell population was divided into subpopulations that have resistance to a subset of the drugs used in the schedule. This is problematic because the number of compartments will grow exponentially. There are 3 drugs used in the study leading to 8 compartments. For treatments involving up to 5 drugs, the approach would require 32 compartments. The main problem here is not computational cost as the number of drugs is unlikely to rise much above 5. However, the approach requires a large number of parameters to determine the rates at which cells move between compartments. These are unlikely to be clinically measurable.

12.10.4 Parallel Evolutionary Algorithm

Tan et al. (2002a,b) applied a parallel EA to a chemotherapy optimization problem from Martin and Teo (1994) with the objective of tumor eradication using a single drug. In this problem, two evaluations were explored. One related to the standard tumor burden subject to the usual dose, toxicity, and tumor size constraints. In the other, point constraints were added to force a 50% reduction in tumor size for every 3 weeks of treatment. Treatments were encoded as a string of numerical pairs. Each pair consisted of a floating-point number representing the drug dose and an integer representing the day during the treatment period on which the dose is delivered. The strings can be of variable length, allowing the treatment to vary in both the number and timing of doses applied. An interesting feature of the best treatments produced is that, once again they favor late-intensity scheduling. This is more pronounced in the optimization without the point constraints as there is no need to reduce the tumor in the early stages of treatment. Tan et al. compared their results with a number of non-EA approaches on the same problem. The comparison is based on a performance index related to tumor burden developed by Martin and Teo (1994). The EA approach came within 0.02% of the best known value, attained by Luus et al. (1995). When point constraints were added, the index of the best treatment found by the EA approach equaled the best known value.

Another interesting feature of Tan et al.'s (2002a,b) work is that their EA algorithm can be configured within an optimization tool, Paladin-DEC, developed by the authors. In particular, the software contains an interactive human decision-maker interface, wherein clinicians can customize the performance index and view the results of the optimization in a graphical display. This is an important step in bridging the gap between clinicians and optimization experts. In the next section we describe a decision-making tool we have developed specifically for use in chemotherapy treatment design.

12.11 ONCOLOGY WORKBENCH

The Oncology Workbench is a system to support the design of novel multidrug cancer chemotherapy treatments. The aim of the system is to allow clinicians to interact with models of tumor response and to use powerful optimization algorithms to search for improved treatments. The system:

- Allows treatment parameters to be set in a straightforward manner.
- Suggests and/or evaluates possible treatment strategies.
- Presents an evaluation of likely treatment outcome in a manner that is readily comprehensible to the user and that has direct medical significance.

It is a significant problem to find the best way of presenting the system to oncologists, who may have some understanding of the tumor models used and their limitations, but who do not understand the algorithms and do not formulate their objectives and constraints in explicit mathematical terms. They have to interact with the system both by inputting treatment scenarios and by interpreting the output of the system.

Figure 12.4 illustrates how different components of the Workbench interact and shows how information flows between them. Using the treatment editor, a clinician may design a treatment by hand or specify general parameters and constraints. The editor communicates hand-designed treatments to the simulation engine. Treatment parameters are sent to the optimization engine and specify an optimization problem. Having received treatment data, both engines call the information repository for additional information concerning the effectiveness and toxicity constraints of the anticancer drugs used. After obtaining this information, the simulation engine evaluates the newly composed treatment and sends the results to the results viewer. The optimization engine determines the best possible treatment and advises the user via the treatment editor. Treatment designs may be stored in the information repository for future reference. The oncologist deals only with the treatment editor and the

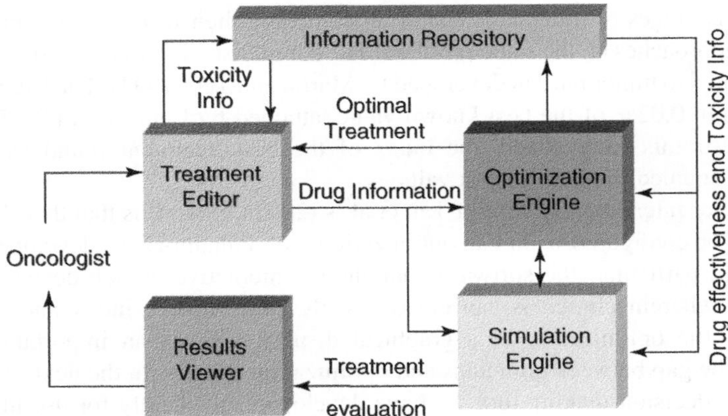

Figure 12.4 Architecture of the Oncology Workbench.

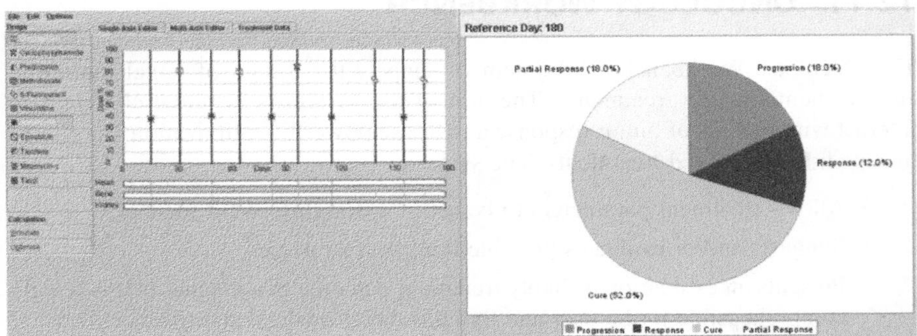

Figure 12.5 Treatment editor and results viewer.

results viewer, which play the role of interface objects between the user and the system. These make significant use of visual representations of the therapy design problem and the decision output, which are "natural" in terms of the problem domain. Figure 12.5 shows screen shots of the treatment editor and results viewer. The treatment editor presents treatments in a range of views on separate tabs. Figure 12.5 shows a graphical multidrug representation of the treatment. Numerical views are also available. A range of drugs is available to the design. These can be dragged and dropped onto the graphical representation. As many treatment designs use a repeating dose pattern, doses can be copied and repeatedly pasted at intervals throughout the treatment period.

Below the treatment representation, the treatment editor displays a number of colored bars that represent the severity of toxic side effects at points during the treatment. The clinician can select which side effects are important to the design. The bars are a useful tool for visually identifying points in the treatment where toxicity is particularly severe. These are the points at which the treatment is at risk of failure because the patient is unable to continue.

The results viewer again offers a number of views to the user, including a graphical representation of the tumor response over time. The pie-chart view shown in Figure 12.5 represents simulation of a treatment applied to a population of tumors with variable tumor response parameters. The chart shows the percentages of tumors regressed to the point of cure, regressed to some extent (response and partial response), and progression at the end of the treatment period. This corresponds to commonly reported statistics from clinical drug trials and so is a good comparison tool from the clinicians' point of view. The interface objects are delivered to users as Java applets via the Internet so that minimal setup is required at the user end. All the simulation and optimization is done on a fast server dedicated to the purpose (Boyle et al., 1997).

In Petrovski and McCall (2005), we proposed the development of a general-problem-solving environment for cancer chemotherapy, based on the Oncology Workbench. The use of Web services and XML formats to encode treatments and model parameters offers the prospect of integrating the diverse approaches to chemotherapy optimization in a distributed solver environment accessible to clinicians globally.

12.12 CONCLUSION

In this chapter we have shown how EAs have been applied to the problem of chemotherapy optimization by ourselves and other researchers. The quantity of work in this area has increased in recent years, indicating a growing interest in the EA community in addressing treatment optimization problems and a growing recognition in the medical community that in silico approaches have much to offer in the search for improved treatment. This growth in activity extends beyond cancer chemotherapy, although it is beyond the scope of this chapter to report on applications of EAs to other areas of medical treatment.

The extent to which optimization can be successfully applied to chemotherapy depends on the extent to which tumor response can be adequately modeled. To date, attention has focused on numerical simulation based on optimal control formulations. The models of Martin and Teo (1994) have been particularly influential in considering non-cycle-specific therapies, though work has been done on cycle-specific response models, notably by Agur (1986).

Evolutionary algorithms have proved highly competitive in comparison to other approaches, performing well on treatment objectives and in search efficiency. EAs are also notably flexible: They do not rely on mathematical features such as the existence of gradients and are adaptable to other models of tumor response that are not necessarily mathematical but are perhaps learned from treatment data.

All of the work discussed in this chapter was carried out by computing science researchers working in collaboration with clinical oncologists. It is important to note that in silico chemotherapy optimization is not yet at the stage where it is having a practical impact on patient treatment or even drug design. There are a number of reasons for this. Most treatments generated by EA optimizers are radically different from conventional treatments. There are two possible causes. Either the models do not adequately capture all essential features of treatment and response or conventional treatments are not successfully optimized by clinical development. It is very likely that both are true. Currently, groups working in this area are isolated and depend on a small number of research oncologists who are open to alternative approaches and willing to provide data and expertise to support the development of models and algorithms. Conversely, the vast body of research oncologists remains unaware or unconvinced of the opportunities that these techniques have to offer. The development of integrated problem-solving environments that give clinicians access to optimization tools is a necessary but not sufficient step to bridging this communication gap.

All of the work presented in this chapter depends on parameters that vary from patient to patient. Some of these are almost impossible to estimate accurately for particular patients, thus rendering remote the possibility of patient-specific in silico chemotherapy design. There is some hope, however, that the intensive work currently in progress in the bioinformatics community to identify biomarkers for patient-specific therapy will come to have application in this area. In the meantime, the notorious difficulty in accessing reliable treatment data in sufficient quantities will remain a significant barrier to realistic modeling and optimization.

REFERENCES

AGUR, Z. (1986). "The effect of drug schedule on responsiveness to chemotherapy," *Ann. NY Acad. Sci.*, Vol. 504, pp. 274–277.

AGUR, Z., R. HASSIN, and S. LEVY (2006). "Optimizing chemotherapy scheduling using local search heuristics," *Operat. Res.*, Vol. 54, pp. 829–846.

BOYLE, J., D. HENDERSON, J. MCCALL, H. MCLEOD, and J. USHER (1997). "Exploring novel chemotherapy treatments using the WWW," *Int. J. Med. Informatics*, Vol. 47, pp. 107–111.

CASSIDY, J. and H. MCLEOD (1995). "Is it possible to design a logical development plan for an anti-cancer drug?" *Pharm. Med.*, Vol. 9, pp. 95–103.

DEARNALEY, D. P., I. JUDSON, and T. ROOT (1995). *Handbook of Adult Cancer Chemotherapy Schedules*. Medicine Group (Education), Oxfordshire.

HENDERSON, D. (1997). *Mathematical Modelling in the Scheduling of Cancer Chemotherapy Treatment When Drug Resistance Is Present*. Ph.D. Dissertation, Robert Gordon University.

HENDERSON, I. C., D. F. HAYES, and R. GELMAN (1988). "Dose-response in the treatment of breast cancer: A critical review," *J. Clin. Oncol.*, Vol. 6, pp. 1501–1515.

HOOS, H. and T. STUTZLE (1999). "Towards a characterisation of stochastic local search algorithms for SAT," *Artif. Intell.*, Vol. 112, pp. 213–232.

JAIN, R. K. (1994). "Barriers to drug delivery in solid tumors," *Sci. Am.*, Vol. 271, pp. 58–65.

LIANG, Y., K.-S. LEUNG, and S.-K. MOK (2006a). "Automating the drug scheduling with different toxicity clearance in cancer chemotherapy via evolutionary computation," Proceedings of the Genetic and Evolutionary Computation Conference 2006. ACM, New York, pp. 1705–1712.

LIANG, Y., K.-S. LEUNG, and S.-K. MOK (2006b). "Optimal control of a cancer chemotherapy problem with different toxic elimination processes," in Proceedings of the 2006 IEEE Congress in Evolutionary Computation. IEEE Press, Piscataway, NJ, pp. 8644–8651.

LUUS R., F. HARTIG, and F. J. KEIL (1995). "Optimal scheduling of cancer chemotherapy by direct search optimization," *Hung. J. Ind. Chem.*, Vol. 23, pp. 55–58.

MARTIN, R. B. and K. L. TEO (1994). *Optimal Control of Drug Administration in Cancer Chemotherapy*. World Scientific, Singapore.

MARUSIC, M., Z. BAJZER, J. P. FREYER, and S. VUK-PAVLOVIC (1994). "Analysis of growth of multicellular tumor spheroids by mathematical models," *Cell Prolife.*, Vol. 27, pp. 73–94.

MCCALL, J. and A. PETROVSKI (1999). "A decision support system for chemotherapy using genetic algorithms," in *Computational Intelligence for Modelling, Control and Automation*, M. Mouhammadian, Ed. IOS Press, Amsterdam, pp. 65–70.

MURRAY, J. M. (1990a). "Optimal control for a cancer chemotherapy problem with general growth and loss functions," *Math. Biosci.*, Vol. 98, pp. 273–287.

MURRAY, J. M. (1990b). "Some optimal control problems in cancer chemotherapy with a toxicity limit," *Math. Biosci.*, Vol. 100, pp. 49–67.

PETROVSKI, A. (1998). An Application of Genetic Algorithms to Chemotherapy Treatment. Ph.D. Dissertation. Robert Gordon University.

PETROVSKI, A., A. BROWNLEE, and J. MCCALL (2005). "Statistical optimization and tuning of GA factors," in Proceedings of the 2005 IEEE Congress in Evolutionary Computation. IEEE Press, Piscataway, NJ, pp. 758–764.

PETROVSKI, A. and J. MCCALL (2000). "Computational optimization of cancer chemotherapies using genetic algorithms," in *Soft Computing Techniques and Applications*, R. John and R. Birkenhead, Eds. Physica, Heidelberg, New York, pp. 117–122.

PETROVSKI, A. and J. MCCALL (2001). "Multi-objective optimization of cancer chemotherapy using evolutionary algorithms," in *Evolutionary Multi-Criterion Optimization, Lecture Notes in Computer Science 1993*, E. Zitzler et al., Eds. Springer, Berlin, Heidelberg, pp. 531–545.

PETROVSKI, A. and J. MCCALL (2005). "Smart problem solving environments for medical decision support," Genetic and Evolutionary Computation Conference (GECCO'05), Proceedings of the 2005 Workshops on Genetic and Evolutionary Computation. ACM Press, New York, pp. 152–158.

PETROVSKI, A., B. SUDHA, and J. MCCALL (2004). "Optimising cancer chemotherapy using particle swarm optimization and genetic algorithms," in *Parallel Problem Solving from Nature VIII, 8th International Conference Proceedings, Lecture Notes in Computer Science Volume 3242*. Springer, Berlin, Heidelberg, pp. 633–641.

SHAKYA, S., J. MCCALL, and D. BROWN (2005). "Using a Markov network model in a univariate EDA: An empirical cost-benefit analysis," in Proceedings of the Genetic and Evolutionary Computation Conference 2005. ACM, New York, pp. 727–734.

SHAKYA, S., J. MCCALL, and A. PETROVSKI (2006). "Optimising cancer chemotherapy using an estimation of distribution algorithm and genetic algorithms," in Proceedings of the Genetic and Evolutionary Computation Conference 2006. ACM Press, New York, pp. 413–418.

SWAN, G. W. (1990). "Role of optimal control theory in cancer chemotherapy," *Math. Biosci.*, Vol. 101, pp. 237–284.

SWAN, G. W. and T. L. VINCENT (1977). "Optimal control analysis in the chemotherapy of IgG multiple myeloma," *Bull. Math. Biol.*, Vol. 39, pp. 317–337.

TAN, K. C., E. F. KHOR, J. CAI, C. M. HENG, and T. H. LEE (2002a). "Automating the drug scheduling of cancer chemotherapy via evolutionary computation," *Artif. Intell. Med.*, Vol. 25, pp. 169–185.

TAN, K. C., T. H. LEE, J. CAI, and Y. H. CHEW (2002b). "Automating the drug scheduling of cancer chemotherapy via evolutionary computation," in Proceedings of the 2002 IEEE Congress on Evolutionary Computation. IEEE Press, Piscataway, NJ, pp. 908–913.

TSE, S., Y. LIANG, K.-S. LEUNG, K.-H. LEE, and S.-K. MOK (2005). "Multiple drug cancer chemotherapy scheduling by a new memetic optimization algorithm," in Proceedings of the 2005 IEEE Congress in Evolutionary Computation. IEEE Press, Piscataway, NJ. pp. 699–706.

VILLASANA, M. and G. OCHOA (2004). "Heuristic design of cancer chemotherapy," *IEEE Trans. Evol. Comput.*, Vol. 8, pp. 513–521.

VILLASANA, M. and A. RADUNSKAYA (2003). "A delay differential equation model for tumor growth," *J. Math. Biol.*, Vol. 47, pp. 270–294.

WHELDON, T. E. (1988). *Mathematical Models in Cancer Research.* Adam Hilger, Bristol.

ZEITZ, S. and C. NICOLINI (1979). "Mathematical approaches to optimization of cancer chemotherapy," *Bull. Mathe. Biol.*, Vol. 41, pp. 305–324.

ZITZLER, E. (1999). Evolutionary Algorithms for Multi-objective Optimization: Methods and Applications. Ph.D. Dissertation. Swiss Federal Institute of Technology (ETH), Zurich.

Chapter 13

Fuzzy Ontology-Based Text Mining System for Knowledge Acquisition, Ontology Enhancement, and Query Answering from Biomedical Texts

Lipika Dey and Muhammad Abulaish

13.1 INTRODUCTION

The field of molecular biology has witnessed phenomenal growth in research activities in the recent past. The number of text documents disseminating knowledge in this area has increased manyfold. Consequently, there is an increased demand for automatic information extraction (IE) methods to extract knowledge from scientific documents and store them in a structured form. Without such approaches, the assimilation of knowledge from this vast repository is becoming practically impossible. However, knowledge that is embedded in natural language texts is difficult to extract using simple pattern-matching techniques. Ontology-guided IE mechanisms can help in the extraction of information stored within unstructured or semistructured text documents effectively. Ontology represents domain knowledge in a structured form and is increasingly being accepted as the key technology wherein key concepts and their interrelationships are stored to provide a shared and common understanding of a domain across applications (Fensel et al., 2001).

Computational Intelligence in Bioinformatics. Edited by G. B. Fogel, D. Corne, and Y. Pan

Figure 13.1 shows the GENIA ontology, which encodes knowledge about molecular biology concepts. In this chapter, we present some examples to elucidate how ontology can facilitate intelligent text information processing. PUBMED (www.pubmedcentral.nih.gov/) is a service of the National Library of Medicine (NLM) that includes over 16 million citations from various life science journals. MEDLINE, the largest component of PUBMED, is NLM's bibliographic database that itself contains over 12 million citations dating back to the mid-1960s, covering all fields related to medicine. It contains bibliographic citations and author abstracts from more than 4800 biomedical journals published in the United States and 70 other countries. Figure 13.2 shows three MEDLINE abstracts. PUBMED is queried by scholars regularly for locating relevant text documents.

We queried PUBMED for retrieving documents containing information on *protein binding sites*, and the top document of Figure 13.2 was returned as a result of this query. It can be seen that this document could be retrieved since the words *protein binding sites* occur in the text. However, the query "NF-kappa B binding sites" returns results not retrieved by the previous, more generalized query. Knowing that NF-kappa B (NF-kB) is a type of protein, a knowledge base (Fig. 13.1) could be used to provide the information that suggests this second document should have been a correct result of the first query.

A third document shown in Figure 13.2 was returned in answer to the query "protein-induced kinase." This is an incorrect response since it talks about seawater-induced Germinal vesicle breakdown (GBVD), which is not a *kinase*. On the other hand, the second document is relevant to this query also, as shown with underlined fragments, but is not retrieved since it requires additional knowledge to infer this.

The problems of query processing highlighted above arise due to the fact that documents are currently retrieved solely on the presence or absence of query words. No additional reasoning or domain knowledge is employed to compute the relevance of a document. The use of an ontology-based query-answering system can facilitate different forms of intelligent query answering. In the earlier set of examples, the first query contained a complex biological concept (e.g., protein binding sites), while the third query contained biological concepts at different levels of specificity. While *protein* is a generic concept, *kinase* represents a specific subclass of *protein*. The queries also restrict the scope of concept occurrence in the text by specifying a relation, which defines the context in which the concepts are required to occur in the document. Ontologies, which store domain knowledge in a structured machine-interpretable format, can play a very important role in text information retrieval by providing additional domain knowledge to the search engine. Ontology consists of classes, attributes, relationships, axioms, and also entities.

There are certain hurdles that need to be overcome before such systems are designed and deployed. First and foremost, ontologies cannot be defined unambiguously. While expert opinion on concepts can differ based on the context in which the ontology is to be used, it is also difficult to draw a list of concepts, entities, and relations that can define a domain exhaustively. Besides, since text documents are

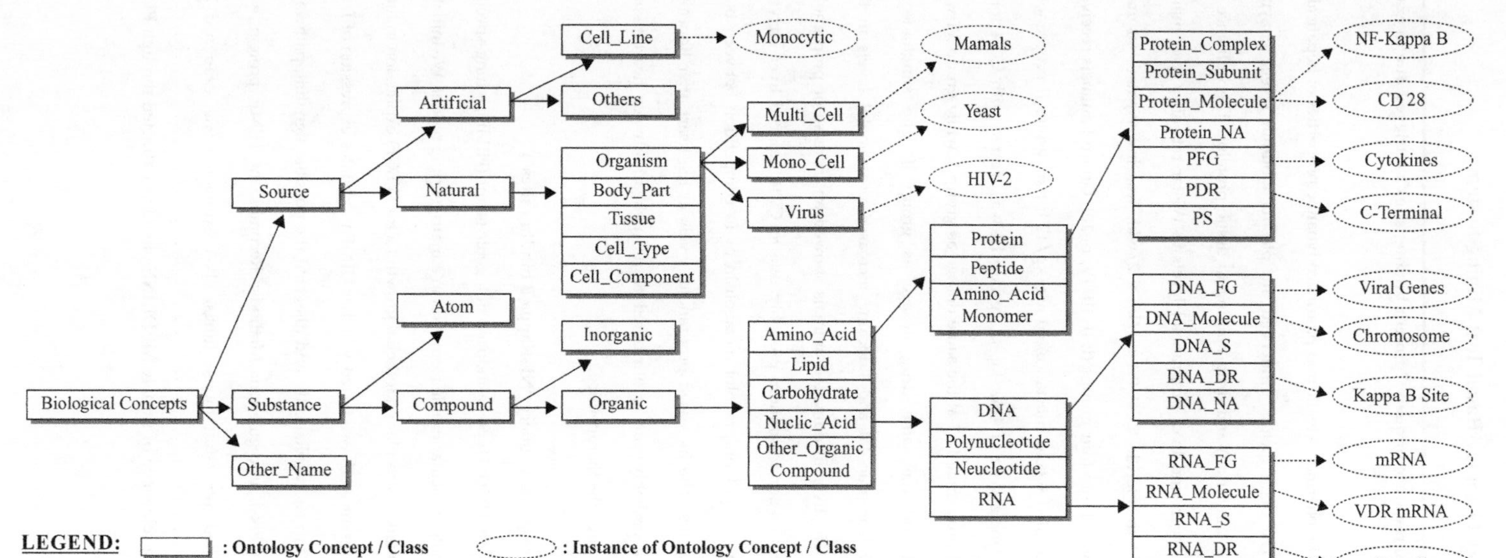

Figure 13.1 GENIA ontology with partial instances.

PMID: 16898077 (in response to query "protein binding sites": retrieved due to exact match)

Liver disease alters the pharmacokinetic and pharmacodynamic properties of hepatically eliminated drugs. The main factors influenced are plasma albumin levels, enzyme balance (induction & inhibition) and drug binding to tissue proteins. The influence of lidocaine on serum, heart and liver propranolol levels in Wistar rats after liver injury induced by carbon tetrachloride CCl4 0.4 ml/kg x 2/wkl, was investigated. 40 male Wistar rats were divided into four groups (I, II, III, IV; n=10), Group I animals received only propranolol (labelled + cold substance) 40 mg/kg/12 h p.o., group II propranolol plus lidocaine in a single dose of 4mg/kg s.c., group III was treated with CCl4 for 6 weeks and received propranolol x2 at the same dosage as group I, while group VI was treated with CCl4 and the same drug dosage as group II. The simultaneous administration of H3-propranolol and lidocaine increased propranolol levels in the serum and tissues. The liver in damaged animals showed an increase of propranolol level under lidocaine co-administration, probably due to CCl4 induced liver enzyme activity, resulting in a rapid propranolol metabolism or to competition between both drug *protein binding sites*. The increased propranolol levels in the heart after lidocaine administration were probably due to attributed to its high affinity for heart tissue. Consequently, as regards the therapeutic ...

PMID: 16872509(in response to query "NF-Kappa B binding sites")

ABSTRACT: BACKGROUND: Hypoxia-induced mitogenic factor (HIMF), a lung-specific growth factor, promotes vascular tubule formation in a matrigel plug model. We initially found that HIMF enhances vascular endothelial growth factor (VEGF) expression in lung epithelial cells. In present work, we tested whether HIMF modulates expression of fetal liver kinase-1 (Flk-1) in endothelial cells, and dissected the possible signaling pathways that link HIMF to Flk-1 upregulation. Methods: Recombinant HIMF protein was intratracheally instilled into adult mouse lungs, Flk-1 expression was examined by

Figure 13.2 Analyzing relevance of sample MEDLINE abstracts extracted through PUBMED.

immunohistochemistry and Western blot. The promoter-luciferase reporter assay and real-time RT-PCR were performed to examine the effects of HIMF on Flk-1 expression in mouse endothelial cell line SVEC 4-10. The activation of NF-kappa B (NF-kB) and phosphorylation of Akt, IKK, and IkBalpha were examined by luciferase assay and Western blot, respectively. Results: Intratracheal instillation of HIMF protein resulted in a significant increase of Flk-1 production in lung tissues. Stimulation of SVEC 4-10 cells by HIMF resulted in increased phosphorylation of IKK and IkBa leading to activation of NF-kB. Blocking NF-kB signaling pathway by dominant-negative mutants of IKK and IkBa suppressed *HIMF-induced Flk-1 upregulation*. Mutation or deletion of *NF-kB binding site* within Flk-1 promoter also abolished HIMF-induced Flk-1 expression in SVEC 4-10 cells. Furthermore, HIMF strongly induced phosphorylation of Akt. A dominant-negative mutant of PI-3K, Deltap85, as well as PI-3K inhibitor LY294002, blocked HIMF-induced NF-kB activation and attenuated Flk-1 production. Conclusion: These results suggest that HIMF upregulates Flk-1 expression in endothelial cells in a PI-3K/Akt-NF-kB signaling pathway-dependent manner, and may play critical roles in pulmonary angiogenesis.

PMID: 16902952 (in response to query "protein induced kinase")

Unlike in most animals, oocytes of marine nemertean worms initiate maturation (=germinal vesicle breakdown, GVBD) following an increase, rather than a decrease, in intraoocytic cAMP. ... Both cAMP elevators and SW triggered GVBD while activating MAPK, its target p90Rsk, and MPF. Similarly, neither cAMP- nor *SW-induced GVBD* was affected by several Ser/Thr phosphatase inhibitors, and both stimuli apparently accelerated GVBD via a MAPK-independent, PI3K-dependent mechanism. However, inhibitors of Raf-1, a kinase that activates MAPK kinase, blocked GVBD and MAPK activation during SW-, but not cAMP-induced maturation. ...

Figure 13.2 *(continued)*

unstructured in nature, to locate the domain concepts within text and reason with them is itself a difficult problem, which needs the involvement of natural language processing (NLP) techniques.

In this chapter, we describe an ontology-based text information processing system that aims at alleviating some of these problems through the following mechanisms:

1. An ontology-guided information extraction system is proposed that exploits NLP techniques to mine interconcept biological relations from text documents. The relations are used to index a document repository and retrieve information intelligently. The mined relations are associated with a fuzzy membership value that indicates the strength of co-occurrence of a particular relation with a pair of biological concepts. Unlike most of the related work, which has described methods for mining a fixed set of biological relations occurring with a set of predefined tags (Sekimizu et al., 1998; Thomas et al., 2000; Ono et al., 2001; Rinaldi et al., 2004), the proposed system identifies all verbs in a document and then identifies the feasible biological relational verbs using contextual analysis. The system has been designed to work with a collection of abstracts in which biological entities have been tagged as ontology concepts, along with the underlying ontology as input.
2. A fuzzy ontology framework is proposed to enhance the underlying ontology to store the mined fuzzy relations. Most of the existing biological ontologies only store taxonomical and partonomical relationships between biological concepts. The novelty of our framework lies in the fact that feasible biological relations mined from an underlying corpus are included in the enhanced ontology.
3. A query-answering system is integrated with the text mining system, which allows users to formulate queries using biological concepts at multiple levels of specificity and also include biological relations in the query.

The remaining sections of the chapter have been arranged as follows. An overview of ontologies in general and biological ontologies in particular is provided in Section 13.2 to provide a perspective of the work discussed later in the chapter. Section 13.3 presents related work on information extraction from biological text documents. Starting with work on biological entity recognition from text documents, we review work on biological relation extraction and then review earlier systems that perform ontology-based information extraction from biological documents. Sections 13.4 through 13.6 present the design of the ontology-based IE system that mines feasible biological relations from text documents and uses them for enhancing the underlying ontology structure. Sample queries using biological relations and the answers generated by the system are presented in Section 13.7. Experimental details along with system performance analysis in terms of the *precision* and *recall* value for the relation mining process is presented in Section 13.8. The extracted relations are generalized using a novel technique that is framed as an optimization problem over an AND-OR concept-pair tree. This is discussed in Section 13.9. In Sections 13.10 and 13.11 we have proposed the use of a fuzzy relations and fuzzy relational ontology to store biological domain knowledge. We present how an existing concept ontology is enhanced to a fuzzy relational ontology. Finally, in Section 13.12 we conclude with a summary and direction for possible future enhancements to the proposed system.

13.2 BRIEF INTRODUCTION TO ONTOLOGIES

The word *ontology* has been borrowed from philosophy, where it means "a systematic explanation of being" (Casely-Hayford, 2005) and has been used in the field of NLP for quite some time now to represent and manipulate meanings of texts. The knowledge engineering community has adopted ontology as a key enabling technology to realize the notion of semantic web (Berners-Lee, 1998). Gruber (1993) defined an ontology as "an explicit specification of a conceptualization," which has become one of the most acceptable definitions to the ontology community. Ontology specifies the key concepts in a domain and their interrelationships to provide an abstract view of an application domain (Fensel et al., 2001).

Usually, a concept in ontology is defined in terms of its mandatory and optional properties along with the value restrictions on those properties. A typical ontology consists of a set of classes, a set of properties to describe these classes, a taxonomy defining the hierarchical organization of these classes, and a set of semantic and structural relations defined among these classes. Ontology is also accompanied by a set of inference rules implementing reasoning functions for reasoning with ontology concepts. Formally, an ontology can be defined as follows:

Definition An ontology Θ is a 4-tuple of the form $\Theta = (C, P, \Re_s, \Re_g)$, where:

- C is a set of concepts. For example, in the GENIA ontology shown in Figure 13.1, C includes Amino_acid, Lipid, Virus, Cell_line, and so on.
- P is a set of properties. A property $p \in P$ is defined as a unary relation of the form $p(c)$, where property p is said to hold for concept $c \in C$. For example, the concept "Virus" may have the properties "virus name," "virus type," and so on.
- $\Re_s$ = {is-a, kind-of, part-of, has-part} is a set of structural semantic relations defined between a pair of concepts. A structural semantic relation $r_s \in \Re_s$ is defined as a binary relation of the form $r_s(C_i, C_j)$, where $C_i, C_j \in C$ are two concepts related through r_s.
- $\Re_g$ is a set of feasible generic relations between concepts. Like structural semantic relations, a generic relation $r_g \in \Re_g$ can be defined as a binary relation of the form $r_g(C_i, C_j)$, where $C_i, C_j \in C$ are the concepts related through r_g. These relations are usually specific to domain and/or application.

13.2.1 Biological Ontologies

Biology is one of the fields where knowledge representation has always followed a structured mechanism and hence lends itself to machine-interpretable ontology building readily. Biological concepts are well defined, and biological ontologies usually store the taxonomical and partonomical relations among these concepts. Some of the well-known ontologies in the field of molecular biology domain are Gene Ontology (GO) (Ashburner et al., 2000), Transparent Access to Multiple Bioinformatics Information Sources (TAMBIS) ontology (also abbreviated "TaO") (Baker et al., 1999), RiboWeb ontology (Altman et al., 1999), EcoCyc ontology (Karp et al., 1999), Ontology for the Molecular Biology (MBO) (Schulze-Kremer, 1998), and GENIA ontology (Tateisi et al., 2000).

The Gene Ontology is mainly used for database annotation. GO has grown up from within a group of databases rather than being proposed from outside. It seeks to capture information about the role of gene products within an organism. GO lacks any upper-level organizing ontology and organizes processes in three levels of hierarchy, representing the function of a gene product, the process in which it takes place, and cellular location and structure. Starting from a few thousand terms describing the genetic workings of three organisms, GO has since grown to a terminology of nearly 16,000 terms and is becoming a de facto standard for describing functional aspects of biological entities in all types of organisms.

The TAMBIS ontology (TaO) describes a wide range of bioinformatics tasks and resources and has a central role within the TAMBIS system that enables biologists to ask questions over multiple external databases using a common query interface. TaO contains knowledge about bioinformatics and molecular biology concepts and their relationships—the instances they represent reside in the external databases. TaO is available in two forms—a small model that concentrates on proteins and a larger scale model that includes nucleic acids. The small TaO, with 250 concepts and 60 relationships, describes proteins and enzymes, as well as their motifs, secondary and tertiary structure, functions, and processes. The larger model, with 1500 concepts, broadens these areas to include concepts pertinent to nucleic acid, its children, and genes.

RiboWeb is a resource whose primary aim is to facilitate the construction of three-dimensional models of ribosomal components and compare the results to existing studies. The knowledge that RiboWeb uses to perform these tasks is captured in four ontologies: the *physical-thing ontology*, the *data ontology*, the *publication ontology*, and the *methods ontology*. The physical-thing ontology describes ribosomal components and associated "physical things." The data ontology captures knowledge about experimental detail as well as data on the structure of physical things. The methods ontology contains information about techniques for analyzing data. It holds knowledge of which techniques can be applied to which data, as well as the input and outputs of each method.

In a manner similar to RiboWeb, EcoCyc creates an ontology to describe the richness and complexity of a domain and the constraints acting within that domain to specify a database schema. EcoCyc is presented to biologists using an encyclopedia metaphor. It covers *Escherichia coli* genes, metabolism, regulation, and signal transduction, which a biologist can explore and use to visualize information. The knowledge base currently describes 4391 *E. coli* genes, 695 enzymes encoded by a subset of these genes, 904 metabolic reactions, and the organization of these reactions into 129 metabolic pathways.

The Ontology for Molecular Biology (MBO) attempts to provide clarity and communication within the molecular biology database community. The MBO contains concepts and relationships that are required to describe biological objects, experimental procedures, and computational aspects of molecular biology. It is very wide ranging and has over 1200 nodes representing both concepts and instances. In the conceptual part of the MBO, the primary relationship used is the "is-a" relationship.

The GENIA ontology provides the base collection of molecular biological entity types and relationships among them. GENIA ontology, shown in Figure 13.1, is a taxonomy of 47 biologically relevant nominal categories. The top three concepts in GENIA ontology are *biological source, biological substance*, and *other_name*. The *other_name* does not refer to any fixed biological concept but is used for describing the terms that are regarded as biological concepts but are not identified with any other known concept in the ontology. The concepts at leaf nodes are the terminal concepts, and they form the actual tags for semantic annotation. The GENIA ontology concepts are related through "is-a" and "part-of" semantic relations. In GENIA ontology, substances are classified according to their chemical characteristics rather than their biological roles since chemical classification of a substance is quite independent of the biological context in which it appears, and therefore more stably defined. The biological role of a substance may vary depending on the biological context. Therefore, in this model substances are classified as proteins, DNA (deoxyribonucleic acid), RNA (ribonucleic acid), and the like. They are further classified into families, complexes, individual molecules, subunits, domains, and regions. The sources are classified into natural and cultured sources that are further classified as an organism (human), a tissue (liver), a cell (leukocyte), or a sublocation of a cell [membrane or a cultured cell line (e.g., HeLa cells)]. Organisms are further classified into multicell organisms, virus, and monocell organisms other than virus. GENIA ontology is encoded in DAML + OIL, an XML-based ontology language and is provided together with the GENIA corpus. GENIA corpus consists of a set of 2000 MEDLINE abstracts in which the biological entities have been manually identified and tagged as a GENIA concept.

13.3 INFORMATION RETRIEVAL FROM BIOLOGICAL TEXT DOCUMENTS: RELATED WORK

In this section we present an overview of some of the recent research that has been directed toward the problems of extraction of biological entities, their interactions, and query answering from unstructured biological documents. The use of ontological models to access and integrate large knowledge repositories in a principled way has an enormous potential to enrich and make accessible unprecedented amounts of knowledge for reasoning (Crow and Shadbolt, 2001). Ontologies are incorporated in information retrieval systems as a tool for the recognition of synonymous expressions and linguistic entities that are semantically similar but superficially distinct (Voorhees, 1994; Andreasen et al., 2000, 2002, 2004; Snoussi et al., 2002; Liddle et al., 2003). Information extraction from biological documents is largely dependent on the correct identification of biological entities in the documents. These entities are then tagged or annotated for more meaningful information extraction. Initially, the process of named entity recognition and tagging was done manually. But the sheer increase in text volume has initiated a significant research effort toward automated identification of biological entities in journal articles and tagging them.

Fukuda et al. (1998) proposed a rule-based system, called protein proper-noun phrase extracting rules (PROPER), to extract material names from sentences using surface clue on character strings in medical and biological documents. PROPER identifies protein names in articles with a recall value of 98.8% and a precision of 94.7%. Proux et al. (1998) used a dictionary-based approach to identify non-English words in a document and identify them as gene terms. Hanisch et al. (2003) proposed a hybrid approach including the use of dictionaries and hand-coded rules in combination with robust linear programming (RLP) based optimization to identify biological entities. The ARBITER system developed by Rindflesch et al. (1999) uses dictionaries and rules to identify binding terms in biomedical texts. Machine-learning-based techniques such as hidden Markov models (Collier et al., 2000), naïve Bayes (Nobata et al., 1999), and support vector machines (SVMs) (Kazama et al., 2002) have been applied successfully for the identification and classification of gene/protein names in text documents. Although named-entity recognition from biological text documents has gained reasonable success, reasoning about contents of a text document requires more than identification of the entities present in it. Context of the entities in a document can be inferred from an analysis of the interentity relations present in the document. Hence, it is important that the relationships among the biological entities present in a text are also extracted and interpreted correctly.

Related work in biological relation extraction can be classified into the following three categories:

1. *Co-occurrence-based approaches* After biological entities are extracted from a document, relations among them are inferred based on the assumption that two entities in the same sentence or abstract are related. Negation in the text is not taken into account. Jenssen et al. (2001) collected a set of almost 14,000 gene names from publicly available databases and used them to search MEDLINE abstracts. Two genes were assumed to be linked if they appeared in the same abstract; the relation received a higher weight if the gene pair appeared in multiple abstracts. For the pairs with high weights (i.e., with five or more occurrences of the pair), it was reported that 71% of the gene pairs were indeed related. However, the primary focus of the work is to extract related gene pairs rather than studying the nature of these relations.

2. *Linguistics-based approaches* Usually, shallow parsing techniques are employed to locate a set of hand-picked verbs or nouns. Shallow parsing is the process of identifying syntactical phrases (e.g., noun phrases, verb phrases) in natural language sentences on the basis of the parts of speech of the words but does not specify their internal structure, nor their role in the main sentence. Rules are specifically developed to extract the surrounding words of these predefined terms and to format them as relations. As with the co-occurrence-based approach, negation in sentences is usually ignored.

Sekimizu et al. (1998) collected the most frequently occurring verbs in a collection of abstracts and developed partial and shallow parsing techniques to find the verb's subject and objects. The estimated precision of inferring relations is about 71%. Thomas et al. (2000) modified a preexisting parser based on cascaded finite state machines to fill templates with information on protein interactions for three verbs—*interact with, associate with,* and *bind to*. They calculated recall and precision in four different manners for three samples of abstracts. The recall values ranged from 24 to 63% and precision from 60 to 81%. The Protein Active Site Template Acquisition (PASTA) system is a more comprehensive system that extracts relations between proteins, species, and residues (Gaizauskas et al., 2003). Text documents are mined to instantiate templates representing relations among these three types of elements. This work reports a precision of 82% and a recall value of 84% for recognition and classification of the terms, and 68% recall and 65% precision for completion of templates. Craven and Kumlien (1999) proposed identification of possible drug interaction relations between protein and chemicals using a "bag of words" approach applied at the sentence level. This produces inferences of the type: *drug-interactions (protein, pharmacologic-agent)*, which specifies an interaction between an agent and a protein. Ono et al. (2001) reported a method for extraction of *protein-protein interactions* based on a combination of syntactic patterns. They employ a dictionary lookup approach to identify proteins in the document. Sentences that contain at least two proteins were selected and parsed with parts-of-speech (POS) matching rules. The rules were triggered by a set of keywords, which are frequently used to name protein interactions (e.g., *associate, bind*).

Rinaldi et al. (2004) proposed an approach toward automatic extraction of a predefined set of seven relations in the domain of molecular biology, based on a complete syntactic analysis of an existing corpus. They extracted relevant relations from a domain corpus based on full parsing of the documents and a set of rules that mapped syntactic structures into the relevant relations. Friedman et al. (2001) developed a natural language processing system, called GENIES (GENomics Information Extraction System), for the extraction of molecular pathways from journal articles. GENIES used the MedLEE parser to retrieve target structures from full-text articles. GENIES identified a predefined set of verbs using templates for each one of these, which are encoded as a set of rules. Friedman et al. (2001) reported a precision of 96% for identifying relations between biological molecules from full-text articles.

3. *Mixed approaches* Ciaramita et al. (2005) reported an unsupervised learning mechanism for extracting semantic relations between molecular biology concepts from tagged MEDLINE abstracts from the GENIA corpus. For each sentence containing two biological entities, a dependency graph highlighting the dependency between the entities was generated based on linguistic analysis. A relation between two entities C_i and C_j was extracted as the shortest path between the pair following the dependency relations. Though a large number of directed paths are initially extracted from the corpus, only the significant ones were retained to be used as biological relations. The approach used is that of hypothesis testing, where the null hypothesis, H_0, is formulated as concepts *A* and *B* do not co-occur more frequently than expected at chance. The probability of co-occurrence of *A* and *B* denoted by *P*(*AB*) is then computed using corpus statistics and H_0 is rejected if *P*(*AB*) is less than a specified level of significance. The major emphasis of this work was to determine the role of a concept in a significant relation and enhance the biological ontology to include these roles and relations. Sentences that contained complex embedded conjunctions/disjunctions or contained more than 100 words were not used for relation extraction. In the presence of nested tags, the system considers only the innermost tag.

It can be observed that most of the systems (other than Ciaramita et al., 2005) were developed to extract a predefined set of relations. The relation set was chosen manually to include a set of frequently occurring relations. Each system was tuned to work with a predetermined set of relations and did not address the problem of relation extraction in a generic way. For example, the method of identification of interaction between genes and gene products cannot work for extraction of enzyme interactions from journal articles or for automatic extraction of protein interactions from scientific abstracts.

The growth of several biological ontologies mentioned in the earlier section has prompted a lot of attention toward ontology-based processing of biomedical texts. The aim is to provide intelligent search mechanisms for extracting relevant information components from a vast collection.

Textpresso (Muller et al., 2004) is an ontology-based biological information retrieval and extraction system. Textpresso analyzes tagged biological documents. Two types of tags are used for tagging text elements manually. The first set of tags defines a collection of biological concepts, and the second set of tags defines a set of relations that can relate two categories of biological concepts. A tag is defined by a collection of terms including nouns, verbs, and the like that can be commonly associated to the concept. Portions of the document containing a relevant subset of terms are marked by the corresponding biological concept or relation tag. The search engine allows the user to search for combinations of concepts, keywords, and relations. With specific relations like commonly occurring gene–gene interactions encoded as a relation tag, Textpresso assists the user to formulate semantic queries. The recall value of the system is reported to vary from 45 to 95%, depending on whether the search is conducted over abstracts or full-text documents.

Uramoto et al. (2004) proposed a text-mining system, MedTAKMI, for knowledge discovery from biomedical documents. The system can be used to dynamically and interactively mine a large collection of documents with biomedically motivated categories to obtain characteristic information from them. The MedTAKMI system performs entity extraction using dictionary lookup from a collection of 2 million biomedical entities, which are then used along with their associated category names to search for documents that contain keywords belonging to specific categories. Users can submit a query and receive a document collection in which each document contains the query keywords or their synonyms. The system also uses syntactic information with a shallow parser to extract binary (a verb and a noun) and ternary (two nouns and a verb) relations that are used as keywords by various MedTAKMI mining functions (e.g., dictionary-based full-text searching, hierarchical category viewer, chronological viewer).

13.4 ONTOLOGY-BASED IE AND KNOWLEDGE ENHANCEMENT SYSTEM

We now present an overview of the ontology-based IE and knowledge enhancement system. The system uses an underlying ontology and processes ontologically tagged biological texts to extract information about biological relations occurring in these documents. A biological relation is assumed to bind two biological entities in a specific way. We have used the GENIA ontology and the GENIA corpus 3.01 (Tateisi et al., 2000), which contains 2000 MEDLINE abstracts, in which the biological entities have been manually tagged as leaf concepts in the GENIA ontology. The system also works with untagged texts. The extracted information is used both for answering user queries intelligently and also to enhance the ontology itself.

A biological relation is assumed to be binary in nature, which defines a specific association between an ordered pair of biological entities. Based on the observation that biological relations can be characterized by verbs and prepositions occurring in the vicinity of biological entities and concepts, the proposed text mining system has the following three key functionalities:

- *Extraction of Information Components from Texts* An *information component* is defined as an ordered triplet $<C_i$, verb/verb accompanied by a preposition, $C_j>$, where C_i and C_j are biological ontology tags occurring in association to biological entities within a single sentence in the text. The information component provides a template for holding biological relations.
- *Compiling a Collection of Feasible Biological Relations* Not all information components extracted from the documents are proper representatives of biological relations. Hence these are subjected to feasibility analysis to identify the ones that can represent feasible biological relations. A feasible fuzzy biological relation is denoted by $<C_i, \mathfrak{R}, C_j, \mu_{(C_i,C_j)}(\mathfrak{R})>$, where C_i and C_j are leaf-level biological concepts, $\mathfrak{R}$ has been established as a *feasible biological relational verb,* and $\mu_{(C_i,C_j)}(\mathfrak{R})$ denotes the strength of association of $\mathfrak{R}$ with ordered concept-pair $<C_i, C_j>$. The strength $\mu_{(C_i,C_j)}(\mathfrak{R})$ is computed as a function of the relative frequency of the relation observed within the corpus.
- *Enhancing the Underlying Ontology into Fuzzy Ontology Structure* Since the GENIA ontology comprises of biological concepts only, our aim is to

enhance this ontology to include biological relations that frequently occur between a pair of these concepts. Since the documents are initially tagged by leaf-level concepts only, all the mined feasible relations are defined between a pair of leaf-level concepts. These relations are, therefore, subjected to specificity analysis to identify the most significant concept level at which these arguments can be defined. Each significant relation is associated with a fuzzy quantifier WEAK, MEDIUM, or STRONG, which indicates the likelihood of observing a particular relation among a pair of concepts. This is also indicative of the appropriateness of including a relation into the ontology without making it superfluous. Since the relations are accompanied by a fuzzy qualifier, we have proposed the extension of a standard ontology structure to a fuzzy relational ontology structure, which is defined later.

Figure 13.3 presents the architecture of the system, which comprises of the following five modules.

- *Document Processor* The function of this module is to filter the meta language (ML) tags from input documents. A parts-of-speech (POS) tagger is used to extract grammatical information from the sentences. Based on its parse structure, each sentence is converted to a semistructured form that is used later to mine biological relations.

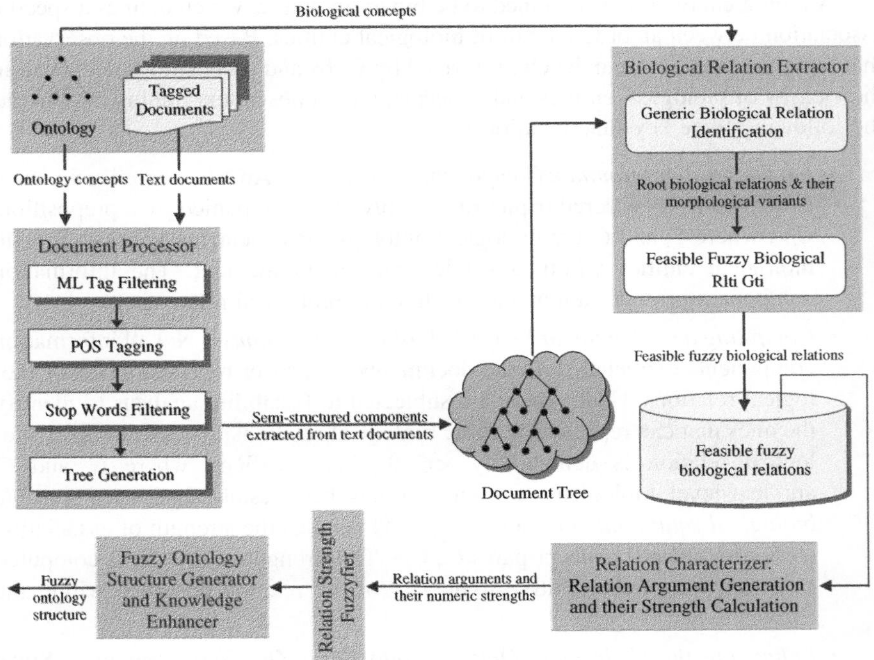

Figure 13.3 Ontology-based IE and knowledge enhancement system architecture.

- *Biological Relation Extractor* This module uses the semistructured records generated by the earlier module in collaboration with the underlying ontology to extract information component triplets <C_i, $\Re/\Re$ accompanied by preposition, C_j>, where $\Re$ is a *verb*. This module is also responsible for conducting a feasibility analysis to identify those verbs that can represent feasible biological relations. For each feasible biological relation, this module also computes its strength $\mu_{(C_i,C_j)}(\Re)$.
- *Relation Characterizer* This module is responsible for identifying the appropriate concept levels for characterizing a relation defined between a pair of biological concepts.
- *Relation Strength Fuzzifier* This module generates fuzzy membership functions to fuzzify the numeric values of the relation strength and maps them into linguistic quantifiers WEAK, MEDIUM, and STRONG to accompany each relation.
- *Fuzzy Ontology Structure Generator and Knowledge Enhancer* The feasible fuzzy relations along with their associated concept pairs and fuzzy strengths are used by this module to enhance the underlying ontology into fuzzy ontology structures. The concept of multiple inheritances as used in object-oriented paradigm is used to design the structure of the fuzzy ontology. We have used Protégé 3.1 to implement the fuzzy ontology structure (http://protege.stanford.edu/).

The design and working principles of these modules are explained in the following sections.

13.5 DOCUMENT PROCESSOR

Figure 13.4 shows a sample tagged MEDLINE abstract from which relevant information components are extracted for further processing. The document processor cleans the tagged abstracts and removes most of the meta language tags (e.g., *<cons>*, *<title>*, *<abstract>*, *<sentence>*) and retains only those tags that provide domain knowledge such as the generic pair <ontology_concept> and </>. The cleaned sentences are subjected to parts-of-speech (POS) analysis. POS analysis assigns parts of speech tags to every word in a sentence, where a tag reflects the syntactic category of the word (Allen, 2004). Since our intention is to identify relevant action verbs along with accompanying prepositions, POS analysis plays an important role in text information extraction. To optimize the memory usage and accuracy of the system, stop words such as *is, was, that, which*, and the like are also removed. We have used a subset of the stop words used by the PubMed database. The entities, verbs, and prepositions retained in a sentence are arranged in a binary tree for further analysis. Each document is converted into a document tree that consists of binary trees for each sentence. The binary tree is built by analyzing the POS tags of each sentence recursively and is defined as follows:

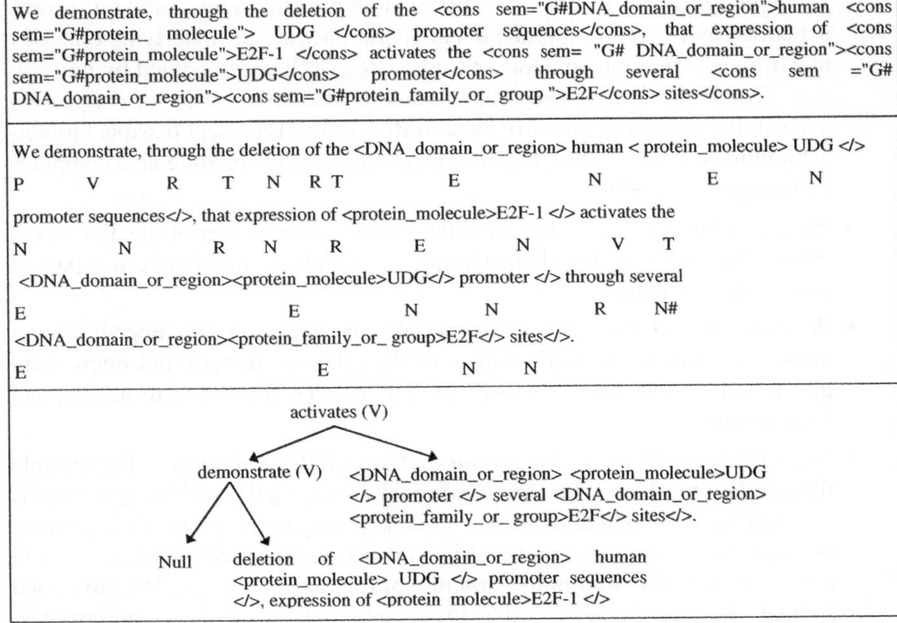
We demonstrate, through the deletion of the <cons sem="G#DNA_domain_or_region">human <cons sem="G#protein_ molecule"> UDG </cons> promoter sequences</cons>, that expression of <cons sem="G#protein_molecule">E2F-1 </cons> activates the <cons sem= "G# DNA_domain_or_region"><cons sem="G#protein_molecule">UDG</cons> promoter</cons> through several <cons sem ="G# DNA_domain_or_region"><cons sem="G#protein_family_or_ group ">E2F</cons> sites</cons>.
We demonstrate, through the deletion of the <DNA_domain_or_region> human < protein_molecule> UDG </>
P V R T N R T E N E N
promoter sequences</>, that expression of <protein_molecule>E2F-1 </> activates the
N N R N R E N V T
<DNA_domain_or_region><protein_molecule>UDG</> promoter </> through several
E E N N R N#
<DNA_domain_or_region><protein_family_or_ group>E2F</> sites</>.
E E N N
activates (V)
demonstrate (V)
<DNA_domain_or_region> <protein_molecule>UDG </> promoter </> several <DNA_domain_or_region> <protein_family_or_ group>E2F</> sites</>.
Null
deletion of <DNA_domain_or_region> human <protein_molecule> UDG </> promoter sequences </>, expression of <protein molecule>E2F-1 </>

Figure 13.4 (*Top*) A Sentence from MEDLINE: 95197524; (*Middle*) its stripped and POS tagged form; and (*Bottom*) the binary tree generated from it.

```
Struct Sentence
Begin
    string Value;
    struct sentence* Lchild;
    struct sentence* Rchild;
End
Struct Sentence* Document [number_of_sentence];
```

A sentence is initially segmented into two portions, one to the left and another to the right of the rightmost verb occurring in the sentence. The process of segmentation continues until no more verbs are found in a segment. The root of each tree contains the rightmost verb in the corresponding sentence. The segments in a sentence are linked to each other as follows:

Root (*R*) Contains the rightmost verb of the sentence

Lchild (L_c) Points to the subtree representing sentence segment that is to the left of the word stored at *R*.

Rchild (R_c) Points to the subtree representing sentence segment that is to the right of the word stored at *R*.

The equivalent context-free grammar for the scheme is as follows.

$$\text{Document (D)} \rightarrow S^*$$
$$S \rightarrow L_C R R_C \mid \epsilon$$
$$L_C \rightarrow L_C R R_C \mid (E+N+A+J+R)^* \mid \epsilon$$
$$R_C \rightarrow (E+N+A+J+R)^* \; ; \quad R \rightarrow V|G|L$$

where S represents a sentence, and N, A, J, R, and V represent noun, adverb, adjective, preposition, and verb, respectively, as assigned by the POS tagger. G and L represent a verb used as a gerund or participle, respectively. The system ignores the POS tag assigned to ontological tags, and replaces it by the symbol E to indicate the positions of entities. A sample tagged MEDLINE abstract from the GENIA corpus, its stripped and POS tagged form, and generated binary tree is shown in Figure 13.4.

13.6 BIOLOGICAL RELATION EXTRACTOR

The process of identifying biological relations is accomplished in two stages. During the first stage, prospective information components that might embed biological relations within them are identified from the binary representation of the sentences. During the second stage, a feasibility analysis is employed to identify correct and feasible biological relations. These stages are elaborated further in the following subsections.

13.6.1 Extraction of Information Components

The aim of this module is to first identify information components characterized by the triplet <C_i, $\Re$/$\Re$ accompanied by preposition, C_j>, where C_i and C_j are biological concepts and $\Re$ represents a verb that may be accompanied by a preposition. A verb may occur in a sentence in its root form or as a variant of it. Different classes of variants of a relational verb are recognized by our system. The first class comprises *morphological variants* of the root verb, which are essentially modifications of the root verb itself. In the English language, the word *morphology* is usually categorized into *inflectional* and *derivational* morphology. *Inflectional morphology* studies the transformation of words for which the root form only changes, keeping the syntactic constraints invariable. For example, the root verb *activate* has three inflectional verb forms—*activates, activated,* and *activating. Derivational morphology* on the other hand deals with the transformation of the stem of a word to generate other words that retain the same concept but may have different syntactic roles. Thus, *activate* and *activation* refer to the concept of "making active," but one is a verb and the other one a noun. Similarly, *inactivate, transactivate, deactivate,* and so on are derived morphological variants created with addition of prefixes. Presently, the system does not consider derivational morphology, and only inflectional variants of a root verb are recognized.

In the context of biological relations, we also observe that the occurrence of a verb in conjunction with a preposition very often changes the nature of the relation. For example, the focus of the relation *activates* may be quite different from the relation *activates in,* in which the verb *activates* is followed by the preposition *in.* Thus our system also considers a third category of biological relations, which are combinations of *root verbs or their morphological variants* and *prepositions* that follow these. Typical examples of biological relations identified in this category include *activated in, binds to, stimulated with,* and so forth. This category of relations can take care of special biological interactions involving substances and sources or localizations.

The relation extractor module identifies all possible information component triplets containing a relational verb, preposition (if any), and the associated concepts by traversing the binary tree built earlier. The working principle of this module is explained by the following steps:

- List of information components, L_{IC}, is initialized to *null.*
- The binary tree generated earlier is traversed in postorder fashion to locate and extract information components. Starting at the leftmost leaf node, if both left and right siblings contain biological tags, the verb represented at the parent of these siblings is assumed to represent a biological relation. If the right child of a node contains a preposition, then the preposition is associated to the verb in the parent node, and the verb–preposition pair is identified as a possible biological relation. If the right child does not contain a preposition, only the verb that may be a root verb or an inflectional variant of it is identified as a possible biological relation. A unique combination of a possible biological relational verb along with the biological tags occurring in the neighborhood of these verbs are added to the list of information components L_{IC}.

By using the above process, certain verbs such as *demonstrate* (appears as a node in the tree shown in Fig. 13.4), which occurred in the text, do not appear in any information component since they do not occur in the vicinity of biological concepts. A large number of irrelevant verbs are thus eliminated from being considered as biological relations. All verbs identified as possible biological relations at this stage are subjected to further feasibility analysis for retention as a biological relational verb.

13.6.2 Identifying Feasible Generic Biological Relations

Although a large number of commonly occurring verbs are eliminated due to the design of the information component itself, it is found that further processing is necessary to consolidate the final list of relations. During the consolidation process, we take care of two things. In the first stage, since various forms of the same verb represent a basic biological relation in different forms, so a feasible collection of unique root verb forms is identified. In the second stage of analysis, however, each morphological variant of a feasible relational verb is considered along with its co-occurring prepositions to obtain a final list of all feasible biological relations.

The core functionalities of the biological relation finding module are summed up in the following steps.

- Let L_{IC} be the collection of all verbs or verb–preposition pairs, which are extracted as part of information components. The biological relation finding module analyzes L_{IC} to determine the set of unique root forms from this collection. The frequency of occurrence of each root verb is the sum total of its occurrence frequencies in each form. All root verbs with frequency less than a user-given threshold are eliminated from further consideration. The surviving verbs are termed as *most frequently occurring* root verbs and represent important *biological relations.*

- Once the frequent root verb list is determined, the morphological variant identification module operates on L_{IC} and identifies the complete list of all biological relation verbs including frequent root verbs, their morphological variants, and their co-occurrence with prepositions.

13.6.3 Feasible Generic Biological Relations Identified from the GENIA Corpus

The methodology described earlier in this section was used to extract the feasible biological relations from the GENIA corpus of tagged MEDLINE abstracts. To check the consistency of the relation set extracted, we have employed the following cross-validation technique. The entire GENIA corpus consisting of 2000 MEDLINE abstracts was divided randomly into three subsets—one containing 640 documents and the other two containing 680 documents each. The relation extraction process was applied on each subset separately. For each subset, the list of frequently occurring root verbs was identified as those verbs that co-occur with two biological concepts on both sides. To retain only frequently occurring interaction verbs, a verb is termed as *frequent* and retained provided it has a minimum occurrence of 2.5% of the total number of documents in the entire collection. It was observed that the root verb collection was fairly stable and the relative frequencies of the various root verbs extracted also remained identical in all the three corpus subsets. Hence, the relative frequencies of various root verbs can be concluded to reflect the actual distribution of various biological interactions in the corpus.

Thereafter we compiled the frequent root verb collection from the entire corpus and computed their relative frequencies of occurrence. Using the same threshold for determining frequent verbs, initially, 15,780 relations were identified as verbs co-occurring with a pair of biological entities in the corpus. For some verbs, there are multiple variants included in the set. On passing this set through a root verb identification module, only 24 root verbs were identified as *frequent.* The root verbs are representative elements for the entire set of frequently occurring biological relations since the entire set will contain morphological variants of these verbs only. Figure 13.5 shows the overall distribution of the root verbs. Figure 13.6 shows the relative frequencies of these root verbs.

On passing the initial set of 15,780 relations through a morphological variant identification module and using the collection of frequent root verbs identified, the overall set of frequent biological relations was found to contain 246 relations that include the morphological variants of the root verbs. The entire set of frequent biological relations is given in the Appendix to this chapter. All unique forms identified

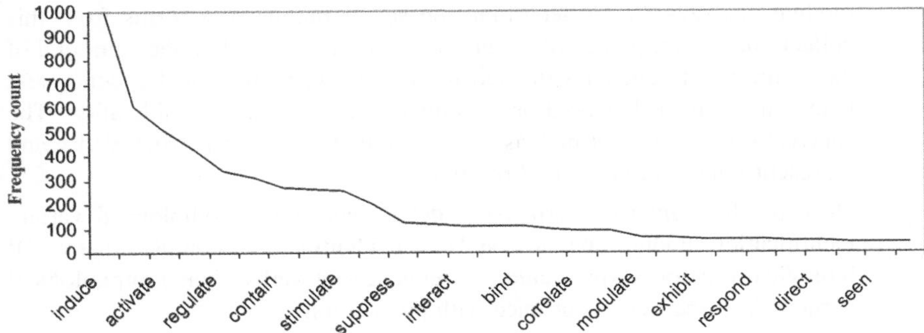

Figure 13.5 Plot of the frequency of relational verbs occurrence and their rank order.

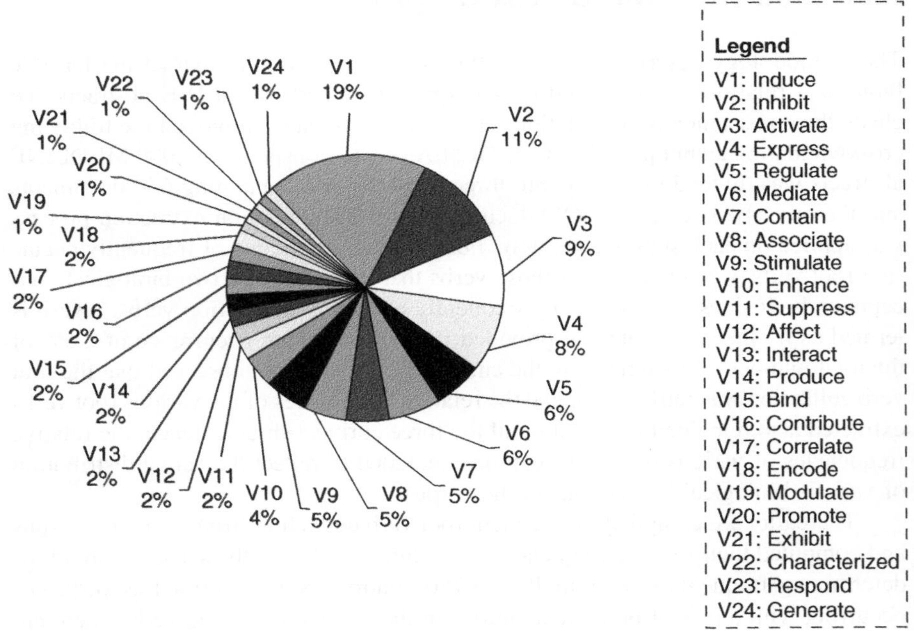

Figure 13.6 Relative frequencies of occurrence of extracted root biological relational verbs.

by the parser have been retained in the list. Equivalence of some of these forms is established during query processing, if needed. For example, if a query is for finding *X that binds to Y*, all relevant sentences containing either *bind to* or *binds to* have to be retrieved. However, if the query is for *X associates with Y*, then the specific relation *associates with* is looked for and not *associated with*, since they are used in very different contexts. While the first form is used in the sense of a biological interaction, the second one is very often used in the sense of assistance or conjunction.

13.6.4 Generating Strengths of Feasible Fuzzy Biological Relations

Since all relation triplets do not occur with equal frequency in the corpus and a particular relation may occur in conjunction with multiple concept pairs and a particular concept pair may be related through multiple relations, therefore each relation is assigned a *strength*, where strength of a relation reflects the frequency of co-occurrence of a relational verb in conjunction with an ordered pair of biological concepts. For example, analysis of the GENIA corpus reveals that the relation *activation* is defined between different concept pairs such as <multi_cell, protein_molecule>, <protein_molecule, protein_molecule>, and, moreover, the concept pair <multi_cell, protein_molecule> is related through different biological relations such as *activation, stimulation*, and the like. A relation triplet is a unique combination of a particular variant of a biological relational verb $\Re$ and an ordered pair of biological concepts denoted by the triplet $<C_i, \Re, C_j>$. These relations are not symmetric. For example, C_i *activates* C_j does not necessarily imply that C_j *activates* C_i.

The strength of the relation $<C_i, \Re, C_j>$ is computed as a fuzzy membership value $\mu_{(C_i,C_j)}(\Re)$ indicating the degree of co-occurrence of the triplet. $\mu_{(C_i,C_j)}(\Re)$ is computed as a mean of the ratio of frequency of the relation $\Re$ occurring in conjunction with the ordered concept pair $<C_i, C_j>$ against all occurrences of $\Re$, and the ratio of frequency of the ordered concept pair $<C_i, C_j>$ occurring in conjunction with $\Re$ against all occurrences of the pair. This is shown in Eq. (13.1), where $|<C_i, \Re, C_j<|$ represents the frequency count of the relation triplet $<C_i, \Re, C_j>$:

$$\mu_{(C_i,C_j)}(\Re) = \frac{1}{2}\left\{ \frac{|<C_i, \Re, C_j>|}{\sum_{a,b}|C_a, \Re, C_b|} + \frac{|<C_i, \Re, C_j>|}{\sum_r |C_i, \Re_r, C_j|} \right\}. \tag{13.1}$$

Continuing with the 246 biological relational verbs or their variants identified earlier, the strength of each relation triplet was computed using Eq. (13.1). Since GENIA has a total of 36 terminal concept tags, the upper bound on the unique relation triplets that can be mined from the GENIA corpus as 36 × 36 × 246 (= 318,816). However, it was found that only 4162 unique biological relation triplets have nonzero strength within the GENIA corpus. Each of these 4162 feasible biological relations is assigned a unique ID and stored in a local knowledge base as a quadruple of the form $<C_i, \Re, C_j, \mu_{(C_i,C_j)}(\Re)>$.

A large number of concept pairs do not co-occur within the neighborhood of any frequent biological relation extracted earlier. Table 13.1 shows a partial listing of concept pairs that do not co-occur in any sentence. This information can be used to restrict the users to formulate queries using valid concept pairs only. Table 13.2 shows some of the *feasible fuzzy biological relations* along with their membership values. This table also illustrates the many-to-many relationship between a pair of biological concepts and a feasible biological relation.

The study of feasibility of various biological relations in the context of a pair of biological concepts is one of our major contributions. The overall statistics of

Table 13.1 Partial List of Non Related Biological Concept Pairs Mined from the GENIA Corpus 3.01

Nonrelated Ordered Concept Pairs	Nonrelated Ordered Concept Pairs
<protein_family_or_group, RNA_domain_or_region>	<protein_complex, RNA_substructure>
<cell_component, RNA_domain_or_region>	<Carbohydrate, Polynucleotide>
<DNA_family_or_group, Carbohydrate>	<Carbohydrate, Atom>
<Lipid, protein_substructure>	<protein_substructure, RNA_substructure>
<Nucleotide, Carbohydrate>	<Polynucleotide, RNA_domain_or_region>
<Tissue, DNA_substructure>	<RNA_substructure, protein_complex>
<amino_acid_monomer, DNA_substructure>	<RNA_substructure, protein_domain_or_region>
<amino_acid_monomer, RNA_domain_or_region>	<Virus, RNA_substructure>

biological interactions extracted from the GENIA corpus 3.01 is summarized in Table 13.3.

13.7 RELATION-BASED QUERY ANSWERING

One of the main uses of the extracted relation instances is in answering user queries intelligently. In this section we illustrate the effectiveness of the proposed approach with some sample queries and the resulting documents extracted from the GENIA corpus. The most important aspects of querying is that in the proposed mechanism queries can accommodate concept names specified at multiple levels of generality and also have relation names contained in them. A query template is of the form <left actor, relation/*, right actor> where the left and right actors could be entity_names, or a concept_name, or represented by a *, where a * indicates a wild-card entry and represents any match. We have presented the results for three representative queries in this section. Each query template presented is followed by a snapshot showing a partial list of the sentences retrieved by the system from GENIA corpus. It can be seen that the proposed system addresses some of the shortcomings presented in Section 13.1. More details about the relation-based query processing system can be obtained in Abulaish and Dey (2005).

Query 1: <Interleukin-10, Inhibits, *>

In query 1, the right actor of relation is left unspecified. Five out of nine answers generated by the system are shown in Table 13.4. It may be noted from the answers generated that our entity recognizer is capable of recognizing "IL-10" as a variation of the protein Interleukin-10.

Table 13.2 Partial List of Feasible Fuzzy Biological Relation Triplets Where Matrix Entries Show the Strength of a Relation Triplet

Relations → Concept pairs ↓	Affect	Induce	Stimulated by	Stimulated with	Activated by	Expressed in	Associated to	Regulated by	…
<AAM, PS>	0	0	0	0	0.26	0	0.42	0	…
<AAM, PFG>	0.11	0	0.11	0.11	0	0	0	0	…
<CL, Tissue>	0	0.13	0	0	0.13	0	0	0	…
<CT, Lipid>	0	0	0.09	0.27	0	0	0	0	…
<PM, CT>	0.14	0.01	0	0	0	0.22	0	0.02	…
<CL, Peptide>	0	0	0	0	0	0	0	0.17	…
<CT, Peptide>	0	0	0.22	0	0	0	0	0.11	…
<DNA_DR, MC>	0	0	0	0	0	0.25	0	0	…
…	…	…	…	…	…	…	…	…	…

Abbreviations: AAM: Amino_Acid_Monomer; DNA_DR: DNA_Domain_or_Region; CL: Cell_Line; CT: Cell_Type; PS: Protein_Subunit; PM: Protein_Molecule; MC: Mono_Cell PFG: Protein_Family_or_Group.

Table 13.3 Fuzzy Relation Extraction Statistics of the Biological Relation Extractor Module

Attribute	Value	Attribute	Value
No. of ontology tags	36	No. of root verbs along with morphological variants as biological relations	246
No. of possible ordered tag pairs	1,296	No. of possible <C_i, $\Re$, C_j> triplets (taking only valid tag pairs)	290,280
No. of related tag pairs	1,180	No. of extracted valid relation triplets having nonzero membership value	4,162
No. of selected root verbs	24		

Table 13.4 Results from Query 1

MEDLINE No.	Sentence No.	Sentence
MEDLINE: 95238477	1	Interleukin (IL)-10 inhibits nuclear factor kappa B (NF kappa B) activation in human monocytes.
MEDLINE: 95238477	3	Our previous studies in human monocytes have demonstrated that interleukin (IL)-10 inhibits lipopolysaccharide (LPS)-stimulated production of inflammatory cytokines, IL-1 beta, IL-6, IL-8, and tumor necrosis factor (TNF)-alpha by blocking gene transcription.
MEDLINE: 95238477	4	Using electrophoretic mobility shift assays (EMSA), we now show that, in monocytes stimulated with LPS or TNF alpha, IL-10 inhibits nuclear stimulation of nuclear factor kappa B (NF kappa B), a transcription factor involved in the expression of inflammatory cytokine genes.
MEDLINE: 99155321	1	Interleukin-10 inhibits expression of both interferon alpha- and interferon gamma-induced genes by suppressing tyrosine phosphorylation of STAT1.
MEDLINE: 97335975	1	Interleukin-10 inhibits interferon-gamma-induced intercellular adhesion molecule-1 gene transcription in human monocytes.

Query 2: <Protein_molecule, Inhibits, *>

Query 2 is a generalized case of query 1 in which the left actor (protein_molecule) is a more generic concept than the earlier instance Interleukin 10. On reviewing the answers for the two queries, it is observed that the set of sentences retrieved in answer to query 1 is contained in the set of sentences retrieved for query 2, which

Table 13.5 Results of Query 2

MEDLINE No.	Sentence No.	Sentence
MEDLINE: 95184007	2	I kappa B-alpha inhibits transcription factor NF-kappa B by retaining it in the cytoplasm.
MEDLINE: 95280909	3	Secreted from activated T cells and macrophages, bone marrow-derived MIP-1 alpha/GOS19 inhibits primitive hematopoietic stem cells and appears to be involved in the homeostatic control of stem cell proliferation.
MEDLINE: 95238477	1	Interleukin (IL)-10 inhibits nuclear factor kappa B (NF kappa B) activation in human monocytes.
MEDLINE: 95238477	3	Our previous studies in human monocytes have demonstrated that interleukin (IL)-10 inhibits lipopolysaccharide (LPS)-stimulated production of inflammatory cytokines, IL-1 beta, IL-6, IL-8, and tumor necrosis factor (TNF)-alpha by blocking gene transcription.
MEDLINE: 95238477	4	Using electrophoretic mobility shift assays (EMSA), we now show that, in monocytes stimulated with LPS or TNF alpha, IL-10 inhibits nuclear stimulation of nuclear factor kappa B (NF kappa B), a transcription factor involved in the expression of inflammatory cytokine genes.
MEDLINE: 99155321	1	Interleukin-10 inhibits expression of both interferon alpha- and interferon gamma-induced genes by suppressing tyrosine phosphorylation of STAT1.
MEDLINE: 97335975	1	Interleukin-10 inhibits interferon-gamma-induced intercellular adhesion molecule-1 gene transcription in human monocytes.

is correct. Seven out of 56 answers retrieved by the system for this query are shown in Table 13.5.

Query 3: <Protein_family_or_group, binds to, DNA_domain_or_region>

Query 3 is very restrictive in that both the left and right actors as well as the relation are specified. For this query there are 16 answers retrieved, out of which 6 are shown in Table 13.6.

13.8 EVALUATION OF THE BIOLOGICAL RELATION EXTRACTION PROCESS

The performance of the whole system was analyzed by taking into account the performance of the biological relation extraction process. Since the feasible relation

Table 13.6 Results of Query 3

MEDLINE No.	Sentence No.	Relevant Sentence
MEDLINE: 95222739	2	Core binding factor (CBF), also known as polyomavirus enhancer-binding protein 2 and SL3 enhancer factor 1, is a mammalian transcription factor that binds to an element termed the core within the enhancers of the murine leukemia virus family of retroviruses.
MEDLINE: 96344715	5	NGFI-B/nur77 binds to the response element by monomer or heterodimer with retinoid X receptor (RXR).
MEDLINE: 97051821	6	An unidentified Ets family protein binds to the EBS overlapping the consensus GAS motif and appears to negatively regulate the human IL-2R alpha promoter.
MEDLINE: 96239482	3	ICSAT is structurally most closely related to the previously cloned ICSBP, a member of the IFN regulatory factor (IRF) family of proteins that binds to interferon consensus sequences (ICSs) found in many promoters of the IFN-regulated genes.
MEDLINE: 99102381	3	NF-Y is a ubiquitous and evolutionarily conserved transcription factor that binds specifically to the CCAAT motif present in the 5 promoter region of a wide variety of genes.
MEDLINE: 97419190	3	Studies of the mechanisms that enable EBV to infect nonactivated, noncycling B cells provide compelling evidence for a sequence of events in which EBV binding to CD21 on purified resting human B cells rapidly activates the NF-kappaB transcription factor, which, in turn, binds to and mediates transcriptional activation of Wp, the initial viral latent gene promoter.

triplets have been extracted from the entire GENIA corpus, to evaluate the correctness of the extraction process, we selected 10 different feasible relation triplets and 500 documents from the corpus randomly for manual verification.

A relation triplet was said to be "correctly identified" if its occurrence within a sentence along with its left and right tags was grammatically correct, and the system was able to locate it in the right context. To judge system performance, it was not enough to judge only the extracted relations; analysis of all the correct relations that were missed by the system was also required. The system was evaluated for its *recall* and *precision* values for each of the selected relation triplets. Recall and precision for this purpose are defined by Eqs. (13.2) and (13.3), respectively.

$$R = \frac{\begin{array}{c}\text{Number of times a relation triplet is correctly identified by the}\\ \text{system (ture positives)}\end{array}}{\begin{array}{c}\text{Number of times the relation triplet actually occurs in the corpus}\\ \text{with the specified tags in the correct sense (true positives +}\\ \text{false negatives)}\end{array}}, \tag{13.2}$$

$$P = \frac{\begin{array}{c}\text{Number of times a relation triplet is correctly indentified}\\ \text{by the systme (true positives)}\end{array}}{\begin{array}{c}\text{Number of times the relation triplet is identified by the system}\\ \text{as correct (true positives + false positives)}\end{array}} \tag{13.3}$$

Table 13.7 summarizes the performance of the system for 10 different relation triplets. The precision value of the system reflects its capability to identify a relational verb along with the correct pair of tags within which it is occurring. The average precision of the proposed system was found to be 95.5%. Recall value reflects the capability of the system to locate all instances of a relation within the corpus. The recall value of this module was 87.23%, which can be improved. On analysis, we determined that most of the errors occur when a biological tag/entity has been used earlier in the sentence or in an earlier sentence, and is referred to in the vicinity of the relation using a pronoun. Since pronouns were not tagged, these were not recognized by the system. Errors also occurred in the case where a relevant tag occurs in conjunction with other tags separated by such operators as "or," ",", and the like, and the tags in the immediate vicinity of the relation did not match the tag given in the relation triplet.

13.9 BIOLOGICAL RELATION CHARACTERIZER

The feasible biological relations mined from the GENIA corpus can be used to enhance the underlying ontology itself. Since the strength of a relation reflects the significance of a particular type of biological association between a concept pair, we propose using this strength to identify relations that are *frequent enough* or have strength greater than a predefined threshold, to be included into the ontology. This prevents the ontology from being converted into a specific instance of knowledge base and helps in retaining the generic nature of knowledge stored in the ontology.

However, since the GENIA corpus is tagged using only leaf-level concepts, all relations that have been mined from the corpus are defined between pairs of leaf-level concepts only. Therefore, the relations should be further generalized before including them in the ontology. The need for this analysis is further clarified through the following example. Out of a total 219 instances of the relation *expressed in* occurring between different leaf concept pairs, it is observed that 170 instances have the left concept as a subclass of *source* and the right concept as a subclass of

Table 13.7 Precision and Recall Values of the Biological Relation Extraction Process

Relation Triplets	Total Number of Times a Relation Triplet Is Identified by the System	Total Number of Times a Relation Triplet Is Correctly Identified by the System	Total Number of Times a Relation Triplet Occurs Correctly in Test Corpus	Precision (%)	Recall (%)
<Protein_molecule, Activates, Protein_molecule>	7	7	8	100.00	87.50
<Protein_molecule, Expressed in, Cell_type>	15	12	15	80.00	80.00
<DNA_domain_or_region, Expressed_in, Cell_type>	5	5	7	100.00	71.43
<Protein_molecule, binds to, <DNA_domain_or_region>	7	7	7	100.00	100.00
<Protein_family_or_group, Interacts with, DNA_domain_or_region>	3	3	3	100.00	100.00
<Protein_molecule, Activated in, Protein_molecule>	2	2	2	100.00	100.00
<DNA_domain_or_region, Regulated by, Protein_family_or_group>	3	3	4	100.00	75.00
<DNA_family_or_group, Associated with, DNA_domain_or_region>	2	2	2	100.00	100.00
<Protein_molecule, Associated with, Protein_molecule>	5	5	6	100.00	83.33
<Protein_molecule, Induces, Protein_family_or_group>	4	3	4	75.00	75.00
Average				**95.50**	**87.23**

substance, where *source* and *substance* are two top-level biological concepts in the GENIA ontology. Studying the distribution of these 170 instances reveals the following:

- 48 instances associate the concept pair <protein_molecule, cell_type>
- 22 instances are defined between concept pair <protein_family_or_group, cell_type>

- 21 instances of the same occur between <protein_molecule, cell_line>
- 10 for <protein_family_or_group, cell_line>
- 9 for <DNA_domain_or_region, cell_type>
- 7 for <RNA_molecule, cell_type>
- 6 for <DNA_family_or_group, cell_type>
- 5 for <RNA_molecule, cell_line>
- 4 each for <RNA_family_or_group, cell_type> and <protein_molecule, tissue>
- 3 each for <protein_molecule, body_part> and <protein_molecule, mono_cell>
- 2 each for <DNA_domain_or_region, cell_line>, <protein_domain_or_region, cell_type>, <DNA_domain_or_region, tissue>, and <DNA_domain_or_region, body_part>
- 1 instance each between the pairs <DNA_family_or_group, cell_line>, <protein_family_or_group, multi_cell>, <protein_complex, cell_line>, <protein_family_or_group, cell_component>, <protein_molecule, cell_component>, <protein_domain_or_region, cell_line>, <DNA_molecule, cell_type>, <protein_molecule, multi_cell>, <RNA_molecule, tissue>, <protein_subunit, cell_component>, <protein_family_or_group, body_part>, <protein_domain_or_region, multi_cell>, <protein_subunit, cell_type>, <DNA_family_or_group, tissue>, <lipid, cell_type>, <peptide, cell_type>, <RNA_molecule, body_part>, <protein_molecule, other_artificial_source>, <DNA_domain_or_region, other_artificial_source>, and <DNA_domain_or_region, mono_cell>

Obviously, rather than storing each unique relation that has low occurrence rates like 1 or 2, it is better to consolidate these occurrences into more generic interactions. It may be noted that cell_type, mono_cell, cell_component and multi_cell are all subclasses of *natural cells*. Hence, a number of relations in the last category may be grouped together to state that their right actors are natural cells. Again all right actors of the instances shown above may be grouped together to state that all these reactions have a *source* as their right actor. Similarly, it is possible to say that 26 of the above instances have a subclass of DNA as the left actor, which is substantially more than the individual numbers of 1, 2, 6, or 9. Similarly, considering subclasses of both DNA and RNA, it is possible to conclude that 44 instances have *nuclic_acid* as its left actor. At the extreme level of generalization, it may be concluded that all instances of *expressed in* shown above occur between the concept pair *<substance, source>*. However, just as overspecialization induces noise, overgeneralization reduces information content. Hence an intermediate level consolidation may be a better characterization of the *expressed in* relation.

Knowing the number of instances at a more specific-level concept pair, it is easy to generate the number of instances that exist at a more generic level with appropriate summations. With an appropriate technique to generalize, which will be discussed later, an appropriate level of generalization arrives at the following distribution of the *expressed in* relation:

- 91 instances occur with the pair <amino acid, natural sources>
- 43 instances occur with <compound, artificial source>
- 35 instances occur with the pair <nuclic acid, natural sources>
- Only 1 relation occurs with the pair <lipid, natural sources>

Thus, instead of defining all feasible relations between most generic-level concept pairs (root concept pairs) or between most specific-level concept pairs (leaf concept pairs), generalization aims at finding the most appropriate specificity level for a relation.

13.9.1 Generating a Concept Pair Tree to Define Biological Relations

In this section we propose a two-way technique to achieve the appropriate generalization of each relation concept pairs $<C_i, C_j>$, where C_i and C_j are concepts in GENIA ontology (not necessarily leaf-level concepts only). A *bottom–up* approach is used to compute the frequencies of the relations at each generic level using the instance numbers mined earlier. A *top–down* approach is thereafter applied in conjunction with an information-theoretic measure to find the most appropriate level of specificity. Each relation deemed feasible is further associated with a fuzzy qualifier that reflects its relative strength in the relation knowledge base.

The GENIA ontology is a taxonomy of 47 biologically relevant nominal categories in which the top three concepts are *biological source, biological substance*, and *other_name*. The *other_name* refers to all biological concepts that are not identified with any other known concept in the ontology. Since it is not possible to characterize relations having one or more participating concept as *other_name*, these instances are ignored. Thus the GENIA ontology tree can be treated as a forest of two significant subtrees one rooted at *source* class containing 13 nominal categories and other at *substance* class containing 34 nominal categories. Taking into account that biological relations are noncommutative, if we propagate the argument concept pairs observed for the relation instances from the leaves toward the root, a relation at the most generic level can be characterized as belonging to one of the following root concept pairs:

- <substance, source>
- <substance, substance>
- <source, source>
- <source, substance>

Starting with a pair of root concept pairs, a *concept pair tree* stores all possible ordered concept pairs that match the root concept pair and can be generated from the underlying concept ontologies. Thus, possible nodes in the tree rooted at <source, source> include <tissue, artificial>, <mono_cell, body_part>, <natural, artificial>, and so forth. Similarly, the node <DNA, cell_line> would be found in the concept tree rooted at <substance, source>, while the node <protein, DNA> will be found in the tree rooted at <substance, substance> and so on. In order to generate the complete list of valid argument pairs for any relation, we map the instances of the relations over concept pair trees rooted at the appropriate concept pair tree.

The concept pair trees are created as follows. Each node N in a concept pair tree has two constituent concepts $<C_i, C_j>$ denoted as the LEFT and the RIGHT concepts. The concept pair tree is represented as an AND-OR tree, where each node consists of two sets of children denoted by L_1 and L_2. L_1 and L_2 each contain a set of nodes that are connected with each other through an AND operator. The two sets L_1 and L_2 are themselves connected by an OR operator. For every node N, the two sets of child nodes L_1 and L_2 are created as follows:

- L_1 consists of concept pairs created by expanding the LEFT concept to consider all its child nodes in the concept ontology, while keeping the RIGHT concept unchanged.
- L_2 is created by keeping the LEFT concept unchanged while considering all children of the RIGHT concept in the concept ontology.

Starting from a root concept pair $<rC_i, rC_j>$, the complete concept pair tree is created recursively as follows:

OR(AND ($<$children of $rC_i, rC_j>$), AND ($<rC_i,$ children of $rC_j>$))

Figure 13.7 shows the resulting AND-OR tree in which the nodes that are part of L_1 or L_2 are ANDed with each other (represented by "∪"), while the set of nodes com-

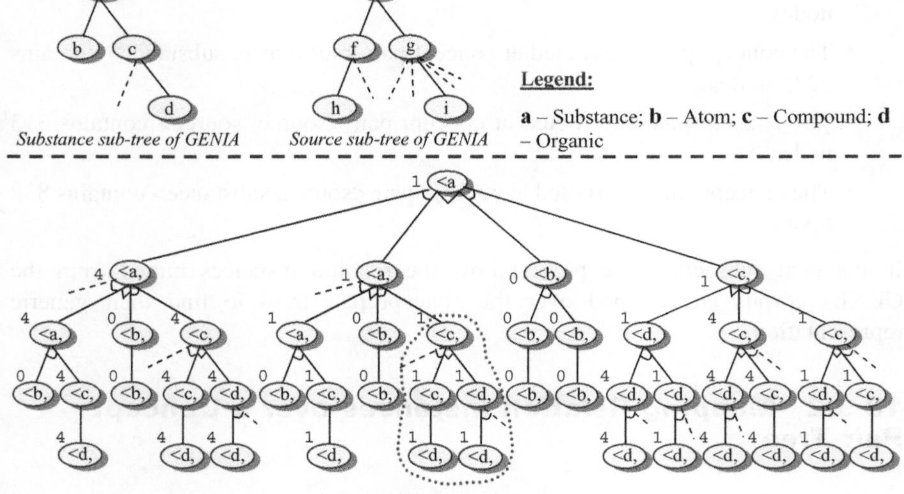

Figure 13.7 AND-OR concept pair tree generated from substance and source subtrees of GENIA ontology tree.

prising L_1 and L_2 are ORed with each other and represented using the symbol "∨." Nodes at each level are expanded in a similar fashion.

It may be observed that nodes are repeated in the concept pair tree since the constituent elements of a node may have been derived through expansion of either the LEFT concept or the RIGHT concept. For example, the node $<b, f>$ occurs both on the LEFT and RIGHT branches of the tree in Figure 13.7 since it arises once by expanding a and once by expanding e. Instances of the same concept pair lie on different subtrees, which are connected with the OR operator at a common ancestor. These subtrees provide alternate paths for generalizing a relation defined at the leaf-level concept pair. All alternate paths are searched before deciding on the best representation for a relation.

The number of times a concept pair $<C_i, C_j>$ occurs in the concept pair tree is obtained using the recursive Eq. (13.4), where i and j represent the level of concepts C_i and C_j, respectively, in the concept ontology tree:

$$N(i, j) = \begin{cases} N(i-1, j) + N(i, j-1) & \forall\, i, j \geq 2, \\ 1 & \text{otherwise.} \end{cases} \tag{13.4}$$

The total number of nodes in a concept pair tree, N, is given as

$$N = \sum_{i=1}^{l_1} \left[n_i \times \sum_{j=1}^{l_2} n_j \times N(i, j) \right], \tag{13.5}$$

where, l_1 and l_2 are depths of the left and right concept ontology trees, respectively, n_i is the number of nodes at level i in the concept ontology tree.

The number of nodes in the concept pair trees constructed from the GENIA ontology are as follows:

- The concept pair tree rooted at concept pair <substance, source> contains 838 nodes.
- The concept pair tree rooted at concept pair <substance, substance> contains 2245 nodes.
- The concept pair tree rooted at concept pair <source, source> contains 313 nodes
- The concept pair tree rooted at concept pair <source, substance> contains 838 nodes.

In the next subsection, we present how the relation instances mined from the GENIA corpus are mapped over the concept pair trees to find their generic representations.

13.9.2 Mapping Relation Instances over a Concept Pair Tree

To identify the most feasible representation of a relation, we consider mapping all instances of that relation to appropriate nodes in the concept pair trees. Suppose there are N instances of a relation r_g observed over the corpus. Each of these instances is defined for a pair of leaf-level concepts. Based on the generic category of the leaf-level concepts, each relation instance can be mapped to a leaf node in one of the four concept pair trees.

For each concept pair tree T^G, all instances that can be mapped to leaf-level nodes of T^G are mapped at the appropriate nodes. Since each leaf-level node has multiple occurrences, a relation instance is mapped to all matching leaf-level nodes. Count at each leaf node determines the number of unique instances of a relation that has been mapped to it. These counts are then propagated up along the tree exploiting its AND-OR property. For each nonleaf node in the concept pair tree, the total number of relations is equal to the number of instances propagated up through all its children in either L_1 or L_2 but not both. Equation (13.6), which is just another representation of the AND-OR principle, computes the number of instances at each nonleaf node:

$$|< \text{LEFT}, r_g, \text{RIGHT}>| = \sum_i |<\text{Child}_i[\text{LEFT}], r_g, \text{RIGHT}>| \\ \text{OR} \sum_i |< \text{LEFT}, r_g, \text{Child}_i[\text{RIGHT}]>|. \tag{13.6}$$

All relations can thus be propagated right up to the root of the concept pair tree. Figure 13.7 shows a part of the <substance, source> concept pair tree generated from the GENIA ontology, with a subset of the instances of the relation *expressed in* mapped over it. Out of a total of 170 instances of the relation, 102 instances of the relation are of the type <organic, expressed in, cell_type>, and these are mapped at all instances of the leaf node <organic (*d*), cell_type (*i*)> in the tree. The portion of the tree enclosed within the dotted blue line has two instances of node <*d*, *i*>. These numbers are propagated up and node <*c*, *i*> also receives 102 instances since the other children of <*c*, *i*> had zero (0) instances mapped to them. Node <*d*, *g*> within the ellipse has 127 instances, 102 of which are propagated up from node <*d*, *i*> and the remaining from its other children (shown in the highlighted subtree of Fig. 13.7 through the dashed line). The nodes <*d*, *g*> and <*c*, *i*> provide alternate paths for propagating the relation instances upward. Upward propagation terminates at root node. It may be observed that the root node receives a total of 170 instances from two alternate paths defined by the two subtrees ORed with each other.

To derive the most appropriate levels for describing a relation, the concept pair tree is now traversed top–down. Starting from the most generic level description at the root level, an information loss function based on set-theoretic approach is applied at each node to determine the appropriateness of defining the relation at that level.

13.9.3 Computing Information Loss at Concept Pair Nodes

The process of determining the most specific concept pairs for relations follows a top–down scanning of the AND-OR tree. Starting from the root node, the aim is to determine those branches and thereby those nodes that can account for sufficiently large numbers of relation instances. When the frequency of a relation drops to an insignificant value at a node, its descendents also need not be considered for the relation conceptualization and may be pruned off without further consideration. Thus, the proposed method is akin to a postpruning method. The lowest unpruned node becomes a leaf and is labeled as the most specific concept pair for defining a relation.

Pruning is done at two levels. First, individual nodes that do not have a significant number of relation instances defined at them are pruned provided the information loss incurred remains within a threshold. Equation (13.7) defines a loss function that is applied at every node N to determine the loss of information incurred if this node is pruned off. The loss function is computed as a symmetric difference between the number of instances that reach the node and the number of relation instances that were defined at its parent.

$$\text{Information loss}\,(N) = \frac{|\text{IC}_P - \text{IC}_N|}{|\text{IC}_P + \text{IC}_N|}, \tag{13.7}$$

where, IC_N = count of instances of relation r_g at N, and IC_P = count of instances of r_g at parent P of N.

Using Eq. (13.7) it is argued that if the information loss at a node N is above a threshold, it is obvious that the node N accounts for a very small percentage of the relation instances that are defined for its parent. Hence any subtree rooted at this node may be pruned off from further consideration while deciding the appropriate level of concept pair association for a relation. For our implementation this threshold has been kept at 10%. Hence any node that accounts for less than 10% of its parent's relations is pruned off along with the subtree rooted at the node. The path marked with dotted lines at node <*a*, *f*> in Figure 13.7 accounts for two instances. Hence information loss for the nodes lying on the dotted path would be 0.91.

The above policy determines the significance of a single node N. However, since there are more ways than one to generalize a relation, a collective estimate of information loss based on individual estimates is applied to decide the appropriate path. Since a parent node has two alternative paths denoted by the expansion of LEFT and RIGHT, respectively, along which a relation may be further specialized, the choice of appropriate level is based on the collective significance of the path composed of retained nodes. For each ANDed set of retained nodes, total information loss for the set is computed as the average information loss for each retained child. The decision to prune off a set of nodes rooted at N is taken as follows:

Let information loss for nodes retained at L_1, be E_1 and information loss for nodes retained at L_2, be E_2.

- If $E_1 = 0$, L_1 is retained and L_2 is pruned off. And, if $E_2 = 0$, L_2 is retained and L_1 is pruned off. Using this rule, the subtree at the left of root node <*a*, *e*> is pruned off since information loss is 0 for the right subtree.
- Otherwise, if $E_1 \approx E_2$, that is, $\text{Min}(E_1, E_2)/\text{Max}(E_1, E_2) \geq 0.995$, then both the subtrees are pruned off, and the node N serves as the appropriate level of specification.

- Otherwise, if $E_1 < E_2$, then L_1 is retained and L_2 is pruned off. If $E_2 < E_1$ then L_2 is retained while L_1 is pruned off.

The set of concept pairs retained are used for conceptualizing the relations. Function AND_OR_Tree_Pruning() implements the tree pruning algorithm formally. The procedure is applied once for each relation over each of the four concept pair trees.

A similar loss function was used by Mena et al. (1998) for computing loss of information when concepts are approximated in an ontology. However, Mena et al. (1998) used *precision* and *recall* of retrieval rather than number of instances for the estimation of the information loss.

13.10 DETERMINING STRENGTHS OF GENERIC BIOLOGICAL RELATIONS

Since all relations are not equally frequent in the corpus, we associate with each relation a strength S that is computed in terms of relative frequency of occurrence of the generic relation in the corpus. Equation (13.8) computes this strength, where G denotes the category of concept pairs: source-substance, source-source, substance-substance, and substance-source. $|T^G|$ denotes the total count of all relations that are defined between ordered concept pairs defined in the tree T^G, and $N^G_{r_g}$ denotes the total number of relation instances of type r_g mapped to T^G:

$$\mu^G_{(C_i,C_j)}(r_g) = \frac{1}{2}\left\{\frac{|<C_i, r_g, C_j>|}{N^G_{r_g}} + \frac{|<C_i, r_g, C_j>|}{|T^G|}\right\}. \tag{13.8}$$

Figure 13.8 shows the distribution of strength values for each of the four basic categories of relations. It can be said from Figure 13.8 that for each category a large number of relations have strength greater than 0.4 and very few have strength around 0.4. It can also be said that most of the substance-substance relations have high

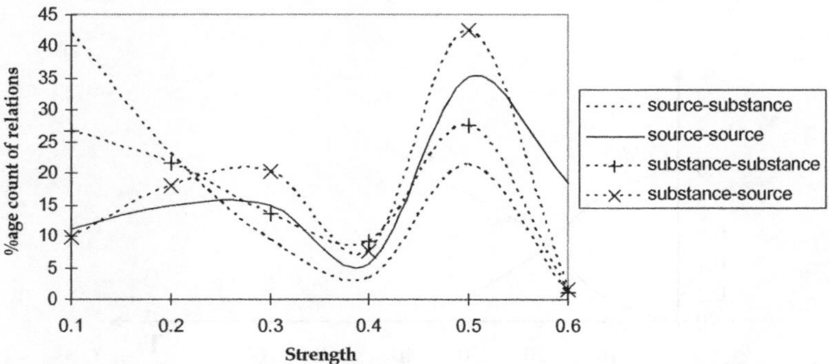

Figure 13.8 Plot of relation strengths and their percentage counts for all four categories of trees.

strength, while most of the source-substance relations have very low strength around 0.1. Since exact numeric values of strength do not convey much information, we choose a fuzzy representation to store the relations. The feasible biological relations are converted into fuzzy relations based on the membership of their strength values to a fuzzy quantifier term set {WEAK, MEDIUM, STRONG}. Since each category has different strength value distributions, the membership functions for determining the values to each of these categories is derived after analyzing the graphs displaying the distributions of strength over a particular tree.

The basic task in designing the fuzzy membership functions is to identify the nature of the membership functions and the parameters for defining those functions. We clarify the notion of fuzzy membership to associate fuzzy qualifiers with concepts using a simple example. Given the exact temperature of a day, one can associate a fuzzy qualifier from the set {HOT, COLD} to a day. However, the single values HOT or COLD may be associated to a potentially infinite set of temperature values each, and given a day's temperature one may like to say that the day is neither HOT nor COLD but rather has some degree of membership to both, which is definitely different from associating a crisp decision. Thus if a day's temperature is 15°C, it could have a membership of 0.6 to HOT and 0.4 to COLD, where as if the temperature on a given day is 25°C, then the day's membership to HOT could be 0.7 and to COLD could be 0.3. The membership values are derived using membership functions. Figure 13.9 shows sample membership functions that can be used for modeling climate as discussed here.

For the proposed system, the chosen classes are fuzzy strength qualifiers {WEAK, MODERATE, STRONG}. The membership functions are derived using the relation strengths obtained earlier. Each curve in Figure 13.8 shows only one valley, and this common valley for all trees is observed at strength 0.4. Hence 0.4 is selected for defining the peak of intermediate strength relations belonging to class MODERATE. The membership functions for the classes WEAK and STRONG for each category are obtained through curve fitting on different sides of the valley. Table 13.8 shows the top 10 relations mined from GENIA corpus and the associated generic concept pairs along with their strength.

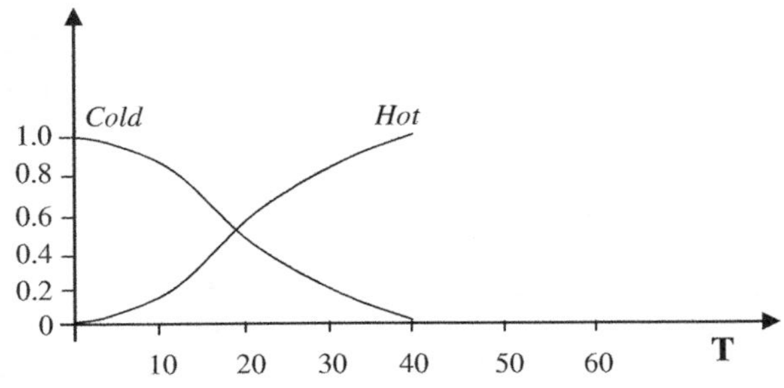

Figure 13.9 Fuzzy membership functions to represent Cold and Hot.

Table 13.8 Biological Relations and Associated Generic Concept Pairs along with Their Fuzzy Strength

Relation	Generic Concept pairs and Their Strengths			
	Substance-Source	Substance-Substance	Source-Source	Source-Substance
Induce	(<OC, Nat>, strong) (<OC, Art>, weak>)	(<OC, AA>, strong) (<OC, NA>, weak)	(<Src, Src>, strong)	—
Inhibits	(<Lip, CT>, weak) (<PFG, CT>, weak) (<PM, CT>, moderate) (<DNADR, CT>, weak)	(<Sbs, Cmp>, strong)	(<CT, Art>, strong) (<CT, Nat>, strong)	(<Nat, AA>, strong) (<Nat, NA>, moderate)
Activate	(<OC, Nat>, strong)	(<Pr, AA>, strong) (<Pr, NA>, weak)	(<CL, CT>, weak) (<CT, CT>, strong) (<MC, CT>, weak)	(<Src, OC>, strong)
Expressed in	(<OC, Src>, strong)	(<DNA, OC>, weak) (<Pr, AA>, moderate) (<Pr, NA>, moderate) (<RNA, OOC>, weak)	(<Nat, Org>, weak) (<Nat, Tis>, weak) (<Nat, CT>, strong)	—
Regulate	(<OC, Art>, weak) (<OC, Nat>, strong)	(<OC, AA>, strong) (<OC, NA>, moderate)	—	(<Nat, AA>, moderate) (<Nat, NA>, strong)
Mediated by	(<DNADR, CT>, strong)	(<OC, AA>, strong) (<OC, NA>, weak)	(<CT, CT>, strong)	(<CL, AA>, weak) (<CL, NA>, weak) (<CT, AA>, strong) (<CT, Lip>, weak) (<CT, NA>, weak)
Contains	(<DNADR, CL>, strong)	(<OC, AA>, moderate) (<OC, NA>, strong)	(<CL, Art>, strong) (<CL, Nat>, strong)	(<CL, DNADR>, weak) (<CT, DNADR>, moderate) (<Vir, DNADR>, weak) (<CT, PM>, moderate) (<Vir, PFG>, weak) (<Vir, PM>, weak)
Associate with	(<PFG, CC>, strong)	(<AA, Pr>, strong) (<OOC, PFG>, weak)	—	(<CC, PM>, strong)
Stimulated by	(<PFG, CT>, strong)	(<OC, OC>, strong)	—	(<CL, AA>, weak) (<CL, OOC>, weak) (<CT, AA>, strong) (<CT, Lip>, weak) (<CT, OOC>, weak)
Enhances	(<OC, Nat>, strong)	(<Lip, AA>, weak) (<PM, AA>, strong) (<PM, NA>, weak) (<PM, OOC>, weak)	(<CT, Tis>, strong)	(<CT, PFG>, strong) (<CT, PM>, moderate)

Abbreviations: OC: Organic compound; AA: Amino_acid; NA: Nuclic_acid; OOC: Other_organic_compound; Sbs: Substance Nat: Natural source; Org: Organism; CT: Cell_type; Pr: Protein; Src: Source; Tis: Tissue; MC: Mono_cell CL: Cell_line; PFG: Protein_family_or_group; Lip: Lipid; DNADR: DNA_domain_or_region; Art: Artificial source; Cmp: Compound; Vir: Virus; PM: Protein_molecule; CC: Cell_component.

13.11 ENHANCING GENIA TO FUZZY RELATIONAL ONTOLOGY

Given that GENIA ontology stores information about biological concepts only, it cannot be exploited for representing biological interactions. Hence, we consider extending this ontology by adding the generic relations to this. Since the relations are accompanied by the fuzzy qualifiers WEAK, MODERATE, and STRONG, we propose an extension of the standard ontology model to a fuzzy relational ontology structure. The proposed fuzzy relational ontology model organizes knowledge in terms of concepts, properties, semantic relations, generic relations, and fuzzy strength and is formally defined as follows.

Definition (Fuzzy Relational Ontology) A fuzzy relational ontology Θ_f is a 5-tuple of the form $\Theta_f = (C, P, \Re_s, \Re_g, S)$, where C, P, $\Re_s$, and $\Re_g$ are the same as defined in Section 13.2. S = {weak, moderate, strong}, is a term set used to represent the strength of the generic biological relations in terms of linguistic qualifiers. A linguistic qualifier $\xi \in S$ is defined as a unary relation of the form $\xi(r_g)$, where $r_g \in \Re_g$ is a feasible generic relation. The strength is determined through fuzzy membership functions (Zadeh, 1965, 1983).

The fuzzy GENIA relational ontology structure is implemented through the introduction of three generic classes: a *ConceptPair* class, a *GenericRelation* class, and a *FuzzyStrength* class. The ConceptPair class consists of *HasLeftConcept* and *HasRightConcept* properties whose values are the instances of the GENIA *concept* classes. FuzzyStrength class has been defined to store the fuzzy quantifiers that can be associated with the generic relations to represent their strength. This class consists of a single property *TermSet*, which is defined as a *symbol* and contains the fuzzy quantifiers *weak*, *moderate*, and *strong*. The GenericRelation class has two properties: *LeftRightActors* and *Strength*. The LeftRightActors property is a kind of owtology web language (OWL) object property whose range is bound to the ConceptPair class. This is also restricted to store exactly one value, an instance of the ConceptPair class, for every instance of a generic relation. The Strength property is also a kind of OWL object property for which the range is bound to the FuzzyStrength class. This property is also restricted to store exactly one value for every instance of the generic relations. All mined generic relations are defined as instances of the class GenericRelation. Figure 13.10 shows a snapshot of a portion of the enhanced fuzzy GENIA relational ontology structure implemented by using Protégé 3.1.

13.12 CONCLUSIONS AND FUTURE WORK

In this chapter, we have proposed a system for ontology-based biological information extraction and ontology enhancement. The proposed system employs text mining to extract information about the likelihood of various biological relation occurrences within tagged biological documents. The set of relations mined from a specific collection is stored as feasible fuzzy biological relations defined between a

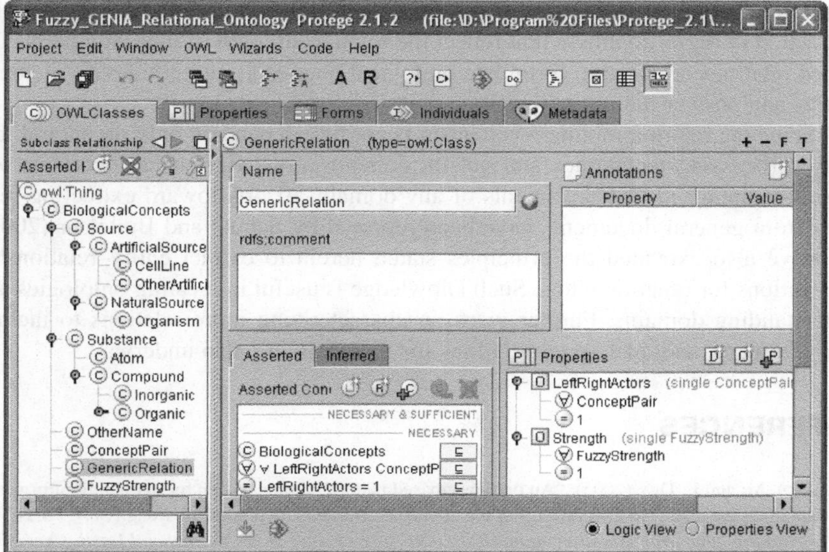

Figure 13.10 Snapshot of the Fuzzy Relational GENIA ontology structure.

pair of biological concepts. Though relations in a text co-occur with entities, the mined relations are characterized at generic concept level since the chemical nature of reactions is better characterized at the concept level than the entity level. Thus, the mined set of relations is not likely to reflect any chance co-occurrences. The strengths of the relations are expressed as fuzzy membership values to categories WEAK, MODERATE, and STRONG, where the membership value reflects likelihood of observing a particular association in a corpus. Coupled with the fact that biological entities can be tagged unambiguously, the characterization of biological relations is also expected to be stable across corpuses.

Performance evaluation of the mining mechanism over the GENIA corpus, which contains 2000 MEDLINE abstracts manually tagged according to the GENIA ontology of molecular biology, shows that the precision of the relation extraction process is high. Reliability of the process is established through the fact that all manually identified relational verbs are extracted correctly. The relation extraction mechanism also works with untagged text documents, though in this case the relations mined are at entity level and not at generic level. The recall value of the entire process may be improved with more rigorous natural language analysis.

In this chapter we have also proposed a methodology to generate generic representation for biological relations mined from a corpus and enhance domain knowledge in terms of a fuzzy relational ontology structure. Since an ontology is not a database, it should not be a storehouse for relation instances. The proposed fuzzy relational ontology adheres to this principle and stores knowledge about the various categories of relations occurring in the corpus at appropriate levels of conceptualization rather than every instance of relation mined. The relations are also associated to fuzzy qualifiers that reflect the relative strength of these relations. The mined relations are used to formulate biological queries at multiple levels of specificities and answer them intelligently.

Since the relation mining mechanism is entirely dependent only on natural language processing techniques and not on domain knowledge, it can work on any tagged or untagged text documents of any domain. Efforts toward extracting relations from general documents have been reported by Schutz and Buitelaar (2005). We have also extended the principles stated herein to extract entity–relationship associations for other domains. Such knowledge is useful in creating ontologies and understanding domains. Further work on characterizing more relations to include unary, ternary, and quaternary relations for a domain are also underway.

REFERENCES

Abulaish, M. and L. Dey (2005). "An ontology-based pattern mining system for extracting information from biological texts," in *Proceedings of the 10th International Conference on Rough Sets, Fuzzy Sets, Data Mining, and Granular Computing (RSFDGrC'05), Lecture Notes in Artificial Intelligence* 3642, Part II. Springer, Berlin/Heidelberg, pp. 420–429.

Allen, J. (2004). *Natural Language Understanding*, 2nd ed., Pearson Education, Singapore.

Altman, R., M. Bada, X. J. Chai, M. Whirl Carillo, R. O. Chen, and N. F. Abernethy (1999). "RiboWeb: An ontology-based system for collaborative molecular biology," *IEEE Intell. Syst.*, Vol. 14, pp. 68–76.

Andreasen, T., P. A. Jensen, J. F. Nilsson, P. Paggio, B. S. Pedersen, and H. E. Thomsen (2002). "ONTOQUERY: Ontology-based querying of texts," AAAI'02 Spring Symposium, Stanford, CA, pp. 28–31.

Andreasen, T., P. A. Jensen, J. F. Nilsson, P. Paggio, B. S. Pedersen, and H. E. Thomsen (2004). "Content-based text querying with ontological descriptors," *Data & Knowledge Eng.*, Vol. 48, pp. 199–219.

Andreasen, T., J. F. Nilsson, and H. E. Thomsen (2000). "Ontology-based querying," in *Flexible Query Answering Systems, Recent Advances*, Larsen et al. Eds. Physica, Heidelberg, New York, pp. 15–26.

Ashburner, M., C. A. Ball, J. A. Blake, D. Botstein, H. Butler, J. M. Cherry, A. P. Davis, K. Dolinski, S. S. Dwight, J. T. Eppig, M. A. Harris, D. P. Hill, L. Issel-Tarver, A. Kasarskis, S. Lewis, J. C. Matese, J. E. Richardson, M. Ringwald, G. M. Rubin, and G. Sherlock (2000). "Gene ontology: Tool for the unification of biology," *Gene Ontol. Consortium, Nat. Genet.*, Vol. 25, pp. 25–29.

Baker, P. G., C. A. Goble, S. Bechhofer, N. W. Paton, R. Stevens, and A. Brass (1999). "An ontology for bioinformatics applications," *Bioinformatics*, Vol. 15, pp. 510–520.

Berners-Lee, T. (1998) "Semantic web road map, W3C design issues," http://www.w3.org/DesignIssues/Semantic.html.

Casely-Hayford, L. B. (2005). "A comparative analysis of methodologies, tools and languages used for building ontologies," http://epubs.cclrc.ac.uk/bitstream/894/OntologvReport.pdf.

Ciaramita, M., A. Gangemi, E. Ratsch, J. Saric, and I. Rojas (2005). "Unsupervised learning of semantic relations between concepts of a molecular biology ontology," in Proceedings of the 19th International Joint Conference on Artificial Intelligence (IJCAI'05), pp. 659–664.

Collier, N., C. Nobata, and J. Tsujii (2000). "Extracting the names of genes and gene products with a hidden Markov model," in Proceedings of the 18th Int'l Conference on Computational Linguistics (COLING'2000), pp. 201–207.

Craven, M. and J. Kumlien (1999). "Constructing biological knowledge bases by extracting information from text sources," in Proceedings of the 8th International Conference on Intelligent Systems for Molecular Biology (ISMB'99), pp. 77–86.

Crow, L. and N. Shadbolt (2001). "Extracting focused knowledge from the semantic web," *Intl. J. Human-Computer Studies*, Vol. 54, pp. 155–184.

Fensel, D., I. Horrocks, F. Harmelen, D. L. van, McGuinness, and P. Patel-Schneider (2001). "OIL: Ontology infrastructure to enable the semantic web," *IEEE Intell. Syst.*, Vol. 16, pp. 38–45.

Friedman, C., P. Kra, H. Yu, M. Krauthammer, and A. Rzhetsky (2001). "GENIES: A natural-language processing system for the extraction of molecular pathways from journal articles," *Bioinformatics*, Vol. 17, pp. s74–s82.

Fukuda, K., T. Tsunoda, A. Tamura, and T. Takagi (1998). "Toward information extraction: Identifying protein names from biological papers," in Proceedings of the Pacific Symposium on Biocomputing, Hawaii, pp. 707–718.

Gaizauskas, R., G. Demetriou, P. J. Artymiuk, and P. Willett (2003). "Protein structures and information extraction from biological texts: The PASTA system," *Bioinformatics* Vol. 19, pp. 135–143.

Gruber, T. R. (1993). "A translation approach to portable ontology specification," *Knowledge Acquisition*, Vol. 5, pp. 199–220.

Hanisch, D., J. Fluck, H. T. Mevissen, and R. Zimmer (2003). "Playing biology's name game: Identifying protein names in scientific text," Pacific Symp. *Biocomput.*, Vol. 8, pp. 403–414.

Jenssen, T.-K., A. Laegreid, J. Komorowski, and E. Hovig (2001). "A literature network of human genes for high-throughput analysis of gene expression," *Nat. Genet.*, Vol. 28, No. 1, pp. 21–28.

Karp, P., M. Riley, S. Paley, A. Pellegrini-Toole, and M. Krummenacker (1999). "EcoCyc: Electronic encyclopedia of *E. coli* genes and metabolism," *Nucleic Acids Res.*, Vol. 27, pp. 55–58.

Kazama, J., T. Makino, Y. Ohta, and J. Tsujii (2002). "Tuning support vector machines for biomedical named entity recognition," in Proceedings of the ACL Workshop of the Natural Language Processing in the Biomedical Domain, Philadelphia, pp. 1–8.

Liddle, S. W., K. A. Hewett, and D. W. Embley (2003). "An integrated ontology development environment for data extraction," in Proceedings of 2nd International Conference on Information Systems Technology and its Applications (ISTA'03), pp. 21–33.

Mena, E., V. Kashyap, A. Illarramendi, and A. Sheth (1998). "Estimating information loss for multi-ontology based query processing," in Proceedings of 2nd International and Interdisciplinary Workshop on Intelligent Information Integration, Brighton Centre, Brighton, UK, pp. 93–108.

Muller, H. M., E. E. Kenny, and P. W. Strenber (2004). "Textpresso: An ontology-based information retrieval and extraction system for biological literature," *PloS Biology*, Vol. 2, p. e309.

Nobata, C., N. Collier, and J. Tsujii (1999). "Automatic term identification and classification in biology texts," in Proceedings of the Natural language Pacific Rim Symposium, pp. 369–375.

Ono, T., H. Hishigaki, A. Tanigami, and T. Takagi (2001). "Automated extraction of information on protein-protein interactions from the biological literature," *Bioinformatics*, Vol. 17, pp. 155–161.

Proux, D., F. Rechenmann, L. Julliard, V. Pillet, and B. Jacq (1998). "Detecting gene symbols and names in biological texts: A first step toward pertinent information extraction," in *Genome Inform Ser Workshop Genome Inform*, Vol. 9, pp. 72–80.

Rinaldi, F., G. Scheider, C. andronis, A. Persidis, and O. Konstani (2004). "Mining relations in the GENIA corpus," in Proceedings of the 2nd European Workshop on Data Mining and Text Mining for Bioinformatics, Pisa, Italy, pp. 61–68.

Rindflesch, T. C., L. Hunter, and A. R. Aronson (1999). "Mining molecular binding terminology from biomedical text," in Proceedings of the AMIA Symposium, pp. 127–131.

Schulze-Kremer, S. (1998). "Ontologies for molecular biology," in Proceedings of the 3rd Pacific Symposium on Biocomputing, pp. 693–704.

Schutz, A. and P. Buitelaar (2005). "RelExt: A tool for relation extraction from text in ontology extension," *9th Int. Semantic Web Conf.*, pp. 593–606.

Sekimizu, T., H. S. Park, and J. Tsujii (1998). "Identifying the interaction between genes and gene products based on frequently seen verbs in Medline abstract," *Genome Inform.*, Vol. 9, pp. 62–71.

Snoussi, H., L. Magnin, and J.-Y. Nie (2002). "Toward an ontology-based web data extraction," in Workshop on Business Agents and the Semantic Web, in Proceedings of the Fifteenth Canadian Conference on Artificial Intelligence (AI'2002). Calgary, Alberta, Canada.

TATEISI, Y., T. OHTA, N. COLLIER, C. NOBATA, and J. TSUJII (2000) "Building annotated corpus in the molecular-biology domain," in Proceedings of the COLING 2000 Workshop on Semantic Annotation and Intelligent Content, pp. 28–34.

THOMAS J., D. MILWARD, C. OUZOUNIS, S. PULMAN, and M. CARROLL (2000). "Automatic extraction of protein interactions from scientific abstracts," in Pacific Symposium on Biocomputing, pp. 538–549.

URAMOTO, N., H. MATSUZAWA, T. NAGANO, A. MURAKAMI, T. TAKEUCHI, and K. TAKEDA (2004). "A text-mining system for knowledge discovery from biomedical documents," *IBM Syst. J.* Vol. 43, pp. 516–533.

VOORHEES, E. M. (1994). "Query expansion using lexical-semantic relations," in Proceedings of the 17th ACM-SIGIR Conference, pp. 61–69.

ZADEH, L. A. (1965). "Fuzzy sets," *J. Info. Control*, Vol. 8, pp. 338–353.

ZADEH, L. A. (1983). "A computational approach to fuzzy quantifiers in natural languages," *Comput. Math. Appl.*, Vol. 9, pp. 149–184.

APPENDIX FEASIBLE BIOLOGICAL RELATIONS

Root Relations	Morphological Variants
Induce	induce, induces, induced, inducers, inducer, induced with, induced to, inducers of, inducer of, induced in, induce in, induces in, induced by
Inhibit	inhibit, inhibits, inhibitory, inhibitor, inhibitors, inhibited, inhibition, inhibiting, inhibited with, inhibition with, inhibitor with, inhibitor to, inhibition to, inhibition of, inhibitors of, inhibitor of, inhibition in, inhibitor in, inhibitors in, inhibited in, inhibits in, inhibitory in, inhibit in, inhibited by, inhibition by, inhibitable by
Activate	activate, activates, activated, activated with, activated to, activated in, activates in, activated by,
Express	express, expresses, expression, expressed, expressing, expressed with, expression with, expression to, expressed to, expression of, expressions of, express in, expression in, expressed in, expression by, expressed by
Regulate	regulate, regulates, regulated, regulated with, regulated to, regulate in, regulated in, regulate by, regulated by
Mediate	mediate, mediates, mediated, mediated in, mediate in, mediated by
Contain	contain, contains, contained, containing, contained in
Associate	associate, associates, associated, associated with, associates with, associate with, associated to, associate to, associate in, associated in
Stimulate	stimulate, stimulates, stimulated, stimulated with, stimulated to, stimulate in, stimulated in, stimulated by

(continued)

APPENDIX *(continued)*

Root Relations	Morphological Variants
Enhance	enhance, enhances, enhancer, enhancement, enhanced, enhancers of, enhancement of, enhancer of, enhanced in, enhancer in, enhancement by, enhancer by, enhanced by
Suppress	suppress, suppresses, suppressive, suppressing, suppressed, suppression, suppressor, suppression of, suppressor of, suppressors of, suppressibility of, suppressed in, suppressed by, suppression by
Affect	affect, affects, affected, affecting, affected with, affected to, affect in, affected in, affected by
Interact	interact, interacts, interacting, interaction, interactions, interactive, interacted, interacts with, interact with, interaction with, interactions with, interacting with, interacted with, interact to interactions of, interaction of, interact in, interaction in, interacts in, interactions in
Produce	produce, produces, produced, producer, producers of, produced in, produced by
Bind	bind, binds, bindings, binding, binding with, binds with, bind with, binds to, binding to, bind to, binding of, binding in, binds in, bind in, binding by
Contribute	contribute, contributes, contributed, contributes to, contribute to, contributed to
Correlate	correlated with, correlates with, correlate with, correlated to, correlates to, correlate to, correlate of, correlate in, correlated in, correlate, correlates, correlated
Encode	encode, encodes, encoded, encoded in, encoded by
Modulate	modulate, modulates, modulated, modulated in, modulated by
Promote	promote, promotes, promoters, promoted, promoter, promoters with, promoter to, promoter of, promoters of, promoter in, promoted in, promoters in, promoter by, promoters by, promoted by
Exhibit	exhibit, exhibits, exhibited, exhibiting
Characterized	characterized, characterized to, characterized in, characterized by
Respond	respond, responds, responded, responding, responders, responds to, responding to, respond to, responded to, responds in, responder in
Generate	generate, generates, generated, generated to, generated in, generated by

Index

Computational Intelligence in Bioinformatics. Edited by G. B. Fogel, D. Corne, and Y. Pan

Printed and bound by CPI Group (UK) Ltd, Croydon, CR0 4YY

16/04/2025

14658416-0004